大数据分析
数据仓库项目实战

尚硅谷IT教育 ◎ 编著

电子工业出版社
Publishing House of Electronics Industry
北京·BEIJING

内 容 简 介

本书按照需求规划、需求实现、需求可视化的流程进行编排，遵循项目开发的实际流程，全面介绍了数据仓库的搭建过程。在整个数据仓库的搭建过程中，本书介绍了主要组件的安装部署过程、需求实现的具体思路、各种问题的解决方案等，并在其中穿插了许多与大数据和数据仓库相关的理论知识，包括大数据概论、数据仓库概论、电商业务概述、数据仓库理论准备、数据仓库建模等。

本书从逻辑上可以分为三部分：第一部分是大数据与数据仓库概论及项目需求描述，主要介绍了数据仓库的概念、应用场景和搭建需求；第二部分是项目部署的环境准备，介绍了如何从零开始搭建一个完整的数据仓库环境；第三部分是需求模块实现，针对不同需求分模块进行实现，是本书的重点部分。

本书适合具有一定的编程基础并对大数据感兴趣的读者阅读。通过阅读本书，读者可以快速了解数据仓库，全面掌握数据仓库的相关技术。

未经许可，不得以任何方式复制或抄袭本书之部分或全部内容。
版权所有，侵权必究。

图书在版编目（CIP）数据

大数据分析：数据仓库项目实战 / 尚硅谷 IT 教育编著. —北京：电子工业出版社，2020.11
ISBN 978-7-121-39600-7

Ⅰ.①大… Ⅱ.①尚… Ⅲ.①数据库系统 Ⅳ.①TP311.13

中国版本图书馆 CIP 数据核字（2020）第 180789 号

责任编辑：李　冰　　　　特约编辑：田学清
印　　刷：北京虎彩文化传播有限公司
装　　订：北京虎彩文化传播有限公司
出版发行：电子工业出版社
　　　　　北京市海淀区万寿路 173 信箱　　邮编：100036
开　　本：787×1092　1/16　　印张：24.75　　字数：650 千字
版　　次：2020 年 11 月第 1 版
印　　次：2024 年 3 月第 6 次印刷
定　　价：100.00 元

凡所购买电子工业出版社图书有缺损问题，请向购买书店调换。若书店售缺，请与本社发行部联系，联系及邮购电话：（010）88254888，88258888。

质量投诉请发邮件至 zlts@phei.com.cn，盗版侵权举报请发邮件到 dbqq@phei.com.cn。
本书咨询联系方式：libing@phei.com.cn。

前 言

大数据发展至今，早已不是一个新兴词语，大数据的应用已经无处不在。在大数据时代，我们面临的不仅是海量的数据，更重要的是海量数据所带来的数据的采集、存储、处理等方方面面的问题。为了更快速、更全面地展示大数据的实践应用，本书以一个数据仓库项目为切入点，带领读者一步步揭开大数据的面纱。

数据仓库项目是学习大数据的重要基石。本书以数据仓库的搭建为主线，从搭建之初的框架选型、数据服务的整体策划到数据的流向，数据的采集、存储和计算，循序渐进，一步步地展开，进行细致剖析。在对数据传输过程的讲解中，穿插了数据仓库的相关理论知识及大数据关键框架组件的讲解，务求让读者对大数据有更深刻的理解，更加全面地了解大数据生态体系。

本书共 9 章，包括大数据与数据仓库概论、项目需求描述、项目部署的环境准备、用户行为数据采集模块、业务数据采集模块、数据仓库搭建模块、数据可视化模块、即席查询模块、元数据管理模块。

本项目采用主流的数据仓库建模方式（确定业务过程、声明粒度、确定维度、确实事实），覆盖当前主流框架——采集，Flume/Kafka/Sqoop；存储，MySQL/Hadoop/HBase；计算，Hive/Tez；查询，Presto/Druid/Kylin；可视化，Superset；任务调度，Azkaban；元数据管理，Atlas；脚本，Shell。

整套项目包含业务指标近 100 个、Shell 脚本 40 多个、用户行为原始表 11 张，业务原始表 24 张、数据仓库总表近 100 张……

阅读本书要求读者具有一定的编程基础，至少掌握一门编程语言（如 Java）及 SQL 查询语言。读者若不具备此项条件，则可以关注 **"尚硅谷教育"公众号（微信号：atguigu）**，在聊天窗口发送**关键字"大数据"**，即可获取尚硅谷大数据学科全套视频教程及学习路线图；发送关键字"数仓项目"，则可获取本书相关学习资料，包括 2760 分钟配套视频及全部的源码、脚本、课件、软件包等。

感谢电子工业出版社的李冰编辑在本书编写过程中给予的指导与支持。

尚硅谷 IT 教育

关于我们

  尚硅谷 IT 教育是一家专业的 IT 教育培训机构，开设了 JavaEE、大数据、HTML5 前端等多门学科，在互联网上发布的 JavaEE、大数据、HTML5 前端、区块链、C 语言、Python 等技术视频教程广受赞誉。

  尚硅谷 IT 教育一直坚持"技术为王"的发展理念，我们专注技术，不断钻研课程，团队中技术型人才占比 60%以上，设有独立的研究院，与多家互联网大型企业的研发团队保持技术交流，保障教学内容始终基于研发一线。

  截至目前，尚硅谷 IT 教育已累计发布视频教程 11 788 集，全长 2742 小时，通过分享优质的教学资源，直接或间接帮助了更多需要帮助的人。让天下没有难学的技术，坚持开源精神，不断打磨优质的教学产品，推出更多进阶的视频教程，同时，为满足更多的学习场景需求，我们会陆续出版系列技术图书，为广大 IT 从业者提供更多优质的学习资源。

  希望通过我们的努力，帮助更多怀揣梦想的年轻人，为中国的软件人才培养尽一点绵薄之力。

  关注"尚硅谷教育"公众号，获取更多视频教程，包括大数据、JavaEE、HTML5 前端、区块链、Go 语言、Linux、Python、C 语言、Android 等，并附赠学习路线图，全部免费获取！

目 录

第1章 大数据与数据仓库概论 1
　1.1 大数据概论 1
　　1.1.1 什么是大数据 1
　　1.1.2 大数据生态圈简介 2
　　1.1.3 大数据应用场景 3
　1.2 数据仓库概论 4
　　1.2.1 什么是数据仓库 4
　　1.2.2 数据仓库能干什么 4
　　1.2.3 数据仓库的特点 5
　1.3 学前导读 6
　　1.3.1 学习的基础要求 6
　　1.3.2 你将学到什么 7
　1.4 本章总结 7

第2章 项目需求描述 8
　2.1 任务概述 8
　　2.1.1 产品描述 9
　　2.1.2 系统目标 9
　　2.1.3 系统功能结构 9
　　2.1.4 系统流程图 10
　2.2 业务描述 10
　　2.2.1 采集模块业务描述 10
　　2.2.2 数据仓库需求业务描述 16
　　2.2.3 数据可视化业务描述 17
　2.3 系统运行环境 17
　　2.3.1 硬件环境 17
　　2.3.2 软件环境 18
　2.4 本章总结 20

第 3 章 项目部署的环境准备 21

3.1 Linux 环境准备 21
3.1.1 VMware 安装 21
3.1.2 CentOS 安装 21
3.1.3 远程终端安装 31

3.2 Linux 环境配置 34
3.2.1 网络配置 34
3.2.2 网络 IP 地址配置 35
3.2.3 主机名配置 36
3.2.4 防火墙配置 37
3.2.5 一般用户设置 38

3.3 Hadoop 环境搭建 38
3.3.1 虚拟机环境准备 39
3.3.2 JDK 安装 45
3.3.3 Hadoop 安装 46
3.3.4 Hadoop 分布式集群部署 47
3.3.5 配置 Hadoop 支持 LZO 压缩 52
3.3.6 配置 Hadoop 支持 Snappy 压缩 53

3.4 本章总结 54

第 4 章 用户行为数据采集模块 55

4.1 日志生成 55
4.2 采集日志的 Flume 57
4.2.1 Flume 组件 58
4.2.2 Flume 安装 58
4.2.3 采集日志 Flume 配置 59
4.2.4 Flume 的 ETL 拦截器和日志类型区分拦截器 61
4.2.5 采集日志 Flume 启动、停止脚本 67

4.3 消息队列 Kafka 68
4.3.1 Zookeeper 安装 68
4.3.2 Zookeeper 集群启动、停止脚本 70
4.3.3 Kafka 安装 71
4.3.4 Kafka 集群启动、停止脚本 73
4.3.5 Kafka Topic 相关操作 74

4.4 消费 Kafka 日志的 Flume 75
4.4.1 消费日志 Flume 配置 75
4.4.2 消费日志 Flume 启动、停止脚本 78

4.5 采集通道启动、停止脚本 79

4.6	本章总结	80

第5章 业务数据采集模块 ... 81

- 5.1 电商业务概述 ... 81
  - 5.1.1 电商业务流程 ... 81
  - 5.1.2 电商常识 ... 82
  - 5.1.3 电商表结构 ... 82
  - 5.1.4 数据同步策略 ... 89
- 5.2 业务数据采集 ... 90
  - 5.2.1 MySQL 安装 ... 90
  - 5.2.2 业务数据生成 ... 92
  - 5.2.3 业务数据建模 ... 94
  - 5.2.4 Sqoop 安装 ... 96
  - 5.2.5 业务数据导入数据仓库 ... 97
- 5.3 本章总结 ... 109

第6章 数据仓库搭建模块 ... 110

- 6.1 数据仓库理论准备 ... 110
  - 6.1.1 范式理论 ... 110
  - 6.1.2 关系模型与维度模型 ... 113
  - 6.1.3 星形模型、雪花模型与星座模型 ... 114
  - 6.1.4 表的分类 ... 116
  - 6.1.5 为什么要分层 ... 117
  - 6.1.6 数据仓库建模 ... 118
  - 6.1.7 业务术语 ... 121
- 6.2 数据仓库搭建环境准备 ... 123
  - 6.2.1 MySQL HA ... 123
  - 6.2.2 Hive 安装 ... 130
  - 6.2.3 Tez 引擎安装 ... 134
- 6.3 数据仓库搭建——ODS 层 ... 138
  - 6.3.1 创建数据库 ... 138
  - 6.3.2 用户行为数据 ... 138
  - 6.3.3 ODS 层用户行为数据导入脚本 ... 141
  - 6.3.4 业务数据 ... 142
  - 6.3.5 ODS 层业务数据导入脚本 ... 151
- 6.4 数据仓库搭建——DWD 层 ... 154
  - 6.4.1 用户行为启动日志表解析 ... 154
  - 6.4.2 用户行为事件表拆分 ... 157
  - 6.4.3 用户行为事件表解析 ... 167

- 6.4.4 业务数据维度表解析 189
- 6.4.5 业务数据事实表解析 195
- 6.4.6 拉链表构建之用户维度表 209
- 6.4.7 DWD 层数据导入脚本 214
- 6.5 数据仓库搭建——DWS 层 223
  - 6.5.1 系统函数 223
  - 6.5.2 用户行为数据聚合 224
  - 6.5.3 业务数据聚合 226
  - 6.5.4 DWS 层数据导入脚本 237
- 6.6 数据仓库搭建——DWT 层 246
  - 6.6.1 设备主题宽表 247
  - 6.6.2 会员主题宽表 249
  - 6.6.3 商品主题宽表 251
  - 6.6.4 优惠券主题宽表 254
  - 6.6.5 活动主题宽表 256
  - 6.6.6 DWT 层数据导入脚本 258
- 6.7 数据仓库搭建——ADS 层 264
  - 6.7.1 设备主题 264
  - 6.7.2 会员主题 272
  - 6.7.3 商品主题 275
  - 6.7.4 营销主题 279
  - 6.7.5 ADS 层数据导入脚本 283
- 6.8 结果数据导出脚本 291
- 6.9 会员主题指标获取的全调度流程 293
  - 6.9.1 Azkaban 安装 293
  - 6.9.2 创建可视化的 MySQL 数据库和表 300
  - 6.9.3 编写指标获取调度流程 301
- 6.10 本章总结 306

第 7 章 数据可视化模块 307

- 7.1 模拟可视化数据 307
  - 7.1.1 会员主题 307
  - 7.1.2 地区主题 308
- 7.2 Superset 部署 310
  - 7.2.1 环境准备 310
  - 7.2.2 Superset 安装 312
- 7.3 Superset 使用 314
  - 7.3.1 对接 MySQL 数据源 314
  - 7.3.2 制作仪表盘 317

7.4 本章总结 ... 322

第8章 即席查询模块 ... 323

8.1 Presto ... 323
8.1.1 Presto 特点 ... 323
8.1.2 Presto 安装 ... 324
8.1.3 Presto 优化之数据存储 ... 328
8.1.4 Presto 优化之查询 SQL ... 329
8.1.5 Presto 注意事项 ... 330

8.2 Druid ... 330
8.2.1 Druid 简介 ... 330
8.2.2 Druid 框架原理 ... 331
8.2.3 Druid 数据结构 ... 332
8.2.4 Druid 安装（单机版）... 333

8.3 Kylin ... 338
8.3.1 Kylin 简介 ... 338
8.3.2 HBase 安装 ... 339
8.3.3 Kylin 安装 ... 341
8.3.4 Kylin 使用 ... 343
8.3.5 Kylin Cube 构建原理 ... 353
8.3.6 Kylin Cube 构建优化 ... 356
8.3.7 Kylin BI 工具集成 ... 360

8.4 即席查询框架对比 ... 367
8.5 本章总结 ... 368

第9章 元数据管理模块 ... 369

9.1 Atlas 入门 ... 369
9.1.1 Atlas 概述 ... 369
9.1.2 Atlas 架构原理 ... 370

9.2 Atlas 安装及使用 ... 371
9.2.1 安装前环境准备 ... 371
9.2.2 集成外部框架 ... 373
9.2.3 集群启动 ... 377
9.2.4 导入 Hive 元数据到 Atlas ... 377

9.3 Atlas 界面查看及使用 ... 378
9.3.1 查看基本信息 ... 378
9.3.2 查看血缘依赖关系 ... 381

9.4 本章总结 ... 386

第1章

大数据与数据仓库概论

知其然知其所以然,在正式开始学习之前,本章先为读者解答一些基本的概念问题。
- 什么是大数据?
- 大数据生态圈的主要构成是什么?
- 大数据应用在哪些行业?
- 什么是数据仓库?
- 数据仓库可以用来做什么?

本书的学习需要读者具备一定的基础,本章会给出说明。同时,对学习后读者可以收获的成果进行简单的介绍。

1.1 大数据概论

1.1.1 什么是大数据

读者首先需要了解什么是数据?什么是大数据?

数据在我们的生活中无处不在,清晨起床,用手机打开新闻资讯,此时就产生了数据;早高峰乘坐地铁,刷二维码进站,又产生了数据;打开购物网站,下单购买商品,还会产生数据……生活在当今这个高度信息化的世界,一切行为几乎都可以用数据来描述,这种情况发生在每个人的身上。每时每刻都有上亿条数据产生,这些海量数据流进了那些提供互联网服务的公司,存储在他们的系统中。如果不对其加以利用,则这些数据只是拖慢系统的沉重负担,但如果善于挖掘,则这些数据就是蕴藏巨大价值的宝藏!

那么大数据究竟是什么?国际顶级权威咨询机构麦肯锡给出定义,大数据指的是所涉及的数据集规模已经超过了传统数据库软件获取、存储、管理和分析的能力。这是一个被故意设计成主观性的定义,并且是一个关于多大的数据集才能被认为是大数据的可变定义,即并不定义大于一个特定数据的 TB 才叫大数据。因为随着技术的不断发展,符合大数据标准的数据集容量也会增长;并且定义随不同行业也有变化,这依赖于在一个特定行业通常何种软件和数据集有多大。因此,大数据在今天不同行业中的范围可以从几十 TB 到几 PB。随着数据量越来越大,大数据的存储和分析计算面临的挑战也越来越大。

1.1.2 大数据生态圈简介

在大数据飞速发展的几年中,已经形成了一个完备多样的大数据生态圈,如图 1-1 所示。从图 1-1 中可以看出,大数据生态圈分为 7 层,这 7 层如果进一步概括,可以归纳为数据采集层、数据计算层和数据应用层 3 层结构。

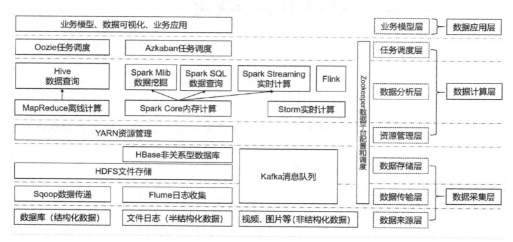

图 1-1 大数据生态圈的层次及其应用举例

1. 数据采集层

数据采集层是整个大数据平台的源头,是整个大数据系统的基石。当前许多公司的业务平台每天都会产生海量的日志数据,收集日志数据供离线和在线的分析系统使用是日志收集系统需要做的事情。除日志数据外,大数据系统的数据来源还包括业务数据库的结构化数据,以及视频、图片等非结构化数据。随着大数据的重要性逐渐突显,大数据采集系统的合理搭建就显得尤为重要。

大数据采集过程中的挑战越来越多,主要来自以下几个方面。

① 数据源多种多样。
② 数据量大且变化快。
③ 如何保证所采集数据的可靠性。
④ 如何避免采集重复的数据。
⑤ 如何保证所采集数据的质量。

针对这些挑战,日志收集系统需要具有高可用性、高可靠性、可扩展性等特征。现在主流的数据传输层的工具有 Sqoop、Flume、DataX 等,通过多种工具的配合使用,可以满足多种数据源的采集传输工作。同时数据传输层通常情况下还需要对数据进行初步的清洗、过滤、汇总、格式化等一系列转换操作,使数据转为适合查询的格式。数据采集完成后,需要选用合适的数据存储系统,考虑到数据存储的可靠性及后续计算的便利性,通常选用分布式文件系统,如 HDFS 和 HBase 等。

2. 数据计算层

大数据仅仅被采集到数据存储系统是远远不够的,只有通过整合计算,数据中的潜在价值才可以被挖掘出来。

数据计算层可以划分为离线数据计算和实时数据计算。离线数据计算主要是指传统的数据仓库概念，数据计算可以以天为单位，还可以细分为小时或者汇总为以周和月为单位，主要以 T+1 的模式进行，即每天凌晨处理上一天的数据。

随着业务的发展，部分业务需求对实时性的要求逐渐提高，实时计算开始占有较大的比重，实时计算的应用场景也越来越广泛，比如，电商实时交易数据更新、设备实时运行状态报告、活跃用户区域分布实时变化等。生活中比较常见的有地图与位置服务应用实时分析路况、天气应用实时分析天气变化趋势等。

大数据的计算需要使用的资源是巨大的，大量的数据计算任务通常需要通过资源管理系统共享一个集群的资源，YARN 便是资源管理系统的一个典型代表。通过资源管理系统可以提高集群的利用率、降低运维成本。大数据的计算通常不是独立的，一个计算任务的运行很大可能依赖于另一个计算任务的结果，使用任务调度系统可以很好地处理任务之间的依赖关系，实现任务的自动化运行。

无论何种数据计算，进行数据计算的前提是规范合理地规划数据，搭建规范统一的数据仓库体系。通过搭建合理的、全面的数据仓库体系，尽量规避数据冗余和重复计算等问题，使数据的价值发挥到最大程度。为此，数据仓库分层理念被逐渐丰富完善，目前应用比较广泛的数据仓库分层理念将数据仓库分为 4 层，分别是原始数据层、明细数据层、汇总数据层和应用数据层。通过数据仓库不同层次之间的分工分类，使数据更加规范化，可以帮助用户需求得到更快实现，并且可以更加清楚明确地管理数据。

3. 数据应用层

当数据被整合计算完成之后，需要最终提供给用户使用，这就是数据应用层。不同的数据平台针对其不同的数据需求有各自相应的数据应用层的规划设计，数据的最终需求计算结果可以构建在不同的数据库上，比如，MySQL、HBase、Redis、Elasticsearch 等。通过这些数据库，用户可以很方便地访问最终的结果数据。

最终的结果数据由于面向的用户不同，可能有不同层级的数据调用量，面临着不同的挑战。如何能更稳定地为用户提供服务、满足各种用户复杂的数据业务需求、保证数据服务接口的高可用性等，都是数据应用层需要考虑的问题。

1.1.3 大数据应用场景

大数据无处不在，包括金融、汽车、互联网、电信、物流、电影娱乐等在内的社会各行各业都已经融入了大数据的印迹。

金融业：大数据在金融行业的应用范围很广，比较典型的金融大数据的应用场景集中在用户经营、数据风控、产品设计和决策支持等，例如，大数据分析为客户推荐合适的产品、预测未来理财产品的受欢迎程度。金融行业的数据大部分是结构化数据，存储在传统关系型数据库中，通过数据挖掘可以分析出隐藏在交易数据中的巨大商业价值。

汽车行业：大数据在汽车行业释放出的巨大价值引发越来越多汽车行业人士的关注，最基本的有利用大数据分析消费者的行为决定汽车营销方向、分析用户维保行为助力二手车真实价值评估、智能导航大数据为智能化交通提供更多的空间和可能,此外利用大数据和物联网技术制造的无人驾驶汽车,在不远的未来将走入我们的日常生活中。

互联网行业:大数据在互联网行业的应用已经渗透到了方方面面,几乎客户的所有行为都会在互联网平台上留下痕迹,所以互联网企业可以方便地获取大量的用户行为信息,通过分析用户行为,可以制定更有针对性的服务策略。除利用大数据提升自己的业务以外,互联网企业已经开始实现数据业务化,利用大数据发现新的商业价值。以阿里巴巴为例,它不仅在加强个性化推荐,开发"千人千面"这种面向消费者的大数据应用,并且还在尝试利用大数据进行智能客户服务。

电信业:电信业大数据的发展仍处于探索阶段,目前国内运营商对大数据的应用主要体现在利用大数据进行基础设施建设优化、网络运营管理优化及市场精准营销,并利用大数据实现客户离网分析,及时掌握客户离网倾向,出台客户挽留措施。

物流业:物流业通过海量的物流数据,以及运输、仓储、包装、运输等环节中涉及的数据信息等,挖掘出新的深层价值,进而提高运输与配送效率、减少物流成本,更有效地满足客户服务要求。具体体现在利用大数据优化物流网络,车货匹配,库存预测,供应链协同管理,从而提高物流效率,降低物流成本。

电影娱乐:大数据在精准营销中所起的作用尤其反映在电影领域,挖掘精准人群的意义更加非同一般,因为不同类型的影片背后是截然不同的受众群体,针对不同的受众,还可以利用大数据去挖掘受众的所思、所想、所好,为内容策划甚至演员选角提供更多的数据支持。

大数据的价值,远远不止于此,大数据对各行各业的渗透,大大推动了社会生产和生活的发展,未来必将产生重大而深远的影响。

1.2 数据仓库概论

1.2.1 什么是数据仓库

数据仓库,英文名称为 Data Warehouse,可简写为 DW 或 DWH。数据仓库是为企业所有级别的决策制定过程,提供所有类型数据支持的资源集合。它出于分析性报告和决策支持目的而创建。

随着技术的飞速发展,经过多年的数据积累,各互联网公司已保存了海量的原始数据和各种业务数据,所以数据仓库技术是各互联网公司目前需要着重发展的技术领域。数据仓库是面向分析的集成化数据环境。通过对数据仓库中的数据进行分析,可以帮助企业改进业务流程、控制成本、提高产品质量等。

1.2.2 数据仓库能干什么

数据仓库系统是一个信息服务和管理平台,它从业务处理系统获得数据,主要以星形模型和雪花模型组织数据,并为用户从数据中获取信息和知识提供各种手段。

按照功能结构划分,数据仓库系统至少应该包含数据获取(Data Acquisition)、数据存储

（Data Storage）和数据访问（Data Access）三个关键部分。

企业数据仓库的建设，是以现有企业业务系统和大量业务数据的积累为基础的。数据仓库不是静态的概念，只有把信息及时交给需要这些信息的使用者，帮助他们做出改善其业务经营的决策，信息才能发挥作用、才有意义。而把信息加以整理归纳和重组，并及时提供给相应的管理决策人员，是数据仓库的根本任务。因此，从企业的角度看，数据仓库的建设是一个工程。

1.2.3 数据仓库的特点

1．数据仓库中的数据是面向主题的

与传统数据库面向应用进行数据组织的特点相对应，数据仓库中的数据是面向主题进行数据组织的。什么是主题呢？首先，主题是一个抽象的概念，是较高层次上企业信息系统中的数据综合、归类并进行分析利用的抽象。在逻辑意义上，它对应企业中某一宏观分析领域所涉及的分析对象。面向主题的数据组织方式，就是在较高层次上对分析对象的数据的一个完整、一致的描述，能完整、统一地刻画各分析对象所涉及企业的各项数据，以及数据之间的联系。所谓较高层次是相对于面向应用的数据组织方式而言的，是指按照主题进行数据组织的方式具有更高的数据抽象级别。

2．数据仓库中的数据是集成的

数据仓库中的数据是从原有的、分散的数据库中抽取来的，抽取的数据可分为操作型数据和分析型数据两大类，两者之间差别甚大。第一，数据仓库的每个主题所对应的源数据在原有的各分散数据库中有许多重复和不一致的地方，且来源于不同联机系统的数据都和不同的应用逻辑捆绑在一起；第二，数据仓库中的数据不是从原有的数据库系统中直接得到的。因此，数据在进入数据仓库之前，必然要经过统一与综合，这一步是数据仓库建设中最关键、最复杂的一步，所要完成的工作如下。

① 要统一源数据中的所有矛盾之处，如字段的同名异义、异名同义、单位不统一、字长不一致等。

② 进行数据综合和计算。数据仓库中的数据综合工作可以在从原有数据库中抽取数据时完成，但大多数是在数据仓库内部完成的，即进入数据仓库以后进行数据综合。

3．数据仓库中的数据是不可更新的

数据仓库中的数据主要供企业管理者决策分析使用，所涉及的数据操作主要是数据查询，一般情况下并不进行修改操作。数据仓库中的数据反映的是相当长的一段时间内历史数据的内容，是不同时间的数据库快照的集合，以及基于这些快照进行统计、综合和重组的导出数据，而不是联机处理的数据。数据库中进行联机处理的数据经过集成输入数据仓库中，一旦数据仓库存放的数据已经超过数据仓库的数据存储期限，这些数据会被删除。因为数据仓库只能进行数据查询操作，所以数据仓库管理系统相比数据库管理系统而言要简单得多。数据库管理系统中的许多技术难点，如完整性保护、并发控制等，在数据仓库管理系统中几乎可以忽略。但是在数据仓库中要查询的数据量往往很大，所以就对数据查询提出了更高的要求，它要求采用各

种复杂的索引技术，同时由于数据仓库面向的是商业企业的高层管理者，他们会对数据查询界面的友好性和数据表示提出更高的要求。

4．数据仓库中的数据是随时间不断变化的

数据仓库中的数据不可更新是针对应用来说的，也就是说，数据仓库的用户在进行数据分析和处理时是不进行数据更新操作的。但并不是说，在从数据集成输入数据仓库开始到最终被删除的整个数据生存周期中，数据仓库中的数据都是永远不变的。

数据仓库中的数据是随时间不断变化的，这是数据仓库的第 4 个特点。这一特点表现在以下 3 个方面。

① 数据仓库随着时间的变化不断增加新的数据内容。数据仓库系统必须不断捕捉 OLTP 数据库中变化的数据，并追加到数据仓库中，也就是要不断地生成 OLTP 数据库的快照，经过统一集成后增加到数据仓库中；但对于确实不再变化的数据库快照，如果捕捉到新的变化数据，则只生成一个新的数据库快照增加进去，而不会对原有的数据库快照进行修改。

② 数据仓库随时间的变化不断删除旧的数据内容。数据仓库中的数据也有存储期限，一旦超过这一期限，就要被删除。只是数据仓库中的数据时限要远远长于操作型环境中的数据时限。在操作型环境中一般只保存 60～90 天的数据，而在数据仓库中则需要保存较长时限的数据（如 5～10 年），以满足 DSS 进行趋势分析的要求。

③ 数据仓库中包含了大量的综合数据，其中很多数据与时间密切相关，如数据经常按照时间段进行综合，或隔一定的时间进行抽样等。这些数据要随着时间的变化不断地进行重新综合。因此，数据仓库的数据特征都包含时间项，以标明数据的历史时期。

1.3 学前导读

在开始学习之前，希望读者仔细阅读以下内容，便于打开大数据学习之门。

1.3.1 学习的基础要求

在学习本书之前，读者需要提前了解一些基础知识，有助于更加轻松、快速地掌握大数据的相关内容，在后续项目的搭建过程中能更加得心应手，为深入学习大数据打下坚实的基础。

首先，学习大数据技术，读者一定要掌握一个操作大数据技术的利器，这个利器就是一门编程语言，如 Java、Scala、Python、R 等。本书以 Java 为基础进行编写，所以学习本书需要读者具备一定的 Java 基础知识和 Java 编程经验。

其次，读者还需要掌握一些数据库知识，如 MySQL、Oracle 等，并熟练使用 SQL，本书将出现大量的 SQL 操作。

最后，读者还需要掌握一门操作系统技术，即在服务器领域占主导地位的 Linux，只要能够熟练使用 Linux 的常用系统命令、文件操作命令和一些基本的 Linux Shell 编程即可。大数据系统需要处理业务系统服务器产生的海量日志数据信息，这些数据通常存储在服务器端，各大互联网公司常用的操作系统是在实际工作中安全性和稳定性很高的 Linux 或者 UNIX。大数

据生态圈的各框架组件也普遍运行在 Linux 上。

如果读者不具备上述基础知识，可以关注尚硅谷教育公众号获取学习资料，读者可根据自身需要选择相应课程进行学习。本书所讲解的项目同时提供了视频课程资料，包括尚硅谷大数据的各种学习视频，读者可在尚硅谷教育公众号回复"数仓项目"免费获取。

1.3.2　你将学到什么

本书将带领读者完成一个完整的数据仓库搭建及需求实现项目，大致可以划分为 3 部分：数据仓库概论及项目需求描述、项目框架搭建和项目需求实现。

在项目需求及框架讲解部分，读者可以全面了解一个数据仓库项目的具体需求，以及根据需求如何完成框架选型的过程。

在项目框架搭建部分，读者将跟随本书从操作系统开始，一步步搭建自己的虚拟机系统，了解各框架的基本知识，完成各框架的基本配置，最终形成一个可以正常运行的大数据虚拟机系统。

在项目需求实现部分，本书将从用户行为数据采集模块、业务数据采集模块、数据仓库搭建模块、即席查询模块、元数据管理模块 5 个方面对需求进行实现，读者通过本部分的学习将会了解一个完整的数据仓库系统从数据源到数据的最终展示是如何实现的，同时还能学到数据仓库相关的理论知识，掌握 Hive、Sqoop、Flume 等日志数据采集工具的工作原理及应用方法。本部分对电商数据仓库的常见实战指标及难点实战指标进行了透彻讲解，具体指标包括每日、每周、每月活跃设备明细，留存用户比例，沉默用户、回流用户、流失用户统计，最近连续 3 周活跃用户统计，最近 7 天内连续 3 天活跃用户统计等。

通过对数据仓库系统的学习，读者能够对数据仓库项目建立起清晰、明确的概念，系统、全面地掌握各项数据仓库项目技术，轻松应对各种数据仓库的难题。

1.4　本章总结

本章主要为读者讲述了大数据的概念及大数据的实际应用场景，使读者对数据仓库建立了初步的认识，还介绍了数据仓库的概念、作用及特点，并为读者明确了学习本书应该具备的技术基础以及最终的学习成果目标，为读者之后的学习做好准备。

第2章 项目需求描述

随着互联网的迅速发展，电子商务变得越来越重要，网上购物已经成为广大消费者主要的消费方式之一。满足用户需求，让用户可以方便购物、快乐购物，使用户可以轻易、方便地找到自己所需的商品，商品能够更加精准地推送到需要的用户眼前，如此，才可以实现一个电子商务系统的真正价值，做到利润最大化。

电子商务系统在满足大量用户的访问和购物需求的同时，还会产生大量的用户行为数据及业务交互数据。用户行为数据是指用户在使用系统的时候进行点击、浏览、添加购物车、收藏等行为时产生的日志数据，在这些日志数据中存储着用户的 id、点击时间、手机型号、渠道来源等信息。业务交互数据是指在系统运行过程中由后端程序记录下来的业务数据。例如，用户的订单表、支付情况表、订单详情表、用户详情表等，这些数据或者存在于前端服务器中，或者存在于后端服务器的传统数据库中，如果加以利用则能挖掘出其中巨大的潜在价值。

为了最大限度地挖掘数据中的潜在价值，我们需要搭建数据仓库系统。数据仓库系统从数据的采集流程开始，将不同来源的数据统一采集进数据仓库中，在数据仓库中对数据进行合理的分析、分类、存储和计算，这个数据仓库将面向所有有数据分析需求的用户，包括企业决策者、运营人员、数据分析师等，为用户提供多样的数据服务，解决用户对数据方面的需求。

本章将重点介绍本书中电商数据项目的需求，主要包括项目的产品描述、功能架构设计、系统流程图设计、各模块的功能业务描述，以及根据需求确定的硬件和软件环境。

本项目主要面向想要了解更多大数据项目相关知识的大数据从业人员、有一定编程基础想要对大数据有所了解的大数据技术的初学者，以及想要更多地了解大数据的开发流程以期能更好地应用大数据的用户。

2.1 任务概述

项目的需求说明非常重要，不仅可以让程序员了解产品的需求，知道开发的目标结果，也可以让系统应用人员了解业务流程的设计是否完善。所以在项目需求说明书中必须有详细的系统功能说明。

2.1.1 产品描述

本数据仓库项目将数据采集、数据同步导入、数据分层搭建、需求分层实现、脚本实现、任务定时调度、元数据管理等功能集合，提供数据展示页面，让用户通过本数据仓库项目，将电子商务系统产生的用户行为数据和业务交互数据及时同步到数据仓库中，并对数据需求进行分析计算，个别需求可以通过 Web 页面得到展示。还对即席查询引擎进行了探索接入，并行考虑了三种即席查询引擎，为用户提供即席查询服务。

2.1.2 系统目标

数据仓库系统需要实现的目标如下：
- 环境搭建完整，技术选型合理，框架服务分配合理；
- 信息流完整，包括数据生成、数据采集、数据仓库建模、数据即席查询；
- 能应对海量数据的分析查询；
- 实现元数据管理。

2.1.3 系统功能结构

如图 2-1 所示，该数据仓库系统主要分为 4 个功能模块，分别是数据采集、数据仓库平台、数据可视化和即席查询。

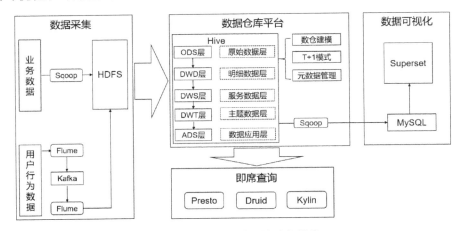

图 2-1　数据仓库系统的功能模块

数据采集模块主要负责将电子商务系统前端的用户行为数据以及业务交互数据采集到大数据存储系统中，所以数据采集模块共分为两大体系：用户行为数据采集体系和业务交互数据采集体系。用户行为数据主要以日志文件的形式落盘（存储在服务器磁盘中，下同），采用 Flume 作为数据采集框架对数据进行实时监控采集；业务交互数据主要存储在 MySQL 中，采用 Sqoop 对其进行 T+1 形式的采集。

数据采集模块负责将原始数据采集到数据仓库中，合理建表，并针对数据进行清洗、转义、分类、重组、合并、拆分、统计等，将数据合理分层，极大地减少数据重复计算的情况。在针对固定长期需求进行数据仓库的合理建设的同时，还应考虑用户的即席查询需求，需对外提供即席查询接口。一方面是为了让用户能够更高效地挖掘和使用数据；另一方面是为了让平台管理人员能够更加有效地做好系统的维护管理工作，对数据仓库的元数据信息建立管理。

数据可视化主要负责将最终需求结果数据导入 MySQL 中，供数据用户使用或者对数据进行 Web 页面展示。

2.1.4 系统流程图

数据仓库系统主要流程如图 2-2 所示。前端埋点（指数据采集的技术方式，下同）用户行为数据经生产层 Flume Agent、Kafka、消费层 Flume Agent 落盘到 HDFS 中，业务交互数据经 Sqoop 采集到 HDFS 中，HDFS 中的数据经过 Hive 的相关操作，将数据进行提取转换，形成合理分层，最终得到需求结果数据，将数据导出 MySQL 中，实现数据可视化，并提供即席查询服务。

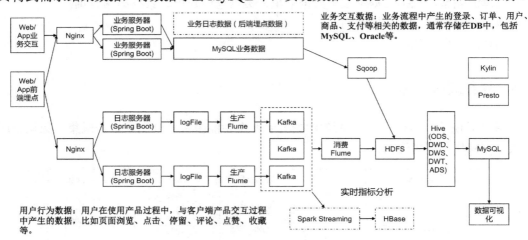

图 2-2　数据仓库系统主要流程

2.2　业务描述

2.2.1　采集模块业务描述

1. 数据生成模块之用户行为数据基本格式

用户执行的一些操作会生成用户行为数据发送到服务器，数据分为公共字段和业务字段。
- 公共字段：基本所有 Android 手机包含的字段。
- 业务字段：前端埋点上报的字段，有具体的业务类型。

如下示例表示业务字段的上传。

```
{
"ap":"xxxxx",                        // 项目数据来源
```

```
"cm": {                                 // 公共字段
    "mid": "",                          // (String) 设备唯一标识
    "uid": "",                          // (String) 用户标识
    "vc": "1",                          // (String) versionCode,程序版本号
    "vn": "1.0",                        // (String) versionName,程序版本名
    "l": "zh",                          // (String) 系统语言
    "sr": "",                           // (String) 渠道号,应用从哪个渠道来
    "os": "7.1.1",                      // (String) Android版本
    "ar": "CN",                         // (String) 区域
    "md": "BBB100-1",                   // (String) 手机型号
    "ba": "blackberry",                 // (String) 手机品牌
    "sv": "V2.2.1",                     // (String) sdkVersion
    "g": "",                            // (String) gmail
    "hw": "1620x1080",                  // (String) heightXwidth,屏幕宽高
    "t": "1506047606608",               // (String) 客户端日志产生的时间
    "nw": "WIFI",                       // (String) 网络模式
    "ln": 0,                            // (double) lng,经度
    "la": 0                             // (double) lat,纬度
},
"et": [                                 //事件
    {
        "ett": "1506047605364",         // 客户端事件产生的时间
        "en": "display",                // 事件名称
        "kv": {                         // 事件结果,以key-value的形式自行定义
            "goodsid": "236",
            "action": "1",
            "extend1": "1",
            "place": "2",
            "category": "75"
        }
    }
]
}
```

示例日志（服务器时间戳 | 日志）如下。

```
1540934156385|{
    "ap": "gmall",
    "cm": {
        "uid": "1234",
        "vc": "2",
        "vn": "1.0",
        "l": "EN",
        "sr": "",
        "os": "7.1.1",
        "ar": "CN",
        "md": "BBB100-1",
        "ba": "blackberry",
        "sv": "V2.2.1",
```

```
        "g": "abc@gmail.com",
        "hw": "1620x1080",
        "t": "1506047606608",
        "nw": "WIFI",
        "ln": 0,
        "la": 0
    },
    "et": [
        {
            "ett": "1506047605364",    // 客户端事件产生的时间
            "en": "display",            // 事件名称
            "kv": {                     // 事件结果,以 key-value 的形式自行定义
                "goodsid": "236",
                "action": "1",
                "extend1": "1",
                "place": "2",
                "category": "75"
            }
        },{
        "ett": "1552352626835",
        "en": "active_background",
        "kv": {
          "active_source": "1"
        }
      }
    ]
}
```

2. 数据生成模块之事件日志数据

(1) 商品列表页加载过程的事件名称为 loading, 产生的日志数据的具体字段名称及字段描述如表 2-1 所示。

表 2-1 商品列表页加载过程 loading

字 段 名 称	字 段 描 述
action	动作:开始加载=1,加载成功=2,加载失败=3
loading_time	加载时长:计算从下拉开始到接口返回数据的时间(开始加载上报 0,加载成功或加载失败上报具体时间)
loading_way	加载类型:读取缓存=1,从接口拉新数据=2(加载成功才会上报加载类型)
extend1	扩展字段
extend2	扩展字段
type	加载方式:自动加载=1,用户下拉页面加载=2,单击底部按钮加载=3
type1	加载失败码:将加载失败状态码报回来(报空为加载成功,没有失败)

(2) 商品点击的事件名称为 display, 产生的日志数据的具体字段名称及字段描述如表 2-2 所示。

表 2-2　商品点击 display

字 段 名 称	字 段 描 述
action	动作：曝光商品=1，点击商品=2
goodsid	商品 id（服务器端下发的商品 id）
place	顺序（第几件商品，第一件为 0，第二件为 1，以此类推）
extend1	曝光类型：首次曝光=1，重复曝光=2
category	品类 id（服务器端定义的品类 id）

（3）商品详情展示的事件名称为 newsdetail，产生的日志数据的具体字段名称及字段描述如表 2-3 所示。

表 2-3　商品详情展示 newsdetail

字 段 名 称	字 段 描 述
entry	页面入口来源：应用首页=1，push=2，详情页相关推荐=3
action	动作：开始加载=1，加载成功=2，加载失败=3，退出页面=4
goodsid	商品 id（服务器端下发的商品 id）
show_style	商品样式：无图=0，一张大图=1，两张大图=2，三张小图=3，一张小图=4，一张大图两张小图=5
news_staytime	页面停留时长：从商品开始加载到用户关闭页面所用的时间。若中途又跳转到了其他页面，则暂停计时，待回到详情页时恢复计时；若中途跳转的时间超过 10 分钟，则本次计时作废，不上报本次数据；若未加载成功就退出，则报空
loading_time	加载时长：从页面开始加载到接口返回数据的时间（开始加载报 0，加载成功或加载失败上报具体时间）
type1	加载失败码：将加载失败状态码报回来（报空为加载成功）
category	品类 id（服务器端定义的品类 id）

（4）广告点击的事件名称为 ad，产生的日志数据的具体字段名称及字段描述如表 2-4 所示。

表 2-4　广告点击 ad

字 段 名 称	字 段 描 述
entry	入口：商品列表页=1，应用首页=2，商品详情页=3
action	动作：请求广告=1，获取缓存广告=2，广告位展示=3，广告展示=4，广告点击=5
content	状态：成功=1，失败=2
detail	失败码（没有则上报空）
source	广告来源：AdMob=1，Facebook=2，ADX（百度）=3，VK（俄罗斯）=4
behavior	用户行为：主动获取广告=1，被动获取广告=2
newstype	Type：图文=1，图集=2，段子=3，GIF=4，视频=5，调查=6，纯文本=7，视频+图文=8，GIF+图文=9，其他=0
show_style	内容样式：无图（纯文本）=6，一张大图=1，三张小图+文=4，一张小图=2，一张大图两张小图+文=3，图集+文=5，一张大图+文=11，GIF（大图）+文=12，视频（大图）+文=13。来源于详情页相关推荐的商品，上报样式都为 0（因为都是左文右图）

（5）消息通知的事件名称为 notification，产生的日志数据的具体字段名称及字段描述如表 2-5 所示。

表 2-5 消息通知 notification

字段名称	字段描述
action	动作：通知产生=1，通知弹出=2，通知点击=3，常驻通知展示（不重复上报，一天之内只报一次）=4
type	通知 id：预警通知=1，天气预报（早=2，晚=3），常驻=4
ap_time	客户端弹出时间
content	备用字段

（6）用户后台活跃的事件名称为 active_background，产生的日志数据的具体字段名称及字段描述如表 2-6 所示。

表 2-6 用户后台活跃 active_background

字段名称	字段描述
active_source	upgrade=1，download（下载）=2，plugin_upgrade=3

（7）用户评价的事件名称为 comment，产生的日志数据的具体字段名称及字段描述如表 2-7 所示。

表 2-7 用户评价 comment

字段名称	字段描述	字段类型	长度	允许空	默认值
comment_id	评价表	int	10,0		
userid	用户 id	int	10,0	√	0
p_comment_id	父级评价 id（为 0 则是一级评价，不为 0 则是回复）	int	10,0	√	
content	评价内容	string	1000	√	
add_time	创建时间	string		√	
other_id	评价的相关 id	int	10,0	√	
praise_count	点赞数量	int	10,0	√	0
reply_count	回复数量	int	10,0	√	0

（8）用户收藏的事件名称为 favorites，产生的日志数据的具体字段名称及字段描述如表 2-8 所示。

表 2-8 用户收藏 favorites

字段名称	字段描述	字段类型	长度	允许空	默认值
id	主键	int	10,0		
course_id	商品 id	int	10,0	√	0
userid	用户 id	int	10,0	√	0
add_time	创建时间	string		√	

（9）用户点赞的事件名称为 praise，产生的日志数据的具体字段名称及字段描述如表 2-9 所示。

表 2-9 用户点赞 praise

字段名称	字段描述	字段类型	长度	允许空	默认值
id	主键	int	10,0		

续表

字段名称	字段描述	字段类型	长度	允许空	默认值
userid	用户 id	int	10,0	√	
target_id	点赞对象的 id	int	10,0	√	
type	点赞类型：问答点赞=1，问答评价点赞=2，文章点赞=3，评价点赞=4	int	10,0	√	
add_time	创建时间	string		√	

（10）产生的错误日志数据的具体字段名称及字段描述如表 2-10 所示。

表 2-10　错误日志

字段名称	字段描述
errorBrief	错误摘要
errorDetail	错误详情

3．数据生成模块之启动日志数据

启动日志，产生的日志数据的具体字段名称及字段描述如表 2-11 所示。

表 2-11　启动日志

字段名称	字段描述
entry	入口： push=1，widget=2，icon=3，notification=4，lockscreen_widget=5
open_ad_type	开屏广告类型：开屏原生广告=1，开屏插屏广告=2
action	状态：成功=1，失败=2
loading_time	加载时长：从下拉开始到接口返回数据的时间（开始加载报 0，加载成功或加载失败上报具体时间）
detail	失败码（没有则上报空）
extend1	失败的 message（没有则上报空）
en	日志类型：en=start

4．数据采集模块之生产数据

通过运行生产日志的 jar 包来模拟数据生成的过程，得到日志数据。

5．数据采集模块

数据采集模块主要采集并落盘到服务器文件夹中的日志数据，需要监控多个日志产生文件夹并能够做到断点续传，实现数据消费"at least once"语义，以及能够根据采集到的日志内容对日志进行分类采集落盘，发往不同的 Kafka topic。Kafka 作为一个消息中间件起到日志缓冲作用，避免同时发生的大量读/写请求造成 HDFS 性能下降，能对 Kafka 的日志生产采集过程进行实时监控，避免消费层 Flume 在落盘 HDFS 过程中产生大量小数据文件，而降低 HDFS 运行性能，并对落盘数据采取适当压缩措施，尽量节省存储空间，降低网络 I/O。

业务数据采集要求按照业务数据库表结构在数据仓库中同步建表，并且根据业务数据库表性质指定对应的同步策略，进行合理的关系建模和维度建模。

2.2.2 数据仓库需求业务描述

1. 数据分层建模

数据仓库被分为 5 层，描述如下。

- ODS（Operation Data Store）层：原始数据层，存放原始数据，直接加载原始日志、数据，数据保持原貌不做处理。
- DWD（Data Warehouse Detail）层：明细数据层，结构和粒度与 ODS 层保持一致，对 ODS 层的数据进行清洗（去除空值、脏数据、超过极限范围的数据）。
- DWS（Data Warehouse Service）层：服务数据层，以 DWD 层的数据为基础，进行轻度汇总。一般聚集到以用户当日、设备当日、商家当日、商品当日等的粒度。在这层通常会以某一个维度为线索，组成跨主题的宽表，比如，由一个用户当日的签到数、收藏数、评价数、抽奖数、订阅数、点赞数、浏览商品数、添加购物车数、下单数、支付数、退款数及点击广告数组成的宽表。
- DWT（Data Warehouse Topic）层：主题数据层，按照主题对 DWS 层数据进行进一步聚合，构建每个主题的全量宽表。
- ADS（Application Data Store）层：数据应用层，也有人把这层称为 APP 层、DAL 层、DM 层等。面向实际的数据需求，以 DWD 层、DWS 层和 DWT 层的数据为基础，组成各种统计报表，统计结果最终同步到关系型数据库，如 MySQL，以供 BI 或应用系统查询使用。

读者需要按照命名规范合理建表。

2. 需求实现

电商业务发展日益成熟，但是如果缺少精细化运营的意识和数据驱动的经验，那么发展将会陷入瓶颈。作为电商数据分析的重要工具——数据仓库的作用就是为运营人员和决策团队提供关键指标的分析数据。电商平台的数据分析主要关注五大关键数据指标，包括活跃用户量、转化、留存、复购、GMV（指成交金额），以及三大关键思路：商品运营、用户运营和产品运营。围绕这一原则，本项目中要求实现的主要需求如下。

- 当日、当周、当月的活跃设备数；
- 每日新增设备数；
- 沉默设备数；
- 本周回流设备数；
- 流失设备数；
- 流存率；
- 最近连续三周活跃设备数；
- 最近七天内连续三天活跃设备数；
- 每日活跃会员数；
- 每日新增会员数；
- 每日新增付费会员数；
- 每日总付费会员数；

- 总会员数；
- 会员活跃率；
- 会员付费率；
- 会员新鲜度；
- 用户行为漏斗分析；
- 商品销量排名；
- 商品收藏排名；
- 商品加入购物车排名；
- 商品退款率排名；
- 商品差评率排名；
- 每日下单信息统计；
- 每日支付信息统计；
- 复购率。

要求将全部需求实现的结果数据存储在 ADS 层，并且完成可用于工作调度的脚本，实现任务自动调度。

2.2.3 数据可视化业务描述

在 MySQL 中根据 ADS 层的结果数据创建对应的表，使用 Sqoop 工具定时将结果数据导出到 MySQL 中，并使用数据可视化工具对数据进行展示。

2.3 系统运行环境

2.3.1 硬件环境

在实际生产环境中，我们需要进行服务器的选型，服务器是选择物理机还是云主机呢？

1．机器成本考虑

物理机，以 128GB 内存、20 核物理 CPU、40 线程、8TB HDD 和 2TB SSD 的戴尔品牌机为例，单台报价约 4 万元，并且还需要考虑托管服务器的费用，一般物理机寿命为 5 年左右。

云主机，以阿里云为例，与上述物理机的配置相似，每年的费用约 5 万元。

2．运维成本考虑

物理机需要由专业运维人员进行维护，云主机的运维工作由服务提供方完成，运维工作相对轻松。

实际上，服务器的选型除了参考上述条件，还应该根据数据量来确定集群规模。

在本项目中，读者可在个人计算机上搭建测试集群，建议将计算机配置为 16GB 内存、8 核物理 CPU、i7 处理器、1TB SSD。测试服务器规划如表 2-12 所示。

表 2-12 测试服务器规划

服务名称	子服务	节点服务器 hadoop102	节点服务器 hadoop103	节点服务器 hadoop104
HDFS	NameNode	√		
	DataNode	√	√	√
	SecondaryNameNode			√
YARN	NodeManager	√	√	√
	ResourceManager		√	
Zookeeper	Zookeeper Server	√	√	√
Flume（采集日志）	Flume	√	√	
Kafka	Kafka	√	√	√
Flume（消费 Kafka）	Flume			√
Hive	Hive	√		
MySQL	MySQL	√	√	√
Keepalived	Keepalived		√	√
Sqoop	Sqoop	√		
Superset	Superset	√		
Presto	Coordinator	√		
	Worker		√	√
Azkaban	AzkabanWebServer	√		
	AzkabanExecutorServer	√		
Druid	Druid	√	√	√
HBase	HRegionServer	√	√	√
Kylin	Kylin	√		
Solr	Solr	√	√	√
Atlas	Atlas	√		
服务数总计		18	12	12

2.3.2 软件环境

1. 技术选型

数据采集运输方面，在本项目中主要完成三个方面的需求：将服务器中的日志数据实时采集到大数据存储系统中，以防止数据丢失及数据堵塞；将业务数据库中的数据采集到数据仓库中；同时将需求计算结果导出到关系型数据库方便进行展示。为此我们选用了 Flume、Kafka 和 Sqoop。

Flume 是一个高可用、高可靠、分布式的海量数据收集系统，可从多种源数据系统采集、聚集和移动大量的数据并集中存储。Flume 提供了丰富多样的组件供用户使用，不同的组件可以自由组合，组合方式基于用户设置的配置文件，非常灵活，可以满足各种数据采集传输需求。

Kafka 是一个提供容错存储、高实时性的分布式消息队列平台。我们可以将它用在应用和处理系统间高实时性和高可靠性的流式数据存储中，也可以实时地为流式应用传送和反馈流式数据。

Sqoop 用于在关系型数据库（RDBMS）和 HDFS 之间传输数据，启用了一个 MapReduce 任务来执行数据采集任务，传输大量结构化或半结构化数据的过程是完全自动化的。其主要通过 JDBC 和关系型数据库进行交互，理论上支持 JDBC 的 Database 都可以使用 Sqoop 和 HDFS 进行数据交互。

数据存储方面，在本项目中主要完成对海量原始数据及转化后各层数据仓库中的数据的存储和对最终结果数据的存储。对海量原始数据的存储，我们选用了 HDFS。HDFS 是 Hadoop 的分布式文件系统，适合应用于大规模的数据集上，将大规模的数据集以分布式文件的方式存储于集群中的各台节点服务器上，提高文件存储的可靠性。对最终结果数据的存储，由于数据体量比较小，且为了方便访问，我们选用了 MySQL。

数据计算方面，我们选用配置了 Tez 运行引擎的 Hive。Hive 是基于 Hadoop 的数据仓库工具，可以将结构化的数据文件映射为一张数据库表，并提供 SQL 查询功能，将 SQL 语句转化为 MapReduce 任务进行运行，可以说在 Hadoop 之上提供了数据查询的功能，主要解决非关系型数据的查询问题。Tez 运行引擎可以将多个有依赖的作业转换为一个作业，这样可以减少中间计算过程产生的数据的落盘次数，从而大大提升作业的计算性能。

即席查询模块，我们对当前比较流行的三种即席查询都进行了探索实验，分别是 Presto、Druid 和 Kylin。三种即席查询各有千秋，Presto 基于内存计算，Druid 是优秀的时序数据处理引擎，Kylin 基于预 Cube 创建计算。

面对海量数据的处理，对元数据的管理会随着数据体量的增大而显得尤为重要。为寻求数据治理的开源解决方案，Hortonworks 公司联合其他厂商与用户于 2015 年发起数据治理倡议，包括数据分类、集中策略引擎、数据血缘、安全、生命周期管理等方面。Apache Atlas 项目就是这个倡议的结果，社区伙伴持续地为该项目提供新的功能和特性。该项目用于管理共享元数据、数据分级、审计、安全性、数据保护等方面。

总结如下。

- 数据采集与传输：Flume、Kafka、Sqoop。
- 数据存储：MySQL、HDFS。
- 数据计算：Hive、Tez。
- 任务调度：Azkaban。
- 即席查询：Presto、Druid、Kylin。
- 元数据管理：Atlas。

2. 框架选型

框架版本的选型要求满足数据仓库平台的几大核心需求：子功能不设局限、国内外资料及社区尽量丰富、组件服务的成熟度和流行度较高。待选择版本如下。

- Apache：运维过程烦琐，组件间的兼容性需要自己调研（本次选用）。
- CDH：国内使用较多，不开源，不用担心组件兼容问题。
- HDP：开源，但没有CDH稳定，使用较少。

笔者经过考量决定选择 Apache 原生版本大数据框架，一方面可以自由定制所需功能组件；另一方面 CDH 和 HDP 版本框架体量较大，对服务器配置要求相对较高。本项目中用到的组件较少，Apache 原生版本即可满足需要。

笔者经过对版本兼容性的调研，确定的版本选型如表 2-13 所示。

表2-13 版本选型

产　品	版　本
Hadoop	2.7.2
Flume	1.7.0
Kafka	0.11.0.2
Hive	2.3.1
Sqoop	1.4.6
MySQL	5.6.24
Azkaban	2.5.0
Java	1.8
Zookeeper	3.4.10
Presto	0.196
Druid	2.7.10
HBase	1.3.1
Kylin	2.5.1
Solr	5.2.1
Atlas	0.8.4

2.4　本章总结

本章主要对本书的项目需求进行了介绍，首先介绍了本项目即将搭建的数据仓库产品需要实现的系统目标、系统功能结构和系统流程图；然后对各主要功能模块进行了重点描述，并对每个模块的重点需求进行了介绍；最后根据项目的整体需求对系统运行的硬件环境和软件环境进行了配置选型。

第3章

项目部署的环境准备

通过上一章的分析，我们已经明确了将要使用的框架类型和实现方式，本章将根据上一章的需求分析，搭建一个完整的项目开发环境，即便读者的计算机中已经具备这些环境，也建议浏览一遍本章内容，因为其对后续开发过程中代码和命令行的理解很有帮助。

3.1 Linux 环境准备

3.1.1 VMware 安装

本节介绍的虚拟机软件是 VMware，VMware 可以使用户在一台计算机上同时运行多个操作系统，还可以像 Windows 应用程序一样来回切换。用户可以如同操作真实安装的系统一样操作虚拟机系统，甚至可以在一台计算机上将几个虚拟机系统连接为一个局域网或者连接到互联网。

在虚拟机系统中，每台虚拟产生的计算机都被称为"虚拟机"，而用来存储所有虚拟机的计算机则被称为"宿主机"。使用 VMware 虚拟机软件安装虚拟机可以减少因安装新系统导致的数据丢失问题，还可以使用户方便地体验各种系统，以进行学习和测试。

VMware 支持多种平台，可以安装在 Windows、Linux 等操作系统上，初学者大多使用 Windows，可下载 VMware Workstation for Windows 版本。VMware 的安装非常简单，与其他 Windows 软件类似，本书不进行详细讲解。值得一提的是，在安装过程中安装的类型包括典型安装或自定义安装，笔者建议初学者选择"典型"安装。

VMware 安装完成启动后，即可进行 Linux 的安装部署。

推荐使用版本：VMware Workstation Pro 或 VMware Workstation Player。其中，Player 版本供个人用户使用，非商业用途，是免费的，其他的 VMware 版本在此不进行过多介绍。

3.1.2 CentOS 安装

在安装 CentOS 之前，用户需要检查本机 BIOS 是否支持虚拟化，开机后进入 BIOS 界面，不同计算机进入 BIOS 界面的操作有所不同，然后进入 Security 下的 Virtualization，选择 Enable 即可。

启动 VMware，进入主界面，依次进行新虚拟机的设置，然后选择配置类型，如图 3-1 所示。

单击"下一步"按钮，进入"安装客户机操作系统"界面，选择"稍后安装操作系统"选项，如图 3-2 所示。

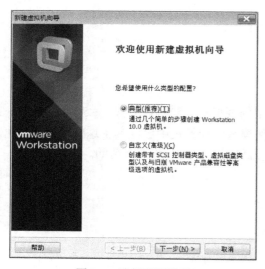 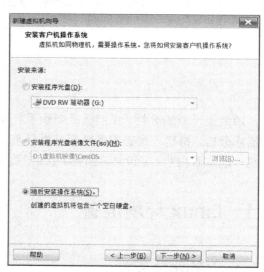

图 3-1　选择配置类型　　　　　　　　　图 3-2　安装客户机操作系统

单击"下一步"按钮，进入"选择客户机操作系统"界面，选择"Linux"选项，然后在"版本"下拉列表中选择要安装的对应的 Linux 版本，此处选择"CentOS"选项，如图 3-3 所示。

图 3-3　选择客户机操作系统

单击"下一步"，进入"命名虚拟机"界面，给虚拟机起一个名字，如"CentOS 6.3"或"PlayBoy"，然后单击"浏览"按钮，选择虚拟机系统安装文件的保存位置，如图 3-4 所示。

单击"下一步"按钮，进入"指定磁盘容量"界面。默认虚拟的最大磁盘大小为20GB（虚拟出来的磁盘会以文件形式存放在虚拟机系统安装目录中），如图3-5所示。

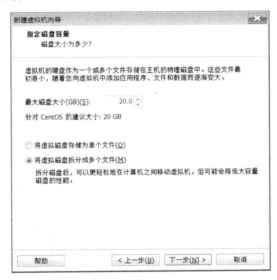

图3-4　命名虚拟机　　　　　　　　　　图3-5　指定磁盘容量

单击"下一步"按钮，进入"已准备好创建虚拟机"界面，确认虚拟机设置，若无须改动，则单击"完成"按钮，开始生成虚拟机，如图3-6所示。

图3-6　准备创建虚拟机

我们可以略做调整，单击"自定义硬件"按钮，打开"硬件"对话框。为使虚拟机中的系统运行速度快一点，我们可以选择"内存"选项来调整虚拟机内存大小，建议调整为4GB，但是虚拟机内存不要超过宿主机内存的一半。CentOS 6.x最少需要628MB的内存，否则会开启简易安装过程，如图3-7所示。

图 3-7 硬件调整

选择"新 CD/DVD(IDE)"选项,可以进行光盘配置。如果选择"使用物理驱动器"选项,则虚拟机会使用宿主机的物理光盘,如果选择"使用 ISO 映像文件"选项,则可以直接加载 ISO 映像文件,单击"浏览"按钮找到 ISO 映像文件的位置即可,如图 3-8 所示。

图 3-8 光盘配置

单击"关闭"按钮即可。如果还想调整虚拟机的硬件配置,则可以选择"虚拟机"下拉菜单中的"设置"命令,重新进入"硬件"对话框,如图 3-9 所示。

第 3 章　项目部署的环境准备

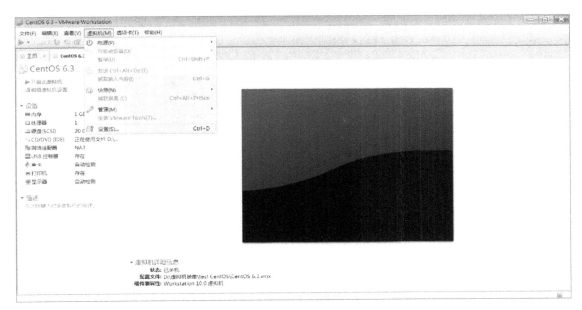

图 3-9　调整虚拟机的硬件配置

选择"电源"→"打开此虚拟机电源"选项，开启虚拟机，就能看到 CentOS6 的安装欢迎界面了，如图 3-10 所示。

图 3-10　CentOS6 安装欢迎页面

选择"Install or upgrade an existing system"选项，安装一个全新的系统。进入安装环境后，向导首先会询问是否检测安装介质的完整性，如图 3-11 所示。这是为了避免因为安装来源不正确，造成无法顺利安装而产生损失，一般情况下，如果下载过程中没有出现问题，则无须检测（检测时间较久），直接单击"Skip"按钮跳过即可。

注意：在虚拟机和宿主机之间，鼠标是不能同时起作用的，如果从宿主机进入虚拟机，则需要把鼠标指针移入虚拟机；如果从虚拟机返回宿主机，则按 Ctrl+Alt 组合键退出。

25

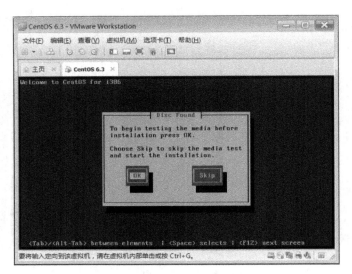

图 3-11　检测安装介质

进入 CentOS 6.3 欢迎界面，单击"Next"按钮，进入选择安装系统的默认语言界面，可以根据需要自行选择，比如，选择"中文（简体）"。选择完成后，单击"Next"按钮，进入键盘布局界面，选择默认的美国式键盘。

单击"下一步"按钮，进入存储设备选择界面，选择"基本存储设备"选项，会弹出存储设备警告。警告安装操作会导致存储设备中的数据丢失，然后单击"是，忽略所有数据"按钮，如图 3-12 所示。

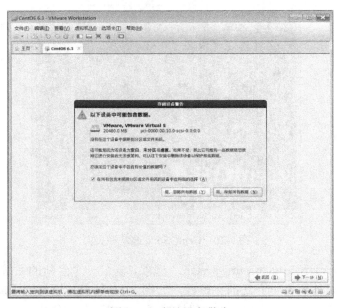

图 3-12　存储设备警告

单击"下一步"按钮，进入主机名配置界面，默认主机名是"localhost.localdomain"，可以自行更改，如图 3-13 所示。在此界面中还可以配置网络，用户也可以在安装完成后执行 setup 或 ifconfig 命令进行网络配置，这里略过。

图 3-13 配置主机名

单击"下一步"按钮,进入时区选择界面,如果住在中国,则选择"亚洲/上海"选项就可以了,建议不勾选"系统时钟使用 UTC 时间"复选框。单击"下一步"按钮,设置管理员密码("根密码"指的是管理员密码,在 Linux 中管理员的名称为"root",翻译为"根用户")。用于学习的系统,密码设置简单是可以接受的,如"123456",但可能会出现如图 3-14 所示[①]的"脆弱密码"提示,单击"无论如何都使用"按钮,依然可以让脆弱密码生效。

图 3-14 设置管理员密码

---

① 图 3-14 中"帐号"的正确写法应为"账号"。

单击"下一步"按钮,进入安装 Linux 中最重要的部分:硬盘分区。在此,笔者推荐选择"创建自定义布局"类型,如图 3-15 所示[①]。

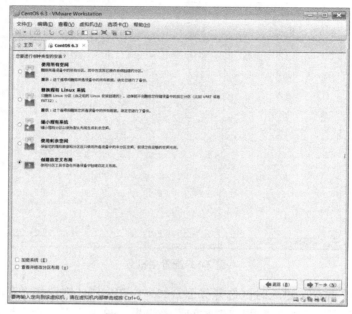

图 3-15 选择硬盘分区类型

单击"下一步"按钮,进入硬盘分区操作界面,如图 3-16 所示。

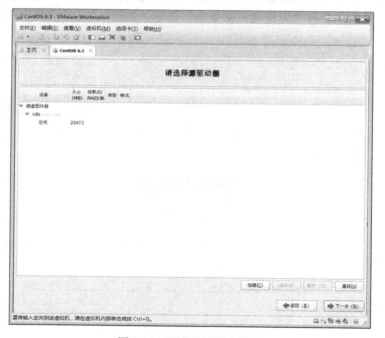

图 3-16 硬盘分区操作界面

---

① 图 3-15 中"其它"的正确写法应为"其他"。

单击"创建"按钮，生成分区，如图 3-17 所示。

图 3-17 生成分区

单击"创建"按钮，进入"添加分区"界面，如图 3-18、图 3-19 和图 3-20 所示。在此界面，我们可以创建/boot 分区、/分区、/home 分区、swap 分区等。

注意：swap 分区是在"文件系统类型"下拉列表中选择的，而不是在"挂载点"下拉列表中选择的。

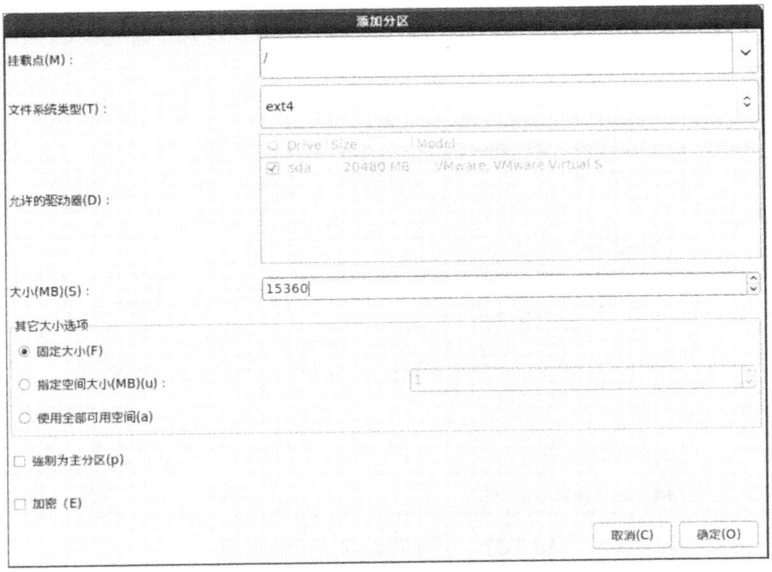

图 3-18 /分区创建

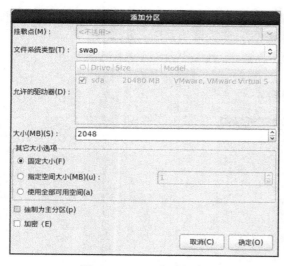

图 3-19 swap 分区创建

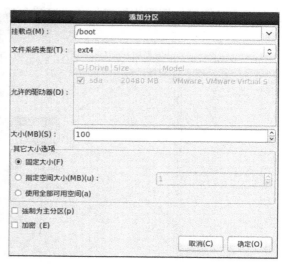

图 3-20 /boot 分区创建

分区创建完成后，单击"确定"按钮，出现格式化警告，单击"格式化"按钮，进入引导装载程序安装界面，如图 3-21 所示。

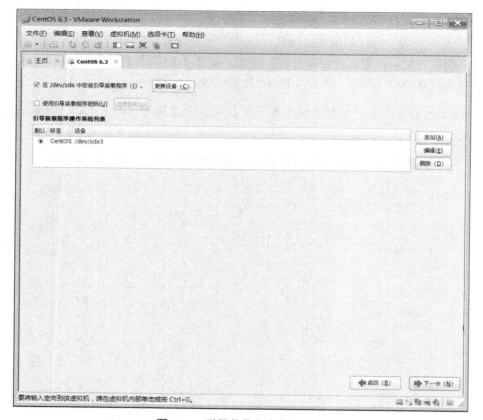

图 3-21 引导装载程序安装界面

单击"下一步"按钮，在弹出的界面中选择"Desktop"选项，并选择"现在自定义"选项，定制系统软件，如图 3-22 所示。

图 3-22　定制系统软件

单击"下一步"按钮，进入系统服务自定义选择界面，建议基本系统部分选择"兼容程序库"和"基本"，应用程序选择"互联网浏览器"，桌面除"KDE 桌面"外全部勾选，语言支持选择"中文支持"，其余部分全部不勾选。

完成配置后，开始安装 CentOS，会等待一段时间，屏幕显示目前安装的软件包及其简介、预估剩余时间以及安装的进度。安装完成后，单击"重新引导"按钮，重启后就可以进入登录界面了。还记得 Linux 的根用户是 root 吗？还记得安装时输入的 root 密码吗？输入正确的用户名和密码就可以登录系统了。

3.1.3　远程终端安装

大多数服务器的日常管理操作，都是通过远程管理工具进行的。常见的远程管理方法包括如 VNC 的图形远程管理、如 Webmin 的基于浏览器的远程管理，不过常用的还是命令行操作。在 Linux 中远程管理使用的是 SSH 协议，本节先介绍两个远程管理工具的使用方法。

1．PuTTY

PuTTY 是一个完全免费的 Windows 远程管理客户端工具，体积小，操作简单，是绿色软件，无须安装，下载后即可使用。对经常到客户公司提供技术支持和维护的用户，相当方便，只要随身带一个 U 盘，即可随处登录。

下载 PuTTY 后双击 putty.exe，弹出如图 3-23 所示的"PuTTY 配置"对话框。

在"主机名称（或 IP 地址）"文本框中输入远程登录主机的 IP 地址，如 192.168.44.8，"端口"根据使用的协议有所区别（选择不同的"连接类型"选项，端口会自动变化，建议选择"SSH"选项）。在"保存的会话"文本框中输入一个名称，单击"保存"按钮即可把本次的连接配置保存起来。设置完成后单击"打开"按钮，即可出现如图 3-24 所示的操作界面。

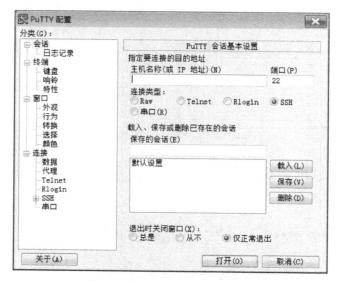

图 3-23　"PuTTY 配置"对话框

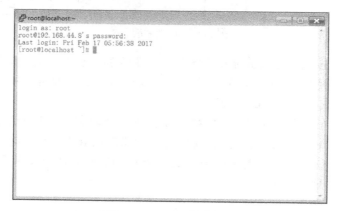

图 3-24　PuTTY 操作界面

2. SecureCRT

SecureCRT 将 SSH（Secure Shell）的安全登录、数据传送性能和 Windows 终端仿真提供的可靠性、可用性、易配置性结合在一起。如果需要管理多台服务器，使用 SecureCRT 可以很方便地记住多个地址，并且还可以通过配置设置自动登录，方便远程管理，效率很高。缺点是 SecureCRT 需要安装，并且是一款共享软件，不付费注册则不能使用。

安装 SecureCRT 并启动后，单击"快速连接"按钮，弹出"快速连接"对话框，如图 3-25 所示，输入"主机名"和"用户名"，单击"连接"按钮，然后按照提示输入密码即可登录。

SecureCRT 默认不支持中文，中文会显示为乱码，解决方法如下。

建立连接后，选择"选项"→"会话选项"命令，在弹出的对话框左侧列表中选择"终端"→"仿真"选项，在右侧"终端"下拉列表中选择"Xterm"选项，勾选"ANSI 颜色"复选框，以支持颜色显示，单击"确定"按钮，如图 3-26 所示[①]。

---

① 图 3-26 中"登陆动作"的正确写法为"登录动作"。

图 3-25 "快速连接"对话框

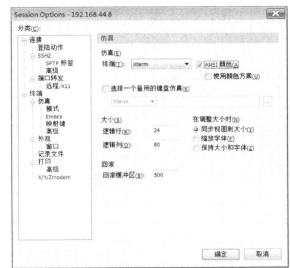

图 3-26 SecureCRT 仿真设置

在左侧列表中选择"终端"→"外观"选项,在右侧"当前颜色方案"下拉列表中选择"Traditional"选项,在"标准字体"和"精确字体"中均选择"新宋体 11pt",并确保"字符编码"选择为"UTF-8"(CentOS 默认使用中文字符集 UTF-8),取消勾选"使用 Unicode 线条绘制字符"复选框,单击"确定"按钮即可,如图 3-27 所示。

图 3-27 SecureCRT 窗口和文本外观设置

至此,我们就搭建好了初步的学习实验环境。

3.2 Linux 环境配置

3.2.1 网络配置

对安装好的 VMware 进行网络配置，方便虚拟机连接网络，本次设置建议选择 NAT（网络地址转换）模式，需要宿主机的 Windows 和虚拟机的 Linux 能够进行网络连接，同时虚拟机的 Linux 可以通过宿主机的 Windows 进入互联网。

选择"编辑"→"虚拟网络编辑器"命令，如图 3-28 所示，对虚拟机进行网络配置。

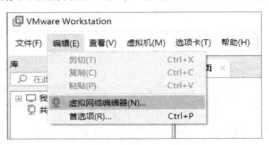

图 3-28 "虚拟网络编辑器"命令

在打开的"虚拟网络编辑器"对话框中，选择 NAT 模式，并修改虚拟机的子网 IP 地址，如图 3-29 所示。

图 3-29 选择 NAT 模式并修改虚拟机的子网 IP 地址

单击"NAT 设置"按钮，在打开的"NAT 设置"对话框中，查看网关设置，如图 3-30 所示。

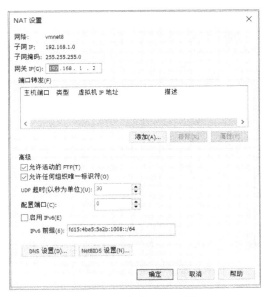

图 3-30　查看网关设置

查看 Windows 环境中的 vmnet8 网络配置，如图 3-31 所示，查看路径为"控制面板"→"网络和 Internet"→"网络连接"。

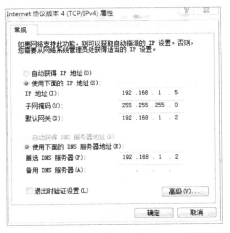

图 3-31　Windows 环境中的 vmnet8 网络配置

3.2.2　网络 IP 地址配置

修改网络 IP 地址为静态 IP 地址，避免 IP 地址经常变化，从而方便节点服务器间的互相通信。

```
[root@hadoop100 桌面]#vim /etc/sysconfig/network-scripts/ifcfg-eth0
```

以下加粗的项必须修改，有值的按照下面的值修改，没有该项的则需要增加。

```
DEVICE=eth0                    #接口名（设备，网卡）
HWADDR=00:0C:2x:6x:0x:xx       #物理 IP 地址
```

```
TYPE=Ethernet                            #网络类型（通常是Ethernet）
UUID=926a57ba-92c6-4231-bacb-f27e5e6a9f44    #随机id
#系统启动的时候网络接口是否有效（yes/no）
ONBOOT=yes
#IP地址的配置方法[none（引导时不使用协议）|static（静态分配IP地址）|bootp（BOOTP协议）|
#dhcp（DHCP协议）]
BOOTPROTO=static
#IP地址
IPADDR=192.168.1.101
#网关
GATEWAY=192.168.1.2
#域名解析器
DNS1=192.168.1.2
```

修改 IP 地址后的结果如图 3-32 所示，执行 ":wq" 命令，保存退出。

图 3-32　修改 IP 地址后的结果

执行 service network restart 命令，重启网络服务，如图 3-33 所示。

```
关闭环回接口：                                                    [确定]
弹出环回接口：                                                    [确定]
弹出界面 eth0：错误：激活连接失败：The connection is not for this device.
                                                                [失败]
```

图 3-33　重启网络服务

如果报错，则执行 "reboot" 命令，重启虚拟机。

3.2.3　主机名配置

修改主机名为一系列有规律的主机名，并修改 hosts 文件添加我们需要的主机名和 IP 地址映射，以便方便管理且方便节点服务器间通过主机名进行通信。

1. 修改 Linux 的主机映射文件（hosts 文件）

（1）进入 Linux 查看本机的主机名。执行 hostname 命令进行查看。

```
[root@hadoop100 桌面]# hostname
hadoop100
```

（2）如果感觉此主机名不合适，则可以进行修改。通过编辑/etc/sysconfig/network 文件进行修改。

```
[root@hadoop100 桌面]# vim /etc/sysconfig/network
NETWORKING=yes
```

```
NETWORKING_IPV6=no
HOSTNAME= hadoop100
```

注意：主机名不要有"_"（下画线）。

（3）打开/etc/sysconfig/network 文件后，可以看到主机名，在此处可以完成对主机名的修改，本例不做修改，仍为 hadoop100。

（4）保存并退出。

（5）打开/etc/hosts 文件。

```
[root@hadoop100 桌面]# vim /etc/hosts
```

添加如下内容。

```
192.168.1.100 hadoop100
192.168.1.101 hadoop101
192.168.1.102 hadoop102
192.168.1.103 hadoop103
192.168.1.104 hadoop104
192.168.1.105 hadoop105
192.168.1.106 hadoop106
192.168.1.107 hadoop107
192.168.1.108 hadoop108
```

（6）重启设备，查看主机名，可以看到已经修改成功。

2．修改 Windows 的主机映射文件（hosts 文件）

（1）进入 C:\Windows\System32\drivers\etc 路径。

（2）复制 hosts 文件到桌面上。

（3）打开桌面上的 hosts 文件并添加如下内容。

```
192.168.1.100 hadoop100
192.168.1.101 hadoop101
192.168.1.102 hadoop102
192.168.1.103 hadoop103
192.168.1.104 hadoop104
192.168.1.105 hadoop105
192.168.1.106 hadoop106
192.168.1.107 hadoop107
192.168.1.108 hadoop108
```

（4）用桌面上的 hosts 文件覆盖 C:\Windows\System32\drivers\etc 路径中的 hosts 文件。

3.2.4 防火墙配置

为了使 Windows 或其他系统可以访问 Linux 虚拟机内的服务，我们有时候需要关闭虚拟机的防火墙服务，以下是常见的防火墙启动/关闭命令。

1. 临时关闭防火墙

（1）查看防火墙状态。

```
[root@hadoop100 桌面]# service iptables status
```

（2）临时关闭防火墙。

```
[root@hadoop100 桌面]# service iptables stop
```

2. 开机启动时关闭防火墙

（1）查看开机启动时防火墙状态。

```
[root@hadoop100 桌面]#chkconfig iptables --list
```

（2）设置开机时关闭防火墙。

```
[root@hadoop100 桌面]#chkconfig iptables off
```

3.2.5 一般用户设置

root 用户具有太大的操作权限，而在实际操作中又需要对用户有所限制，所以我们需要创建一般用户。

（1）创建 atguigu 用户。

（2）配置 atguigu 用户具有 root 权限，接下来的所有操作都将在一般用户身份下完成。

① 添加 atguigu 用户，并对其设置密码。

```
[root@hadoop100 ~]#useradd atguigu
[root@hadoop100 ~]#passwd atguigu
```

② 修改配置文件。

```
[root@hadoop100 ~]#vim /etc/sudoers
```

修改/etc/sudoers 文件，找到第 91 行，在 root 下面添加一行。

```
## Allow root to run any commands anywhere
root      ALL=(ALL)     ALL
atguigu   ALL=(ALL)     ALL
```

或者配置成执行 sudo 命令时，不需要输入密码。

```
## Allow root to run any commands anywhere
root      ALL=(ALL)     ALL
atguigu   ALL=(ALL)     NOPASSWD:ALL
```

修改完毕后，用户使用 atguigu 账号或执行 sudo 命令进行登录，即可获得 root 操作权限。

3.3 Hadoop 环境搭建

在搭建完 Linux 环境之后，我们正式开始搭建 Hadoop 分布式集群环境。

3.3.1 虚拟机环境准备

1. 克隆虚拟机

关闭要被克隆的虚拟机，右击虚拟机名称，在弹出的快捷菜单中选择"管理"→"克隆"命令，如图 3-34 所示。

图 3-34　开始克隆

在欢迎界面单击"下一步"按钮，打开"克隆虚拟机向导"对话框，选择"虚拟机中的当前状态"选项，克隆虚拟机，如图 3-35 所示。

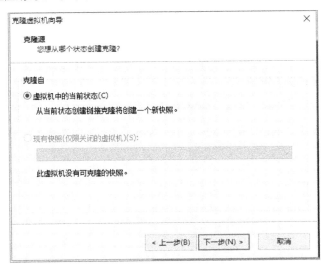

图 3-35　克隆虚拟机

设置"克隆方法"为"创建完整克隆",如图 3-36 所示。

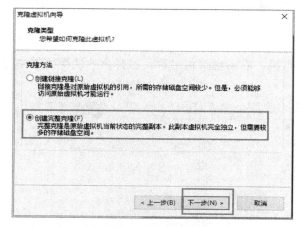

图 3-36 设置"克隆方法"为"创建完整克隆"

设置克隆的"虚拟机名称"和"位置",如图 3-37 所示。

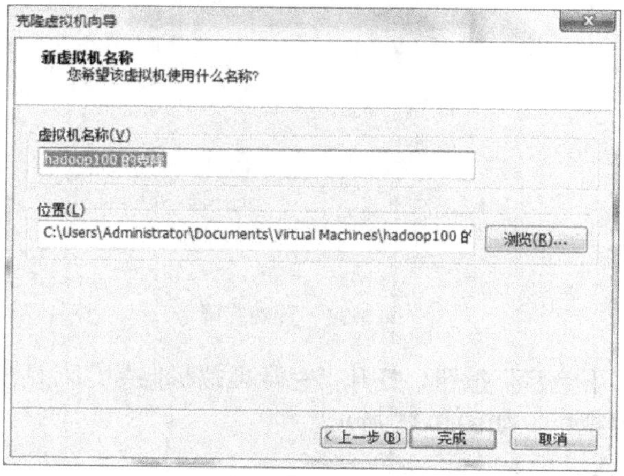

图 3-37 设置克隆的"虚拟机名称"和"位置"

单击"完成"按钮,开始克隆,需要等待一段时间,克隆完成后,单击"关闭"按钮。修改克隆后的虚拟机的 IP 地址。

```
[root@hadoop110 /]#vim /etc/udev/rules.d/70-persistent-net.rules
```

进入如下页面,删除"eth0"所在的行,将"eth1"修改为"eth0",同时复制物理 IP 地址,如图 3-38 所示。

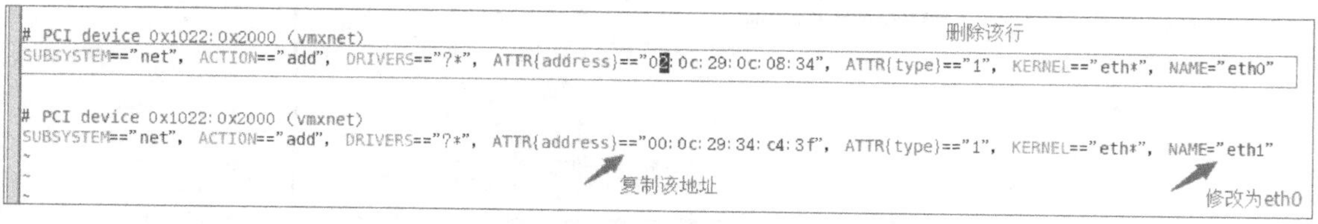

图 3-38 修改网卡

修改 eth0 网卡中的物理 IP 地址。
```
[root@hadoop110 /]#vim /etc/sysconfig/network-scripts/ifcfg-eth0
```
把复制的物理 IP 地址进行更新。
```
HWADDR=00:0c:29:34:c4:3f       #物理IP地址
```
修改为想要设置的 IP 地址。
```
IPADDR=192.168.1.102           #IP地址
```

按照 3.2.3 节中主机名的配置方法修改主机名。

重新启动服务器，按照上述操作分别克隆 3 台虚拟机，命名为 hadoop102、hadoop103、hadoop104，主机名和 IP 地址分别与 3.2.3 节中的 hosts 文件设置一一对应。

2．创建安装目录

（1）在/opt 目录下创建 module、software 文件夹。
```
[atguigu@hadoop102 opt]$ sudo mkdir module
[atguigu@hadoop102 opt]$ sudo mkdir software
```
（2）修改 module、software 文件夹的所有者。
```
[atguigu@hadoop102 opt]$ sudo chown atguigu:atguigu module/ software/
[atguigu@hadoop102 opt]$ ll
总用量 8
drwxr-xr-x. 2 atguigu atguigu 4096 1月  17 14:37 module
drwxr-xr-x. 2 atguigu atguigu 4096 1月  17 14:38 software
```
之后，所有的软件安装操作将在 module 和 software 文件夹中进行。

3．配置三台虚拟机免密登录

为什么需要配置免密登录呢？这与 Hadoop 分布式集群的架构有关。我们搭建的 Hadoop 分布式集群是"主从架构"，配置了节点服务器间免密登录之后，就可以方便地通过主节点服务器启动从节点服务器，而不用手动输入用户名和密码。

第一步：配置 SSH。

（1）基本语法：假设要以用户名 user 登录远程主机 host，只需要输入 ssh user@host，如 ssh atguigu@192.168.1.100，若本地用户名与远程用户名一致，登录时则可以省略用户名，如 ssh host。

（2）SSH 连接时出现"Host key verification failed"的错误提示，直接输入 yes 即可。
```
[atguigu@hadoop102 opt] $ ssh 192.168.1.103
The authenticity of host '192.168.1.103 (192.168.1.103)' can't be established.
RSA key fingerprint is cf:1e:de:d7:d0:4c:2d:98:60:b4:fd:ae:b1:2d:ad:06.
Are you sure you want to continue connecting (yes/no)?
Host key verification failed.
```

第二步：无密钥配置。

（1）免密登录原理如图 3-39 所示。

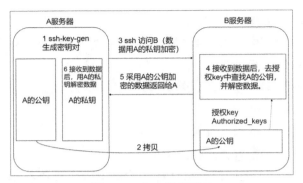

图 3-39　免密登录原理

（2）生成公钥和私钥。

```
[atguigu@hadoop102 .ssh]$ ssh-key-gen -t rsa
```

连续按三次 Enter 键，就会生成两个文件：id_rsa（私钥）、id_rsa.pub（公钥）。

（3）将公钥复制到要免密登录的目标服务器上。

```
[atguigu@hadoop102 .ssh]$ ssh-copy-id hadoop102
[atguigu@hadoop102 .ssh]$ ssh-copy-id hadoop103
[atguigu@hadoop102 .ssh]$ ssh-copy-id hadoop104
```

.ssh 文件夹下的文件功能解释如下。

- known_hosts ：记录 SSH 访问过计算机的公钥。
- id_rsa ：生成的私钥。
- id_rsa.pub ：生成的公钥。
- authorized_keys ：存放授权过的免密登录服务器公钥。

4．配置时间同步

为什么要配置节点服务器间的时间同步呢？

即将搭建的 Hadoop 分布式集群需要解决两个问题：数据的存储和数据的计算。

Hadoop 对大型文件的存储采用分块的方法，将文件切分成多块，以块为单位，分发到各台节点服务器上进行存储。当这个大型文件再次被访问到的时候，需要从 3 台节点服务器上分别拿出数据，然后进行计算。由于计算机之间的通信和数据的传输一般是以时间为约定条件的，如果 3 台节点服务器的时间不一致，就会导致在读取块数据的时候出现时间延迟，可能会导致访问文件时间过长，甚至失败，所以配置节点服务器间的时间同步非常重要。

第一步：配置时间服务器（必须是 root 用户）。

（1）检查计算机中是否安装了 ntp。

```
[root@hadoop102 桌面]# rpm -qa|grep ntp
ntp-4.2.6p5-10.el6.centos.x86_64
fontpackages-filesystem-1.41-1.1.el6.noarch
ntpdate-4.2.6p5-10.el6.centos.x86_64
```

（2）修改 ntp 配置文件。

```
[root@hadoop102 桌面]# vim /etc/ntp.conf
```

修改内容如下。

① 修改 1（设置本地网络上的主机不受限制），将以下配置前的#删除，解开此行注释。

```
#restrict 192.168.1.0 mask 255.255.255.0 nomodify notrap
```

② 修改 2（设置为不采用公共的服务器）。

```
server 0.centos.pool.ntp.org iburst
server 1.centos.pool.ntp.org iburst
server 2.centos.pool.ntp.org iburst
server 3.centos.pool.ntp.org iburst
```

将上述内容修改为：

```
#server 0.centos.pool.ntp.org iburst
#server 1.centos.pool.ntp.org iburst
#server 2.centos.pool.ntp.org iburst
#server 3.centos.pool.ntp.org iburst
```

③ 修改 3（添加一个默认的内部时钟数据，使用它为局域网用户提供服务）。

```
server 127.127.1.0
fudge 127.127.1.0 stratum 10
```

（3）修改/etc/sysconfig/ntpd 文件。

```
[root@hadoop102 桌面]# vim /etc/sysconfig/ntpd
```

增加如下内容（让硬件时间与系统时间一起同步）。

```
SYNC_HWCLOCK=yes
```

重新启动 ntpd 文件。

```
[root@hadoop102 桌面]# service ntpd status
ntpd 已停
[root@hadoop102 桌面]# service ntpd start
正在启动 ntpd:                                              [确定]
```

执行：

```
[root@hadoop102 桌面]# chkconfig ntpd on
```

第二步：配置其他服务器（必须是 root 用户）。

配置其他服务器 10 分钟与时间服务器同步一次。

```
[root@hadoop103 hadoop-2.7.2]# crontab -e
```

编写脚本。

```
*/10 * * * * /usr/sbin/ntpdate hadoop102
```

修改 hadoop103 的节点服务器时间，使其与另外两台节点服务器时间不同步。

```
[root@hadoop103 hadoop]# date -s "2017-9-11 11:11:11"
```

10 分钟后查看该服务器是否与时间服务器同步。

```
[root@hadoop103 hadoop]# date
```

5. 编写集群分发脚本

集群间数据的复制通用的两个命令是 scp 和 rsync，其中，rsync 命令可以只对差异文件进行更新，非常方便，但是使用时需要操作者频繁输入各种命令参数，为了能够更方便地使用该命令，我们编写一个集群分发脚本，主要实现目前集群间的数据分发。

第一步：脚本需求分析。循环复制文件到所有节点服务器的相同目录下。

（1）原始复制。

```
rsync -rv /opt/module root@hadoop103:/opt/
```

（2）期望脚本效果。

```
xsync path/filename #要同步的文件路径或文件名
```

（3）在/home/atguigu/bin 目录下存放的脚本，atguigu 用户可以在系统任何地方直接执行。

第二步：脚本实现。

（1）在/home/atguigu 目录下创建 bin 目录，并在 bin 目录下使用 vim 命令创建文件 xsync，文件内容如下。

```
[atguigu@hadoop102 ~]$ mkdir bin
[atguigu@hadoop102 ~]$ cd bin/
[atguigu@hadoop102 bin]$ touch xsync
[atguigu@hadoop102 bin]$ vim xsync
#!/bin/bash
#获取输入参数个数，如果没有参数，则直接退出
pcount=$#
if((pcount==0)); then
echo no args;
exit;
fi

#获取文件名称
p1=$1
fname=`basename $p1`
echo fname=$fname

#获取上级目录到绝对路径
pdir=`cd -P $(dirname $p1); pwd`
echo pdir=$pdir

#获取当前用户名称
user=`whoami`

#循环
for((host=103; host<105; host++)); do
        echo ---------------------- hadoop$host ------------------
        rsync -rvl $pdir/$fname $user@hadoop$host:$pdir
done
```

（2）修改脚本 xsync，使其具有执行权限。

```
[atguigu@hadoop102 bin]$ chmod 777 xsync
```

（3）调用脚本的形式：xsync 文件名称。

```
[atguigu@hadoop102 bin]$ xsync /home/atguigu/bin
```

3.3.2 JDK 安装

JDK 是 Java 的开发工具箱，是整个 Java 的核心，包括 Java 运行环境、Java 工具和 Java 基础类库，JDK 是学习大数据的基础工具之一。即将搭建的 Hadoop 分布式集群的安装程序就是用 Java 开发的，所有 Hadoop 分布式集群想要正常运行，必须安装 JDK。

（1）在 3 台虚拟机上分别卸载现有的 JDK。

① 检查计算机中是否已安装 Java 软件。

```
[atguigu@hadoop102 opt]$ rpm -qa | grep java
```

② 如果安装的版本低于 1.7，则卸载该 JDK。

```
[atguigu@hadoop102 opt]$ sudo rpm -e 具体软件包名
```

（2）将 JDK 导入 opt 目录下的 software 文件夹中。

① 在 Linux 下的 opt 目录中查看软件包是否导入成功。

```
[atguigu@hadoop102 opt]$ cd software/
[atguigu@hadoop102 software]$ ls
hadoop-2.7.2.tar.gz  jdk-8u144-linux-x64.tar.gz
```

② 解压 JDK 到/opt/module 目录下，tar 命令用来解压.tar 或者.tar.gz 格式的压缩包，通过-z 选项指定解压.tar.gz 格式的压缩包。-f 选项用于指定解压文件，-x 选项用于指定解包操作，-v 选项用于显示解压过程，-C 选项用于指定解压路径。

```
[atguigu@hadoop102 software]$ tar -zxvf jdk-8u144-linux-x64.tar.gz -C /opt/module/
```

（3）配置 JDK 环境变量，方便使用到 JDK 的程序能正常调用 JDK。

① 先获取 JDK 路径。

```
[atgui@hadoop102 jdk1.8.0_144]$ pwd
/opt/module/jdk1.8.0_144
```

② 打开/etc/profile 文件，需要注意的是，/etc/profile 文件属于 root 用户，需要使用 sudo vim 命令才可以对它进行编辑。

```
[atguigu@hadoop102 software]$ sudo vim /etc/profile
```

在 profile 文件末尾添加 JDK 路径，添加的内容如下。

```
#JAVA_HOME
export JAVA_HOME=/opt/module/jdk1.8.0_144
export PATH=$PATH:$JAVA_HOME/bin
```

保存后退出。

```
:wq
```

③ 修改/etc/profile 文件后，需要执行 source 命令使修改后的文件生效。

```
[atguigu@hadoop102 jdk1.8.0_144]$ source /etc/profile
```

（4）通过执行 java -version 命令，测试 JDK 是否安装成功。

```
[atguigu@hadoop102 jdk1.8.0_144]# java -version
java version "1.8.0_144"
```

重启（如果执行 java -version 命令可以正常查看 Java 版本，说明 JDK 安装成功，则不用重启）。

```
[atguigu@hadoop102 jdk1.8.0_144]$ sync
[atguigu@hadoop102 jdk1.8.0_144]$ sudo reboot
```

（5）分发 JDK 给所有节点服务器。

```
[atguigu@hadoop102 jdk1.8.0_144]$ xsync /opt/module/jdk1.8.0_144
```

（6）分发环境变量。

```
[atguigu@hadoop102 jdk1.8.0_144]$ xsync /etc/profile
```

（7）执行 source 命令，使环境变量在每台虚拟机上生效。

```
[atguigu@hadoop103 jdk1.8.0_144]$ source /etc/profile
[atguigu@hadoop104 jdk1.8.0_144]$ source /etc/profile
```

3.3.3 Hadoop 安装

在搭建 Hadoop 分布式集群时，每个节点服务器上的 Hadoop 配置基本相同，所以只需要在 hadoop102 节点服务器上进行操作，配置完成之后同步到另外两个节点服务器上即可。

（1）将 Hadoop 的安装包 hadoop-2.7.2.tar.gz 导入 opt 目录下的 software 文件夹中，该文件夹被指定用来存储各软件的安装包。

① 进入 Hadoop 安装包路径。

```
[atguigu@hadoop102 ~]$ cd /opt/software/
```

② 解压安装包到/opt/module 文件中。

```
[atguigu@hadoop102 software]$ tar -zxvf hadoop-2.7.2.tar.gz -C /opt/module/
```

③ 查看是否解压成功。

```
[atguigu@hadoop102 software]$ ls /opt/module/
hadoop-2.7.2
```

（2）将 Hadoop 添加到环境变量，可以直接使用 Hadoop 的相关指令进行操作，而不用指定 Hadoop 的目录。

① 获取 Hadoop 安装路径。

```
[atguigu@ hadoop102 hadoop-2.7.2]$ pwd
/opt/module/hadoop-2.7.2
```

② 打开/etc/profile 文件。

```
[atguigu@ hadoop102 hadoop-2.7.2]$ sudo vim /etc/profile
```

在 profile 文件末尾添加 Hadoop 路径，添加的内容如下。

```
##HADOOP_HOME
export HADOOP_HOME=/opt/module/hadoop-2.7.2
export PATH=$PATH:$HADOOP_HOME/bin
export PATH=$PATH:$HADOOP_HOME/sbin
```

③ 保存后退出。

```
:wq
```

④ 执行 source 命令，使修改后的文件生效。

```
[atguigu@ hadoop102 hadoop-2.7.2]$ source /etc/profile
```

（3）测试是否安装成功。

```
[atguigu@hadoop102 ~]$ hadoop version
Hadoop 2.7.2
```

（4）重启（如果 hadoop 命令可以用，则不用重启）。

```
[atguigu@ hadoop101 hadoop-2.7.2]$ sync
[atguigu@ hadoop101 hadoop-2.7.2]$ sudo reboot
```

（5）分发 Hadoop 给所有节点服务器。

```
[atguigu@hadoop100 hadoop-2.7.2]$ xsync /opt/module/hadoop-2.7.2
```

（6）分发环境变量。

```
[atguigu@hadoop100 hadoop-2.7.2]$ xsync /etc/profile
```

（7）执行 source 命令，使环境变量在每台虚拟机上生效。

```
[atguigu@hadoop103 hadoop-2.7.2]$ source /etc/profile
[atguigu@hadoop104 hadoop-2.7.2]$ source /etc/profile
```

3.3.4 Hadoop 分布式集群部署

Hadoop 的运行模式包括本地式、伪分布式及完全分布式三种模式。本次主要搭建实际生产环境中比较常用的完全分布式模式，搭建完全分布式模式之前需要对集群部署进行提前规划，不要将过多的服务集中到一台节点服务器上。我们将负责管理工作的 NameNode 和 ResourceManager 分别部署在两台节点服务器上，另一台节点服务器上部署 SecondaryNameNode，所有节点服务器均承担 DataNode 和 NodeManager 角色，并且 DataNode 和 NodeManager 通常存储在同一台节点服务器上，所有角色尽量做到均衡分配。

（1）集群部署规划如表 3-1 所示。

表 3-1　集群部署规划

	hadoop102	hadoop103	hadoop104
HDFS	NameNode DataNode	DataNode	SecondaryNameNode DataNode
YARN	NodeManager	ResourceManager NodeManager	NodeManager

（2）对集群角色的分配主要依靠配置文件，配置集群文件的细节如下。

① 核心配置文件为 core-site.xml，该配置文件属于 Hadoop 的全局配置文件，我们主要对分布式文件系统 NameNode 的入口地址和分布式文件系统中数据落地到服务器本地磁盘的位置进行配置，代码如下。

```
[atguigu@hadoop102 hadoop]$ vim core-site.xml
<!-- 指定 HDFS 中 NameNode 的地址 -->
<property>
  <name>fs.defaultFS</name>
<!-- 其中，hdfs 为协议名称，hadoop102 为 NameNode 的节点服务器主机名称，9000 为端口-->
  <value>hdfs://hadoop102:9000</value>
</property>

<!-- 指定 Hadoop 运行时产生的文件的存储目录，该目录需要单独创建 -->
<property>
  <name>hadoop.tmp.dir</name>
  <value>/opt/module/hadoop-2.7.2/data/tmp</value>
</property>
```

② Hadoop 的环境配置文件为 hadoop-env.sh，在这个配置文件中我们主要需要指定 JDK 的路径 JAVA_HOME，避免程序运行中出现 JAVA_HOME 找不到的异常。

```
[atguigu@hadoop102 hadoop]$ vim hadoop-env.sh
export JAVA_HOME=/opt/module/jdk1.8.0_144
```

③ HDFS 的配置文件为 hdfs-site.xml，在这个配置文件中我们主要对 HDFS 文件系统的属性进行配置。

```
[atguigu@hadoop102 hadoop]$ vim hdfs-site.xml
<!-- 指定 HDFS 存储内容的副本个数 -->
<!-- Hadoop 通过使用文件的冗余来确保文件存储的可靠性，由于有 3 个 DataNode，所以我们可以将副本数量设置为 3 -->
<property>
  <name>dfs.replication</name>
  <value>3</value>
</property>
<!-- 配置 Hadoop 分布式集群的 SecondaryNameNode -->
<!-- SecondaryNameNode 主要作为 NameNode 的辅助，端口为 50090 -->
<property>
      <name>dfs.namenode.secondary.http-address</name>
      <value>hadoop104:50090</value>
```

　　　　</property>

　　④ YARN 的环境配置文件为 yarn-env.sh，同样指定 JDK 的路径 JAVA_HOME。

```
[atguigu@hadoop102 hadoop]$ vim yarn-env.sh
export JAVA_HOME=/opt/module/jdk1.8.0_144
```

　　⑤ 关于 YARN 的配置文件 yarn-site.xml，主要配置如下两个参数。

```
[atguigu@hadoop102 hadoop]$ vim yarn-site.xml
<!-- reducer 获取数据的方式 -->
<!-- yarn.nodemanager.aux-services 是 NodeManager 上运行的附属服务，其值需要配置成
mapreduce_shuffle 才可以运行 MapReduce 程序 -->

<property>
  <name>yarn.nodemanager.aux-services</name>
  <value>mapreduce_shuffle</value>
</property>

<!-- 指定 YARN 的 ResourceManager 的地址 -->
<property>
  <name>yarn.resourcemanager.hostname</name>
  <value>hadoop103</value>
</property>
```

　　⑥ MapReduce 的环境配置文件为 mapred-env.sh，同样指定 JDK 的路径 JAVA_HOME。

```
[atguigu@hadoop102 hadoop]$ vim mapred-env.sh
export JAVA_HOME=/opt/module/jdk1.8.0_144
```

　　⑦ 关于 MapReduce 的配置文件 mapred-site.xml，主要配置一个参数，指明 MapReduce 的运行框架为 YARN。

```
[atguigu@hadoop102 hadoop]$ cp mapred-site.xml.template mapred-site.xml
[atguigu@hadoop102 hadoop]$ vim mapred-site.xml
<!-- MapReduce 计算框架的资源交给 YARN 来管理 -->
<property>
  <name>mapreduce.framework.name</name>
  <value>yarn</value>
</property>
```

　　⑧ 主节点服务器 NameNode 和 ResourceManager 的角色在配置文件中已经进行了配置，还需指定从节点服务器的角色，配置文件 slaves 就是用来配置 Hadoop 分布式集群中各台从节点服务器的角色的。如下所示，对 slaves 文件进行修改，将 3 台节点服务器全部指定为从节点服务器，启动 DataNode 和 NodeManager 进程。

```
/opt/module/hadoop-2.7.2/etc/hadoop/slaves
[atguigu@hadoop102 hadoop]$ vim slaves
hadoop102
hadoop103
hadoop104
```

⑨ 在集群上分发配置好的Hadoop配置文件,这样3台节点服务器都可享有相同的Hadoop的配置,接下来即可通过不同的进程启动命令了。

```
[atguigu@hadoop102 hadoop]$ xsync /opt/module/hadoop-2.7.2/
```

⑩ 查看文件分发情况。

```
[atguigu@hadoop103 hadoop]$ cat /opt/module/hadoop-2.7.2/etc/hadoop/core-site.xml
```

(3) 创建数据目录。

根据在 core-site.xml 文件中配置的分布式文件系统最终落地到各数据节点上的本地磁盘位置信息/opt/module/hadoop-2.7.2/data/tmp,自行创建该目录。

```
[atguigu@hadoop102 hadoop-2.7.2]$ mkdir /opt/module/hadoop-2.7.2/data/tmp
[atguigu@hadoop103 hadoop-2.7.2]$ mkdir /opt/module/hadoop-2.7.2/data/tmp
[atguigu@hadoop104 hadoop-2.7.2]$ mkdir /opt/module/hadoop-2.7.2/data/tmp
```

(4) 启动 Hadoop 分布式集群。

① 如果第一次启动集群,则需要格式化 NameNode。

```
[atguigu@hadoop102 hadoop-2.7.2]$ hadoop namenode -format
```

② 在配置了 NameNode 的节点服务器后,通过执行 start-dfs.sh 命令启动 HDFS,即可同时启动所有的 DataNode 和 SecondaryNameNode。

```
[atguigu@hadoop102 hadoop-2.7.2]$ sbin/start-dfs.sh
[atguigu@hadoop102 hadoop-2.7.2]$ jps
4166 NameNode
4482 Jps
4263 DataNode
[atguigu@hadoop103 hadoop-2.7.2]$ jps
3218 DataNode
3288 Jps
[atguigu@hadoop104 hadoop-2.7.2]$ jps
3221 DataNode
3283 SecondaryNameNode
3364 Jps
```

③ 通过执行 start-yarn.sh 命令启动 YARN,即可同时启动 ResourceManager 和所有的 NodeManager。需要注意的是,NameNode 和 ResourceManager 如果不在同一台服务器上,则不能在 NameNode 上启动 YARN,应该在 ResourceManager 所在的服务器上启动 YARN。

```
[atguigu@hadoop103 hadoop-2.7.2]$ sbin/start-yarn.sh
```

通过执行 jps 命令可在各台节点服务器上查看进程启动情况,若显示如下内容,则表示启动成功。

```
[atguigu@hadoop103 hadoop-2.7.2]$ sbin/start-yarn.sh
[atguigu@hadoop102 hadoop-2.7.2]$ jps
4166 NameNode
4482 Jps
4263 DataNode
```

```
4485 NodeManager
[atguigu@hadoop103 hadoop-2.7.2]$ jps
3218 DataNode
3288 Jps
3290 ResourceManager
3299 NodeManager
[atguigu@hadoop104 hadoop-2.7.2]$ jps
3221 DataNode
3283 SecondaryNameNode
3364 Jps
3389 NodeManager
```

（5）通过 Web UI 查看集群是否启动成功。

① 在 Web 端输入之前配置的 NameNode 的节点服务器地址和端口 50070，即可查看 HDFS 文件系统。例如，在浏览器中输入 http://hadoop102:50070，可以检查 NameNode 和 DataNode 是否正常。NameNode 的 Web 端如图 3-40 所示。

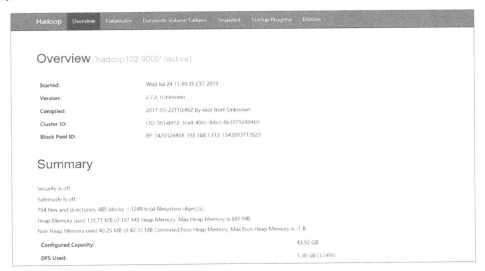

图 3-40 NameNode 的 Web 端

② 通过在 Web 端输入 ResourceManager 的地址和端口 8088，可以查看 YARN 上任务的运行情况。例如，在浏览器输入 http://hadoop103:8088 ，即可查看本集群 YARN 的运行情况。YARN 的 Web 端如图 3-41 所示。

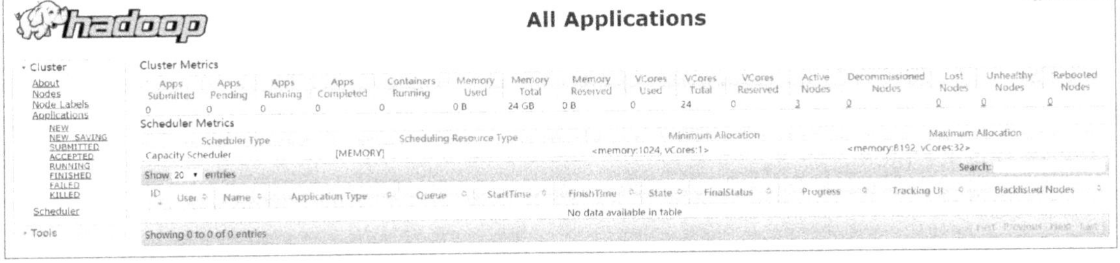

图 3-41 YARN 的 Web 端

（6）运行 PI 实例，检查集群是否启动成功。

在集群任意节点服务器上执行下面的命令，如果看到如图 3-42 所示的运行结果，则说明集群启动成功。

```
[atguigu@hadoop102 hadoop]$ cd /opt/module/hadoop-2.7.2/share/hadoop/mapreduce/
[atguigu@hadoop102 mapreduce]$ hadoop jar hadoop-mapreduce-examples-2.7.2.jar pi 10 10
```

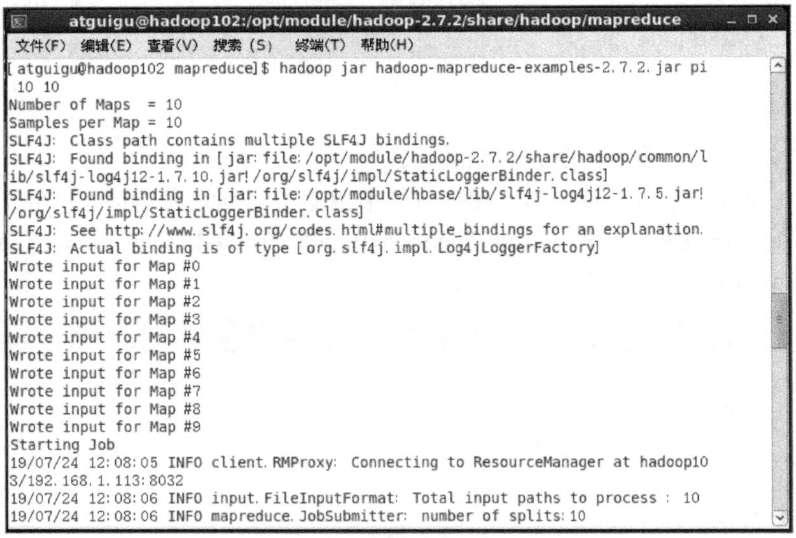

图 3-42　PI 实例运行结果

最后输出为 Estimated value of Pi is 3.20000000000000000000。

3.3.5　配置 Hadoop 支持 LZO 压缩

数据的压缩对海量数据的存储至关重要，合理选用压缩格式可以大大缩小内存占用空间，提高 I/O 传输效率，在 Flume 进行数据传输的过程中，我们需要将 Flume 落盘到 HDFS 文件并保存为压缩格式。

HDFS 支持的压缩格式很多，且各有优点，采集到 HDFS 存储的用户行为日志通常体量很大，落盘的单个文件甚至会超过 HDFS 的文件切片大小，当对这样的大文件进行处理时就会涉及切片操作，支持切片同时压缩率较高的压缩格式为 LZO 压缩，为此，需要配置 Hadoop 支持 LZO 压缩，具体操作如下。

（1）先下载 LZO 的 jar 项目。

（2）下载后的文件名是 hadoop-lzo-master，它是一个.zip 格式的压缩包，先进行解压，然后用 maven 进行编译，生成 hadoop-lzo-0.4.20.jar。在通过尚硅谷教育公众号获取的项目资料中可以找到该 jar 包。

（3）将编译好后的 hadoop-lzo-0.4.20.jar 放入 hadoop-2.7.2/share/hadoop/common 中。

```
[atguigu@hadoop102 common]$ pwd
/opt/module/hadoop-2.7.2/share/hadoop/common
```

```
[atguigu@hadoop102 common]$ ls
hadoop-lzo-0.4.20.jar
```

（4）同步 hadoop-lzo-0.4.20.jar 到 hadoop103、hadoop104。

```
[atguigu@hadoop102 common]$ xsync hadoop-lzo-0.4.20.jar
```

（5）打开 Hadoop 的配置文件 core-site.xml，增加支持 LZO 压缩的配置，代码如下。

```xml
<property>
<name>io.compression.codecs</name>
<value>
org.apache.hadoop.io.compress.GzipCodec,
org.apache.hadoop.io.compress.DefaultCodec,
org.apache.hadoop.io.compress.BZip2Codec,
org.apache.hadoop.io.compress.SnappyCodec,
com.hadoop.compression.lzo.LzoCodec,
com.hadoop.compression.lzo.LzopCodec
</value>
</property>

<property>
    <name>io.compression.codec.lzo.class</name>
    <value>com.hadoop.compression.lzo.LzoCodec</value>
</property>
```

（6）同步 core-site.xml 文件到 hadoop103、hadoop104。

```
[atguigu@hadoop102 hadoop]$ xsync core-site.xml
```

（7）启动并查看集群。

```
[atguigu@hadoop102 hadoop-2.7.2]$ sbin/start-dfs.sh
[atguigu@hadoop103 hadoop-2.7.2]$ sbin/start-yarn.sh
```

① Web 和进程查看。
- Web 查看：http://hadoop102:50070。
- 进程查看：执行 jps 命令，查看各台节点服务器的状态。

② 当启动发生错误时，采取如下措施。
- 查看日志：/opt/module/hadoop-2.7.2/logs。
- 如果进入安全模式，则可以通过执行 hdfs dfsadmin-safemode leave 命令强制离开安全模式。
- 停止所有进程，删除 data 和 log 文件夹，然后执行 hdfs namenode-format 命令进行格式化（在集群没有重要数据的前提下）。

3.3.6 配置 Hadoop 支持 Snappy 压缩

Hadoop 集群本身不支持 Snappy 压缩，若想使 Hadoop 支持 Snappy 压缩，需要对 Hadoop 进行编译，具体编译步骤此处不再赘述，读者可在尚硅谷教育公众号后台回复"snappy"获取编译后的 Hadoop 压缩包，编译后的操作步骤如下。

（1）将编译后支持 Snappy 压缩的 Hadoop jar 包解压，将 lib/native 目录下的所有文件上传到 hadoop102 的/opt/module/hadoop-2.7.2/lib/native 目录下，并分发到 hadoop103 和 hadoop104。

（2）重新启动 Hadoop。

（3）检查支持的压缩方式。

参考代码如下。

```
[atguigu@hadoop102 native]$ hadoop checknative
hadoop:  true /opt/module/hadoop-2.7.2/lib/native/libhadoop.so
zlib:    true /lib64/libz.so.1
snappy:  true /opt/module/hadoop-2.7.2/lib/native/libsnappy.so.1
lz4:     true revision:99
bzip2:   false
```

3.4 本章总结

本章主要对项目运行所需的环境进行了安装和部署，从安装虚拟机和 CentOS 开始，到最终 JDK 和 Hadoop 的安装，对每一步的安装部署进行了详细介绍。本章是整个项目的基础，重点在于 Hadoop 集群的搭建和配置，读者务必掌握。

第4章
用户行为数据采集模块

根据第 2 章中对采集模块的整体分析，在本章中，我们将带领读者完成数据采集模块的搭建。

4.1 日志生成

本项目需要读者模仿前端日志数据落盘过程自行生成模拟日志数据，这部分代码读者可通过尚硅谷教育公众号的项目资料获取，可同时获取完整 jar 包。通过后续内容中日志生成的操作，可以在虚拟机的/tmp/logs 目录下生成每天的日志数据。

1. 日志启动

（1）将获取的 jar 包 log-collector-1.0-SNAPSHOT-jar-with-dependencies.jar 复制到 hadoop102 上，并同步到 hadoop103 的/opt/module 目录下。

```
[atguigu@hadoop102 module]$ xsync log-collector-1.0-SNAPSHOT-jar-with-dependencies.jar
```

（2）在 hadoop102 上执行 jar 程序。

```
[atguigu@hadoop102 module]$ java -classpath log-collector-1.0-SNAPSHOT-jar-with-dependencies.jar com.atguigu.appclient.AppMain >/opt/module/test.log
```

（3）根据程序中配置文件的设置，系统会在虚拟机的/tmp/logs 目录下生成日志文件，生成的用户行为日志默认为服务器的当前系统时间，若想生成不同时间的用户行为日志，则可以通过修改服务器的时间来实现。

```
[atguigu@hadoop102 module]$ cd /tmp/logs/
[atguigu@hadoop102 logs]$ ls
app-2020-03-10.log
```

（4）在/home/atguigu/bin 目录下创建脚本 dt.sh，用于统一修改服务器的时间，以生成不同时间下的用户行为日志。

```
[atguigu@hadoop102 bin]$ vim dt.sh
```

（5）在脚本中编写如下内容，分别在 3 台节点服务器下使用 date 命令修改服务器时间。

```
#!/bin/bash

for i in hadoop102 hadoop103 hadoop104
do
        echo "========== $i =========="
        ssh -t $i "sudo date -s $1"
done
```

(6) 增加脚本执行权限。

`[atguigu@hadoop102 bin]$ chmod 777 dt.sh`

(7) 启动脚本。

`[atguigu@hadoop102 bin]$ dt.sh 2020-03-10`

2. 集群日志生成启动脚本

将日志生成的命令封装成脚本可以方便用户调用执行，具体操作步骤如下。

(1) 在/home/atguigu/bin 目录下创建脚本 lg.sh。

`[atguigu@hadoop102 bin]$ vim lg.sh`

(2) 脚本思路：通过 i 变量在 hadoop102 和 hadoop103 节点服务器间遍历，分别通过 ssh 命令进入两台节点服务器，执行 java 命令，运行日志生成 jar 包，在两台节点服务器的/tmp/logs 目录下生成模拟日志文件。

在脚本中编写如下内容。

```
#! /bin/bash

 for i in hadoop102 hadoop103
 do
  ssh $i "java -classpath /opt/module/log-collector-1.0-SNAPSHOT-jar-with-dependencies.jar com.atguigu.appclient.AppMain $1 $2 >/opt/module/test.log &"
 done
```

(3) 增加脚本执行权限。

`[atguigu@hadoop102 bin]$ chmod 777 lg.sh`

(4) 启动脚本。

`[atguigu@hadoop102 module]$ lg.sh`

(5) 分别在 hadoop102 和 hadoop103 的/tmp/logs 目录下查看生成的数据，判断脚本是否生效。

```
[atguigu@hadoop102 logs]$ ls
app-2020-03-10.log
[atguigu@hadoop103 logs]$ ls
app-2020-03-10.log
```

3. 集群所有进程查看脚本

启动集群后，用户需要通过 jps 命令查看各台节点服务器进程的启动情况，操作起来比较麻烦，所以我们通过写一个集群所有进程查看脚本来实现使用一个脚本查看所有节点服务器的所有进程的目的。

（1）在/home/atguigu/bin 目录下创建脚本 xcall.sh。

[atguigu@hadoop102 bin]$ vim xcall.sh

（2）脚本思路：通过 i 变量在 hadoop102、hadoop103 和 hadoop104 节点服务器间遍历，分别通过 ssh 命令进入 3 台节点服务器，执行传入参数指定命令。

在脚本中编写如下内容。

```
#! /bin/bash

for i in hadoop102 hadoop103 hadoop104
do
        echo --------- $i ----------
        ssh $i "$*"
done
```

（3）增加脚本执行权限。

[atguigu@hadoop102 bin]$ chmod 777 xcall.sh

（4）启动脚本。

[atguigu@hadoop102 bin]$ xcall.sh jps

4.2 采集日志的 Flume

如图 4-1 所示，采集日志层 Flume 主要需要完成的任务为将日志从落盘文件中采集出来，传输给消息中间件 Kafka 集群，这期间要保证数据不丢失，程序出现故障死机后可以快速重启，对日志进行初步分类，分别发往不同的 Kafka Topic，方便后续对日志数据进行分别处理。

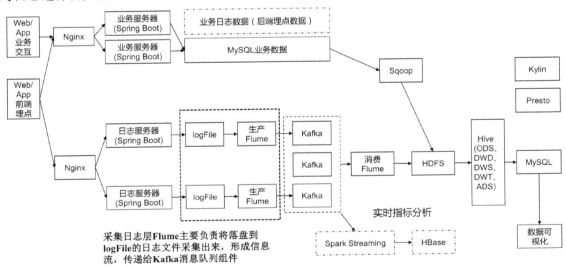

图 4-1 采集日志层 Flume 的流向

4.2.1 Flume 组件

Flume 整体上是 Source-Channel-Sink 的三层架构，其中，Source 层完成对日志的收集，将日志封装成 event 传入 Channel 层中；Channel 层主要提供队列的功能，对 Source 层中传入的数据提供简单的缓存功能；Sink 层取出 Channel 层中的数据，将数据送入存储文件系统中，或者对接其他的 Source 层。

Flume 以 Agent 为最小独立运行单位，一个 Agent 就是一个 JVM，单个 Agent 由 Source、Sink 和 Channel 三大组件构成。

Flume 将数据表示为 event（事件），event 由一字节数组的主体 body 和一个 key-value 结构的报头 header 构成。其中，主体 body 中封装了 Flume 传送的数据，报头 header 中容纳的 key-value 信息则是为了给数据增加标识，用于跟踪发送事件的优先级和重要性，用户可通过拦截器（Interceptor）进行修改。

Flume 的数据流由 event 贯穿始终，这些 event 由 Agent 外部的 Source 生成，当 Source 捕获事件后会进行特定的格式化，然后 Source 会把事件推入 Channel 中，Channel 中的 event 会由 Sink 来拉取，Sink 拉取 event 后可以将 event 持久化或者推向另一个 Source。

除此之外，Flume 还有一些使其应用更加灵活的组件：拦截器、Channel 选择器（Selector）、Sink 组和 Sink 处理器。其功能如下。

- 拦截器可以部署在 Source 和 Channel 之间，用于对事件进行预处理或者过滤，Flume 内置了很多类型的拦截器，用户也可以自定义自己的拦截器。
- Channel 选择器可以决定 Source 接收的一个特定事件写入哪些 Channel 组件中。
- Sink 组和 Sink 处理器可以帮助用户实现负载均衡和故障转移。

4.2.2 Flume 安装

在进行采集日志层的 Flume Agent 配置之前，我们首先需要安装 Flume，Flume 需要安装部署到每台节点服务器上，具体安装步骤如下。

（1）将 apache-flume-1.7.0-bin.tar.gz 上传到 Linux 的/opt/software 目录下。

（2）解压 apache-flume-1.7.0-bin.tar.gz 到/opt/module/目录下。

```
[atguigu@hadoop102 software]$ tar -zxf apache-flume-1.7.0-bin.tar.gz -C /opt/module/
```

（3）修改 apache-flume-1.7.0-bin 的名称为 flume。

```
[atguigu@hadoop102 module]$ mv apache-flume-1.7.0-bin flume
```

（4）将 flume/conf 目录下的 flume-env.sh.template 文件的名称修改为 flume-env.sh，并配置 flume-env.sh 文件，在配置文件中增加 JAVA_HOME 路径，如下所示。

```
[atguigu@hadoop102 conf]$ mv flume-env.sh.template flume-env.sh
[atguigu@hadoop102 conf]$ vim flume-env.sh
export JAVA_HOME=/opt/module/jdk1.8.0_144
```

（5）将配置好的 Flume 分发到集群中其他节点服务器上。

4.2.3 采集日志 Flume 配置

1. Flume 配置分析

针对本项目，在编写 Flume Agent 配置文件之前，首先需要进行组件选型。

1）Source

本项目主要从一个实时写入数据的文件夹中读取数据，Source 可以选择 Spooling Directory Source、Exec Source 和 Taildir Source。Taildir Source 相比 Exec Source、Spooling Directory Source 具有很多优势。Taildir Source 可以实现断点续传、多目录监控配置。而在 Flume 1.6 以前需要用户自定义 Source，记录每次读取文件的位置，从而实现断点续传。Exec Source 可以实时搜集数据，但是在 Flume 不运行或者 Shell 命令出错的情况下，数据将会丢失，从而不能记录数据读取位置、实现断点续传。Spooling Directory Source 可以实现目录监控配置，但是不能实时采集数据。

2）Channel

由于采集日志层 Flume 在读取数据后主要将数据送往 Kafka 消息队列中，所以使用 Kafka Channel 是很好的选择，同时选择 Kafka Channel 可以不配置 Sink，提高了效率。

3）拦截器

本项目中主要部署两个拦截器，一个用来过滤格式不正确的非法数据，这在实际生产环境中也是必不可少的，另一个用来分辨日志类型，根据日志类型给 event 添加 header 信息，可以帮助 Channel 选择器选择日志应该发往的 Channel。

4）Channel 选择器

采集日志层 Flume 主要部署两个 Kafka Channel，分别将数据发往不同的 Kafka Topic，两个 Topic 存储的数据不同，所以需要配置 Channel 选择器决定日志去向，并且配置选择器类型为 multiplexing，在该模式下，会将 event 发送至特定的 Channel，而不会发送至所有 Channel，实现了日志的分类分流。

2. Flume 的具体配置

在/opt/module/flume/conf 目录下创建 file-flume-kafka.conf 文件。

```
[atguigu@hadoop102 conf]$ vim file-flume-kafka.conf
```

在文件中配置如下内容。

```
#定义 Agent 必需的组件名称，同时指定本配置文件的 Agent 名称为 a1
a1.sources=r1
a1.channels=c1 c2

#定义 Source 组件相关配置
#使用 Taildir Source
a1.sources.r1.type = TAILDIR
#配置 Taildir Source，保存断点位置文件的目录
a1.sources.r1.positionFile = /opt/module/flume/test/log_position.json
#配置监控目录组
a1.sources.r1.filegroups = f1
#配置目录组下的目录，可配置多个目录
```

```
a1.sources.r1.filegroups.f1 = /tmp/logs/app.+

#配置Source发送数据的目标Channel
a1.sources.r1.channels = c1 c2

#拦截器
#配置拦截器名称
a1.sources.r1.interceptors = i1 i2
#配置拦截器名称,需要写明全类名
a1.sources.r1.interceptors.i1.type =
com.atguigu.flume.interceptor.LogETLInterceptor$Builder
a1.sources.r1.interceptors.i2.type =
com.atguigu.flume.interceptor.LogTypeInterceptor$Builder

#配置Channel选择器
#配置选择器类型
a1.sources.r1.selector.type = multiplexing
#配置选择器识别header中的key
a1.sources.r1.selector.header = topic
#配置不同的header信息,发往不同的Channel
a1.sources.r1.selector.mapping.topic_start = c1
a1.sources.r1.selector.mapping.topic_event = c2

# configure channel 配置Channel
#配置Channel类型为Kafka Channel
a1.channels.c1.type = org.apache.flume.channel.kafka.KafkaChannel
#配置Kafka集群节点服务器列表
a1.channels.c1.kafka.bootstrap.servers =
hadoop102:9092,hadoop103:9092,hadoop104:9092
#配置该Channel发往Kafka的Topic,该Topic需要在Kafka中提前创建
a1.channels.c1.kafka.topic = topic_start
#配置不将header信息解析为event内容
a1.channels.c1.parseAsFlumeEvent = false
#配置该Kafka Channel所属的消费者组名,为实现multiplexing类型的Channel选择器,应将2
个Kafka Channel配置相同的消费者组
a1.channels.c1.kafka.consumer.group.id = flume-consumer

#配置同上
a1.channels.c2.type = org.apache.flume.channel.kafka.KafkaChannel
a1.channels.c2.kafka.bootstrap.servers =
hadoop102:9092,hadoop103:9092,hadoop104:9092
a1.channels.c2.kafka.topic = topic_event
a1.channels.c2.parseAsFlumeEvent = false
a1.channels.c2.kafka.consumer.group.id = flume-consumer
```

注意:com.atguigu.flume.interceptor.LogETLInterceptor 和 com.atguigu.flume.interceptor.LogTypeInterceptor 是笔者自定义的拦截器的全类名。读者需要根据自己自定义的拦截器进行相应修改。

4.2.4　Flume 的 ETL 拦截器和日志类型区分拦截器

在本项目中自定义了两个拦截器，分别是 ETL 拦截器、日志类型区分拦截器。

ETL 是指将业务系统的数据经过抽取、清洗转换之后加载到数据仓库的过程，目的是将企业中分散、零乱、标准不统一的数据整合到一起，为企业的管理者决策提供分析依据。在这里可以简单地理解为数据清洗。

ETL 拦截器主要用于过滤时间戳不合法和 JSON 数据不完整的日志。

日志类型区分拦截器主要用于将启动日志和事件日志区分开，并增加 event 的 header 信息，方便将不同类型的 event 发往 Kafka 的不同 Topic。

拦截器的定义步骤如下。

（1）创建 Maven 工程 flume-interceptor。

（2）创建包名：com.atguigu.flume.interceptor。

（3）在 pom.xml 文件中添加如下依赖。

```xml
<dependencies>
    <dependency>
        <groupId>org.apache.flume</groupId>
        <artifactId>flume-ng-core</artifactId>
        <version>1.7.0</version>
    </dependency>
</dependencies>

<build>
    <plugins>
        <plugin>
            <artifactId>maven-compiler-plugin</artifactId>
            <version>2.3.2</version>
            <configuration>
                <source>1.8</source>
                <target>1.8</target>
            </configuration>
        </plugin>
        <plugin>
            <artifactId>maven-assembly-plugin</artifactId>
            <configuration>
                <descriptorRefs>
                    <descriptorRef>jar-with-dependencies</descriptorRef>
                </descriptorRefs>
            </configuration>
            <executions>
                <execution>
                    <id>make-assembly</id>
                    <phase>package</phase>
                    <goals>
                        <goal>single</goal>
```

```
            </goals>
          </execution>
        </executions>
      </plugin>
    </plugins>
</build>
```

（4）在 com.atguigu.flume.interceptor 包中创建 LogETLInterceptor 类名。
Flume 的 ETL 拦截器 LogETLInterceptor。

```java
package com.atguigu.flume.interceptor;

import org.apache.flume.Context;
import org.apache.flume.Event;
import org.apache.flume.interceptor.Interceptor;

import java.nio.charset.Charset;
import java.util.ArrayList;
import java.util.List;

public class LogETLInterceptor implements Interceptor {

    @Override
    public void initialize() {

    }

    @Override
    public Event intercept(Event event) {

        // 1 获取数据
        byte[] body = event.getBody();
        String log = new String(body, Charset.forName("UTF-8"));

        // 2 判断数据是否合法
        if (log.contains("start")) {
            if (LogUtils.validateStart(log)){
                return event;
            }
        }else {
            if (LogUtils.validateEvent(log)){
                return event;
            }
        }

        // 3 返回校验结果
        return null;
    }
```

```java
    @Override
    public List<Event> intercept(List<Event> events) {

        ArrayList<Event> interceptors = new ArrayList<>();

        for (Event event : events) {
            Event intercept1 = intercept(event);

            if (intercept1 != null){
                interceptors.add(intercept1);
            }
        }

        return interceptors;
    }

    @Override
    public void close() {

    }

    public static class Builder implements Interceptor.Builder{

        @Override
        public Interceptor build() {
            return new LogETLInterceptor();
        }

        @Override
        public void configure(Context context) {

        }
    }
}
```

（5）编写 Flume 日志过滤工具类 LogUtils，方便 ETL 拦截器 LogETLInterceptor 调用。

```java
package com.atguigu.flume.interceptor;
import org.apache.commons.lang.math.NumberUtils;

public class LogUtils {

    public static boolean validateEvent(String log) {
        // 服务器时间 | JSON
        // 1549696569054 | {"cm":{"ln":"-89.2","sv":"V2.0.4","os":"8.2.0","g":"M67B4QYU@gmail.com","nw":"4G","l":"en","vc":"18","hw":"1080*1920","ar":"MX","uid":"u8678","t":"1549679122062","la":"
```

```
-27.4","md":"sumsung-12","vn":"1.1.3","ba":"Sumsung","sr":"Y"},"ap":"weather","et":[]}

        // 1 切割
        String[] logContents = log.split("\\|");

        // 2 校验
        if(logContents.length != 2){
            return false;
        }

        //3 校验服务器时间
        if (logContents[0].length()!=13
|| !NumberUtils.isDigits(logContents[0])){
            return false;
        }

        // 4 校验JSON
        if (!logContents[1].trim().startsWith("{")
|| !logContents[1].trim().endsWith("}")){
            return false;
        }

        return true;
    }

    public static boolean validateStart(String log) {

        if (log == null){
            return false;
        }

        // 校验JSON
        if (!log.trim().startsWith("{") || !log.trim().endsWith("}")){
            return false;
        }

        return true;
    }
}
```

（6）Flume 日志类型区分拦截器 LogTypeInterceptor。

```
package com.atguigu.flume.interceptor;

import org.apache.flume.Context;
import org.apache.flume.Event;
import org.apache.flume.interceptor.Interceptor;
```

```java
import java.nio.charset.Charset;
import java.util.ArrayList;
import java.util.List;
import java.util.Map;

public class LogTypeInterceptor implements Interceptor {
    @Override
    public void initialize() {

    }

    @Override
    public Event intercept(Event event) {

        // 区分日志类型: body 或 header
        // 1 获取 body 数据
        byte[] body = event.getBody();
        String log = new String(body, Charset.forName("UTF-8"));

        // 2 获取 header 数据
        Map<String, String> headers = event.getHeaders();

        // 3 判断数据类型并向 header 中赋值
        if (log.contains("start")) {
            headers.put("topic","topic_start");
        }else {
            headers.put("topic","topic_event");
        }

        return event;
    }

    @Override
    public List<Event> intercept(List<Event> events) {

        ArrayList<Event> interceptors = new ArrayList<>();

        for (Event event : events) {
            Event intercept1 = intercept(event);

            interceptors.add(intercept1);
        }

        return interceptors;
    }
```

```
    @Override
    public void close() {

    }

    public static class Builder implements Interceptor.Builder{

        @Override
        public Interceptor build() {
            return new LogTypeInterceptor();
        }

        @Override
        public void configure(Context context) {

        }
    }
}
```

(7) 打包。

拦截器打包之后,只需要单独的压缩包,不需要将依赖包上传。打包之后要放入 Flume 的 lib 目录下,如图 4-2 所示。

图 4-2　拦截器压缩包

注意:为什么不需要依赖包?因为依赖包在 Flume 的 lib 目录下已经存在。

(8) 需要先将打好的包放入 hadoop102 的/opt/module/flume/lib 目录下。

```
[atguigu@hadoop102 lib]$ ls | grep interceptor
flume-interceptor-1.0-SNAPSHOT.jar
```

(9) 分发 Flume 到 hadoop103 和 hadoop104。

```
[atguigu@hadoop102 module]$ xsync flume/
```

(10) 执行 flume-ng agent 命令,将上述配置文件启动,其中,--name 选项用于指定本次命令执行的 Agent 名字,本配置文件中为 a1;--conf-file 选项用于指定配置文件的存储路径。

```
[atguigu@hadoop102 flume]$ bin/flume-ng agent --name a1 --conf-file conf/file-flume-kafka.conf
```

该 Flume Agent 的数据流向是 Kafka,由于我们还没有安装 Kafka,所以启动后不能形成完整的数据流,若想看到数据的消费情况,读者可以使用监控工具 Gangalia 进行查看,此处不再赘述。

建议读者后续学习了 Kafka 后再对该部分配置命令进行测试。

4.2.5 采集日志 Flume 启动、停止脚本

同日志生成一样，我们也将采集日志层 Flume 的启动、停止命令封装成脚本，以方便后续调用执行。

（1）在 /home/atguigu/bin 目录下创建脚本 f1.sh。

[atguigu@hadoop102 bin]$ vim f1.sh

脚本思路：通过匹配输入参数的值选择是否启动采集程序，启动采集程序后，设置日志不打印且程序在后台运行。

若停止程序，则通过管道符切割等操作获取程序的编号，并通过 kill 命令停止程序。在脚本中编写如下内容。

```bash
#! /bin/bash

case $1 in
"start"){
        for i in hadoop102 hadoop103
        do
                echo " --------启动 $i 采集Flume-------"
                ssh $i "source /etc/profile ; nohup /opt/module/flume/bin/flume-ng agent --conf-file /opt/module/flume/conf/file-flume-kafka.conf --name a1 -Dflume.root.logger=INFO,LOGFILE >/dev/null 2>&1 &"
        done
};;
"stop"){
        for i in hadoop102 hadoop103
        do
                echo " --------停止 $i 采集Flume-------"
                ssh $i "ps -ef | grep file-flume-kafka | grep -v grep |awk '{print \$2}' | xargs kill"
        done

};;
esac
```

脚本说明如下。

说明 1：nohup 命令可以在用户退出账户或关闭终端之后继续运行相应的进程。nohup 命令就是不挂起的意思，不间断地运行命令。

说明 2：/dev/null 代表 Linux 的空设备文件，所有往这个文件里面写入的内容都会丢失，俗称"黑洞"。企业在进行开发时，如果不想在控制台显示大量的启动过程日志，就可以把日志写入"黑洞"，以减少磁盘存储空间。

标准输入 0：从键盘获得输入 /proc/self/fd/0。

标准输出 1：输出到控制台 /proc/self/fd/1。

错误输出 2：输出到控制台 /proc/self/fd/2。

说明 3：

① "ps -ef | grep file-flume-kafka"用于获取 Flume 进程，查看结果可以发现存在两个进程 id，但是我们只想获取第一个进程 id 21319。

```
atguigu    21319      1 57 15:14 ?        00:00:03
……
atguigu    21428  11422  0 15:14 pts/1    00:00:00 grep file-flume-kafka
```

② "ps -ef | grep file-flume-kafka | grep -v grep"用于过滤包含 grep 信息的进程。

```
atguigu    21319      1 57 15:14 ?        00:00:03
……
```

③ "ps -ef | grep file-flume-kafka | grep -v grep |awk '{print \$2}'"，采用 awk，默认用空格分隔后，取第二个字段，获取到 21319 进程 id。

④ "ps -ef | grep file-flume-kafka | grep -v grep |awk '{print \$2}' | xargs kill"，xargs 表示获取前一阶段的运行结果，即 21319，作为下一个命令 kill 的输入参数。实际执行的是 kill 21319。

（2）增加脚本执行权限。

```
[atguigu@hadoop102 bin]$ chmod 777 f1.sh
```

（3）f1 集群启动脚本。

```
[atguigu@hadoop102 module]$ f1.sh start
```

（4）f1 集群停止脚本。

```
[atguigu@hadoop102 module]$ f1.sh stop
```

4.3 消息队列 Kafka

通过 Flume Agent 程序将日志从落盘文件夹采集出来之后，需要发送到 Kafka，Kafka 在这里起到数据缓冲和负载均衡的作用，大大减轻数据存储系统的压力。在向 Kafka 发送日志之前，需要先安装 Kafka，而在安装 Kafka 之前需要先安装 Zookeeper，为之提供分布式服务。本节主要带领读者完成 Zookeeper 和 Kafka 的安装部署。

4.3.1 Zookeeper 安装

Zookeeper 是一个能够高效开发和维护分布式应用的协调服务，主要用于为分布式应用提供一致性服务，提供的功能包括维护配置信息、名字服务、分布式同步、组服务等。

Zookeeper 的安装步骤如下。

1. 集群规划

在 hadoop102、hadoop103 和 hadoop104 三台节点服务器上部署 Zookeeper。

2. 解压安装

（1）解压 Zookeeper 安装包到/opt/module/目录下。

```
[atguigu@hadoop102 software]$ tar -zxvf zookeeper-3.4.10.tar.gz -C /opt/module/
```

（2）在/opt/module/zookeeper-3.4.10 目录下创建 zkData 文件夹，用于保存 Zookeeper 的相关数据。
```
mkdir -p zkData
```

3. 配置 zoo.cfg 文件

（1）重命名/opt/module/zookeeper-3.4.10/conf 目录下的 zoo_sample.cfg 为 zoo.cfg，我们可以对其中的配置进行自定义设置。
```
mv zoo_sample.cfg zoo.cfg
```

（2）具体配置，在配置文件中找到如下内容，将数据存储目录 dataDir 设置为上文中自行创建的 zkData 文件夹。
```
dataDir=/opt/module/zookeeper-3.4.10/zkData
```

增加如下配置，如下配置指出了 Zookeeper 集群的 3 台节点服务器信息。
```
#######################cluster#########################
server.2=hadoop102:2888:3888
server.3=hadoop103:2888:3888
server.4=hadoop104:2888:3888
```

（3）配置参数解读。
```
Server.A=B:C:D。
```
- A 是一个数字，表示第几台服务器；
- B 是这台服务器的 IP 地址；
- C 是这台服务器与集群中的 Leader 服务器交换信息的端口；
- D 表示当集群中的 Leader 服务器无法正常运行时，需要一个端口来重新进行选举，选出一个新的 Leader 服务器，而这个端口就是用来执行选举时服务器相互通信的端口。

在集群模式下配置一个文件 myid，这个文件在 dataDir 目录下，其中有一个数据就是 A 的值，Zookeeper 启动时读取此文件，并将里面的数据与 zoo.cfg 文件里面的配置信息进行比较，从而判断到底是哪台服务器。

4. 集群操作

（1）在/opt/module/zookeeper-3.4.10/zkData 目录下创建一个 myid 文件，当集群启动时由 Zookeeper 读取此文件。
```
touch myid
```

注意：当添加 myid 文件时，一定要在 Linux 中创建，在文本编辑工具中创建有可能出现乱码。

（2）编辑 myid 文件。
```
vim myid
```

在文件中添加与 Server 对应的编号，根据在 zoo.cfg 文件中配置的 Server id 与节点服务器的 IP 地址对应关系添加，如在 hadoop102 节点服务器中添加 2。

（3）复制配置好的 Zookeeper 到其他服务器上。

```
xsync /opt/module/zookeeper-3.4.10
```

分别修改 hadoop103 和 hadoop104 节点服务器中 myid 文件的内容为 3 和 4。

（4）在 3 台节点服务器中分别启动 Zookeeper。

```
[root@hadoop102 zookeeper-3.4.10]# bin/zkServer.sh start
[root@hadoop103 zookeeper-3.4.10]# bin/zkServer.sh start
[root@hadoop104 zookeeper-3.4.10]# bin/zkServer.sh start
```

（5）执行如下命令，在 3 台节点服务器中查看 Zookeeper 的服务状态。

```
[root@hadoop102 zookeeper-3.4.10]# bin/zkServer.sh status
JMX enabled by default
Using config: /opt/module/zookeeper-3.4.10/bin/../conf/zoo.cfg
Mode: follower
[root@hadoop103 zookeeper-3.4.10]# bin/zkServer.sh status
JMX enabled by default
Using config: /opt/module/zookeeper-3.4.10/bin/../conf/zoo.cfg
Mode: leader
[root@hadoop104 zookeeper-3.4.5]# bin/zkServer.sh status
JMX enabled by default
Using config: /opt/module/zookeeper-3.4.10/bin/../conf/zoo.cfg
Mode: follower
```

4.3.2 Zookeeper 集群启动、停止脚本

由于 Zookeeper 没有提供多台服务器同时启动、停止的脚本，使用单台节点服务器执行服务器启动、停止命令显然操作烦琐，所以可将 Zookeeper 启动、停止命令封装成脚本。具体操作步骤如下。

（1）在 hadoop102 的 /home/atguigu/bin 目录下创建脚本 zk.sh。

```
[atguigu@hadoop102 bin]$ vim zk.sh
```

脚本思路：通过执行 ssh 命令，分别登录集群节点服务器，然后执行启动、停止或者查看服务状态的命令。在脚本中编写如下内容。

```
#! /bin/bash

case $1 in
"start"){
 for i in hadoop102 hadoop103 hadoop104
 do
  ssh $i "source /etc/profile ; /opt/module/zookeeper-3.4.10/bin/zkServer.sh start"
 done
};;
"stop"){
 for i in hadoop102 hadoop103 hadoop104
 do
  ssh $i "source /etc/profile ; /opt/module/zookeeper-3.4.10/bin/zkServer.sh stop"
 done
```

```
};;
"status"){
 for i in hadoop102 hadoop103 hadoop104
 do
  ssh $i "source /etc/profile ; /opt/module/zookeeper-3.4.10/bin/zkServer.sh status"
 done
};;
esac
```

（2）增加脚本执行权限。

```
[atguigu@hadoop102 bin]$ chmod 777 zk.sh
```

（3）Zookeeper 集群启动脚本。

```
[atguigu@hadoop102 module]$ zk.sh start
```

（4）Zookeeper 集群停止脚本。

```
[atguigu@hadoop102 module]$ zk.sh stop
```

4.3.3 Kafka 安装

Kafka 是一个优秀的分布式消息队列系统，通过将日志消息先发送至 Kafka，可以规避数据丢失的风险，增加数据处理的可扩展性，提高数据处理的灵活性和峰值处理能力，提高系统可用性，为消息消费提供顺序保证，并且可以控制优化数据流经系统的速度，解决消息生产和消息消费速度不一致的问题。

Kafka 集群需要依赖 Zookeeper 提供服务来保存一些元数据信息，以保证系统可用性。在完成 Zookeeper 的安装之后，就可以安装 Kafka 了，具体安装步骤如下。

（1）Kafka 集群规划如表 4-1 所示。

表 4-1 Kafka 集群规划

hadoop102	hadoop103	hadoop104
Zookeeper	Zookeeper	Zookeeper
Kafka	Kafka	Kafka

（2）下载安装包。

下载 Kafka 的安装包。

（3）解压安装包。

```
[atguigu@hadoop102 software]$ tar -zxvf kafka_2.11-0.11.0.0.tgz -C /opt/module/
```

（4）修改解压后的文件名称。

```
[atguigu@hadoop102 module]$ mv kafka_2.11-0.11.0.0/ kafka
```

（5）在/opt/module/kafka 目录下创建 logs 文件夹，用于保存 Kafka 运行过程中产生的日志文件。

```
[atguigu@hadoop102 kafka]$ mkdir logs
```

（6）进入 Kafka 的配置目录，打开 server.properties，修改配置文件，Kafka 的配置文件都是以键值对的形式存在的，主要需要修改的内容如下。

```
[atguigu@hadoop102 kafka]$ cd config/
[atguigu@hadoop102 config]$ vim server.properties
```

找到对应的配置并按照如下内容进行修改。

```
#broker 的全局唯一编号，不能重复
broker.id=0
#配置删除 Topic 功能为 true，即在 Kafka 中删除 Topic 为真正删除，而不是标记删除
delete.topic.enable=true
#处理网络请求的线程数量
num.network.threads=3
#用来处理磁盘 I/O 的线程数量
num.io.threads=8
#发送套接字的缓冲区大小
socket.send.buffer.bytes=102400
#接收套接字的缓冲区大小
socket.receive.buffer.bytes=102400
#请求套接字的缓冲区大小
socket.request.max.bytes=104857600
#Kafka 运行日志存放的路径，配置为自行创建的 logs 文件夹
log.dirs=/opt/module/kafka/logs
#Topic 在当前 broker 上的分区个数
num.partitions=1
#用来恢复和清理 data 下数据的线程数量
num.recovery.threads.per.data.dir=1
#数据文件保留的最长时间，超时则被删除
log.retention.hours=168
#配置连接 Zookeeper 集群的地址
zookeeper.connect=hadoop102:2181,hadoop103:2181,hadoop104:2181
```

（7）配置环境变量，将 Kafka 的安装目录配置到系统环境变量中，可以更加方便用户执行 Kafka 的相关命令。在配置完环境变量后，需要执行 source 命令使环境变量生效。

```
[root@hadoop102 module]# vim /etc/profile
#KAFKA_HOME
export KAFKA_HOME=/opt/module/kafka
export PATH=$PATH:$KAFKA_HOME/bin
[root@hadoop102 module]# source /etc/profile
```

（8）安装配置全部修改完成后，分发安装包和环境变量到集群其他节点服务器，并使环境变量生效。

```
[root@hadoop102 etc]# xsync profile
[atguigu@hadoop102 module]$ xsync kafka/
```

（9）修改 broker.id。

分别在 hadoop103 和 hadoop104 上修改配置文件/opt/module/kafka/config/server.operties 中的 broker.id=1、broker.id=2。

注意：broker.id 为识别 Kafka 集群不同节点服务器的标识，不可重复。

（10）启动集群。

依次在 hadoop102、hadoop103 和 hadoop104 上启动 Kafka。

```
[atguigu@hadoop102 kafka]$ bin/kafka-server-start.sh config/server.properties &
[atguigu@hadoop103 kafka]$ bin/kafka-server-start.sh config/server.properties &
[atguigu@hadoop104 kafka]$ bin/kafka-server-start.sh config/server.properties &
```

（11）关闭集群。

```
[atguigu@hadoop102 kafka]$ bin/kafka-server-stop.sh stop
[atguigu@hadoop103 kafka]$ bin/kafka-server-stop.sh stop
[atguigu@hadoop104 kafka]$ bin/kafka-server-stop.sh stop
```

4.3.4 Kafka 集群启动、停止脚本

同 Zookeeper 一样，将 Kafka 集群的启动、停止命令写成脚本，方便以后调用执行。

（1）在/home/atguigu/bin 目录下创建脚本 kf.sh。

```
[atguigu@hadoop102 bin]$ vim kf.sh
```

在脚本中编写如下内容。

```
#! /bin/bash

case $1 in
"start"){
        for i in hadoop102 hadoop103 hadoop104
        do
                echo " --------启动 $i Kafka-------"
                ssh $i "source /etc/profile ; /opt/module/kafka/bin/kafka-server-start.sh -daemon /opt/module/kafka/config/server.properties "
        done
};;
"stop"){
        for i in hadoop102 hadoop103 hadoop104
        do
                echo " --------停止 $i Kafka-------"
                ssh $i " source /etc/profile ; /opt/module/kafka/bin/kafka-server-stop.sh stop"
        done
};;
esac
```

（2）增加脚本执行权限。

```
[atguigu@hadoop102 bin]$ chmod 777 kf.sh
```

（3）Kafka 集群启动脚本。

```
[atguigu@hadoop102 module]$ kf.sh start
```

（4）Kafka 集群停止脚本。

```
[atguigu@hadoop102 module]$ kf.sh stop
```

4.3.5 Kafka Topic 相关操作

本节主要带领读者熟悉 Kafka 的常用命令行操作。在本项目中，学会使用命令行操作 Kafka 已经足够，若想更加深入地了解 Kafka，体验 Kafka 其余的优秀特性，读者可以通过尚硅谷教育公众号获取 Kafka 的相关视频资料，自行学习。

（1）查看 Kafka Topic 列表。

```
[atguigu@hadoop102 kafka]$ bin/kafka-topics.sh --zookeeper hadoop102:2181 --list
```

（2）创建 Kafka Topic。

进入 /opt/module/kafka/ 目录下，分别创建启动日志主题、事件日志主题，创建的日志主题名称应该与采集日志层 Flume 的配置文件中的名称相同。

① 创建启动日志主题。

```
[atguigu@hadoop102 kafka]$ bin/kafka-topics.sh --zookeeper hadoop102:2181,hadoop103:2181, hadoop104:2181  --create --replication-factor 1 --partitions 1 --topic topic_start
```

② 创建事件日志主题。

```
[atguigu@hadoop102 kafka]$ bin/kafka-topics.sh --zookeeper hadoop102:2181,hadoop103:2181,hadoop104:2181  --create --replication-factor 1 --partitions 1 --topic topic_event
```

（3）删除 Kafka Topic 命令。

若在创建主题时出现错误，则可以使用删除主题命令对主题进行删除。

① 删除启动日志主题。

```
[atguigu@hadoop102 kafka]$ bin/kafka-topics.sh --delete --zookeeper hadoop102:2181,hadoop103:2181,hadoop104:2181 --topic topic_start
```

② 删除事件日志主题。

```
[atguigu@hadoop102 kafka]$ bin/kafka-topics.sh --delete --zookeeper hadoop102:2181,hadoop103:2181,hadoop104:2181 --topic topic_event
```

（4）Kafka 控制台生产消息测试。

```
 [atguigu@hadoop102 kafka]$ bin/kafka-console-producer.sh \
--broker-list hadoop102:9092 --topic topic_start
>hello world
>atguigu  atguigu
```

（5）Kafka 控制台消费消息测试。

```
[atguigu@hadoop102 kafka]$ bin/kafka-console-consumer.sh \
--bootstrap-server hadoop102:9092 --from-beginning --topic topic_start
```

其中，--from-beginning 表示将主题中以往所有的数据都读取出来。用户可根据业务场景选择是否增加该配置。

（6）查看 Kafka Topic 详情。

```
[atguigu@hadoop102 kafka]$ bin/kafka-topics.sh --zookeeper hadoop102:2181 \
--describe --topic topic_start
```

（7）开启 Kafka 控制台消费消息测试，执行采集日志 Flume 启动脚本，查看控制台是否有日志打印，以验证日志是否采集成功。

```
[atguigu@hadoop102 kafka]$ bin/kafka-console-consumer.sh \
--bootstrap-server hadoop102:9092 --from-beginning --topic topic_start
[atguigu@hadoop102 module]$ f1.sh start
```

4.4 消费 Kafka 日志的 Flume

将日志从采集日志层 Flume 发送到 Kafka 集群后，接下来的工作需要将日志数据进行落盘存储，我们依然将这部分工作交给 Flume 完成，如图 4-3 所示。

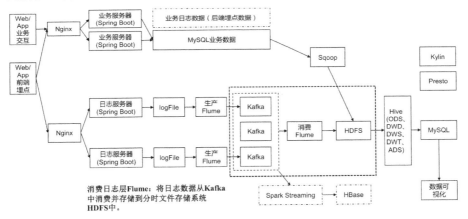

图 4-3 消费日志层 Flume

将消费日志层 Flume Agent 程序部署在 hadoop104 上，实现 hadoop102、hadoop103 负责日志的生成和采集，hadoop104 负责日志的消费存储。在实际生产环境中应尽量做到将不同的任务部署在不同的节点服务器上。消费日志层 Flume 集群规划如表 4-2 所示。

表 4-2 消费日志层 Flume 集群规划

	节点服务器 hadoop102	节点服务器 hadoop103	节点服务器 hadoop104
Flume（消费 Kafka）			Flume

4.4.1 消费日志 Flume 配置

1. Flume 配置分析

消费日志层 Flume 主要从 Kafka 中读取消息，所以选用 Kafka Source。Channel 选用 File

Channel，能最大限度避免数据丢失。Sink 选用 HDFS Sink，可以将日志直接落盘到 HDFS 中。

消费日志层 Flume 配置分析如图 4-4 所示，该层 Flume 需要从不同的 Kafka Topic 消费读取消息，再将日志落盘到 HDFS 的不同目录中，所以我们可以简单搭建两个消息通道，分别进行组装。

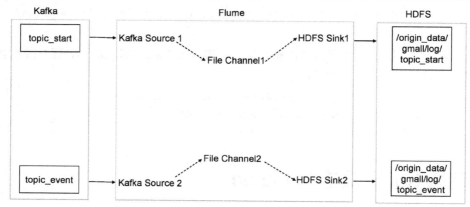

图 4-4 消费日志层 Flume 配置分析

2．Flume 具体配置

（1）在 hadoop104 的 /opt/module/flume/conf 目录下创建 kafka-flume-hdfs.conf 文件。

```
[atguigu@hadoop104 conf]$ vim kafka-flume-hdfs.conf
```

（2）在文件中配置如下内容。

```
## Flume Agent 组件声明
a1.sources=r1 r2
a1.channels=c1 c2
a1.sinks=k1 k2

## Source1 属性配置
#配置 Source 类型为 Kafka Source
a1.sources.r1.type = org.apache.flume.source.kafka.KafkaSource
#配置 Kafka Source 每次从 Kafka Topic 中拉取的 event 个数
a1.sources.r1.batchSize = 5000
#配置拉取数据批次间隔为 2000 毫秒
a1.sources.r1.batchDurationMillis = 2000
#配置 Kafka 集群地址
a1.sources.r1.kafka.bootstrap.servers =
hadoop102:9092,hadoop103:9092,hadoop104:9092
#配置 Source 对接 Kafka 主题
a1.sources.r1.kafka.topics=topic_start

## source2 属性配置，与 Source1 配置类似，只是消费主题不同
a1.sources.r2.type = org.apache.flume.source.kafka.KafkaSource
a1.sources.r2.batchSize = 5000
a1.sources.r2.batchDurationMillis = 2000
a1.sources.r2.kafka.bootstrap.servers =
hadoop102:9092,hadoop103:9092,hadoop104:9092
```

```
a1.sources.r2.kafka.topics=topic_event

## Channel1 属性配置
#配置 Channel 类型为 File Channel
a1.channels.c1.type = file
#配置存储 File Channel 传输数据的断点信息目录
a1.channels.c1.checkpointDir = /opt/module/flume/checkpoint/behavior1
#配置 File Channel 传输数据的存储位置
a1.channels.c1.dataDirs = /opt/module/flume/data/behavior1/
#配置 File Channel 的最大存储容量
a1.channels.c1.maxFileSize = 2146435071
#配置 File Channel 最多存储 event 的个数
a1.channels.c1.capacity = 1000000
#配置 Channel 满时 put 事务的超时时间
a1.channels.c1.keep-alive = 6

## Channel2 属性配置同 Channel1，注意需要配置不同的目录路径
a1.channels.c2.type = file
a1.channels.c2.checkpointDir = /opt/module/flume/checkpoint/behavior2
a1.channels.c2.dataDirs = /opt/module/flume/data/behavior2/
a1.channels.c2.maxFileSize = 2146435071
a1.channels.c2.capacity = 1000000
a1.channels.c2.keep-alive = 6

## Sink1 属性配置
#配置 Sink1 类型为 HDFS Sink
a1.sinks.k1.type = hdfs
#配置发到 HDFS 的存储路径
a1.sinks.k1.hdfs.path = /origin_data/gmall/log/topic_start/%Y-%m-%d
#配置 HDFS 落盘文件的文件名前缀
a1.sinks.k1.hdfs.filePrefix = logstart-

##Sink2 属性配置同 Sink1
a1.sinks.k2.type = hdfs
a1.sinks.k2.hdfs.path = /origin_data/gmall/log/topic_event/%Y-%m-%d
a1.sinks.k2.hdfs.filePrefix = logevent-

## 避免产生大量小文件的相关属性配置
a1.sinks.k1.hdfs.rollInterval = 10
a1.sinks.k1.hdfs.rollSize = 134217728
a1.sinks.k1.hdfs.rollCount = 0

a1.sinks.k2.hdfs.rollInterval = 10
a1.sinks.k2.hdfs.rollSize = 134217728
a1.sinks.k2.hdfs.rollCount = 0

## 控制输出文件是压缩文件
a1.sinks.k1.hdfs.fileType = CompressedStream
```

```
a1.sinks.k2.hdfs.fileType = CompressedStream

a1.sinks.k1.hdfs.codeC = lzop
a1.sinks.k2.hdfs.codeC = lzop

## 拼装
a1.sources.r1.channels = c1
a1.sinks.k1.channel= c1

a1.sources.r2.channels = c2
a1.sinks.k2.channel= c2
```

4.4.2 消费日志 Flume 启动、停止脚本

将消费日志层 Flume 程序的启动、停止命令编写成脚本，方便后续调用执行，脚本包括启动消费层 Flume 程序和根据 Flume 的任务编号停止其运行，与采集日志层 Flume 启动、停止脚本类似，编写步骤如下。

（1）在/home/atguigu/bin 目录下创建脚本 f2.sh。

```
[atguigu@hadoop102 bin]$ vim f2.sh
```

在脚本中编写如下内容。

```
#! /bin/bash

case $1 in
"start"){
      for i in hadoop104
      do
            echo " --------启动 $i 消费flume-------"
            ssh $i " source /etc/profile ; nohup /opt/module/flume/bin/flume-ng agent --conf-file /opt/module/flume/conf/kafka-flume-hdfs.conf --name a1 -Dflume.root.logger=INFO,LOGFILE >/opt/module/flume/log.txt   2>&1 &"
      done
};;
"stop"){
      for i in hadoop104
      do
            echo " --------停止 $i 消费flume-------"
            ssh $i "ps -ef | grep kafka-flume-hdfs | grep -v grep |awk '{print \$2}' | xargs kill"
      done

};;
esac
```

（2）增加脚本执行权限。

```
[atguigu@hadoop102 bin]$ chmod 777 f2.sh
```

(3) f2集群启动脚本。

`[atguigu@hadoop102 module]$ f2.sh start`

(4) f2集群停止脚本。

`[atguigu@hadoop102 module]$ f2.sh stop`

4.5 采集通道启动、停止脚本

在完成所有的采集日志落盘工作后，我们需要将本章涉及的所有命令和脚本统一封装成采集通道启动、停止脚本，否则一项一项开启采集通道的进程也是非常耗时的，编写步骤如下。

（1）在/home/atguigu/bin 目录下创建脚本 cluster.sh。

`[atguigu@hadoop102 bin]$ vim cluster.sh`

在脚本中编写如下内容。

```
#! /bin/bash

case $1 in
"start"){
 echo " -------- 启动 集群 -------"

 echo " -------- 启动 hadoop集群 -------"
 /opt/module/hadoop-2.7.2/sbin/start-dfs.sh
 ssh hadoop103 "source /etc/profile ; /opt/module/hadoop-2.7.2/sbin/start-yarn.sh"

#启动 Zookeeper集群
zk.sh start

    #Zookeeper的启动需要一定时间，此时间根据用户计算机的性能而定，可适当调整
    sleep 4s;

#启动 Flume采集集群
f1.sh start

#启动 Kafka采集集群
    #Kafka的启动需要一定时间，此时间根据用户计算机的性能而定，可适当调整
    kf.sh start

    sleep 6s;

#启动 Flume消费集群
 f2.sh start
};;
"stop"){
```

```
echo " -------- 停止 集群 -------"

#停止 Flume 消费集群
f2.sh stop

#停止 Kafka 采集集群
kf.sh stop

sleep 6s;

#停止 Flume 采集集群
f1.sh stop

#停止 Zookeeper 集群
zk.sh stop

echo " -------- 停止 hadoop 集群 -------"
ssh hadoop103 "/opt/module/hadoop-2.7.2/sbin/stop-yarn.sh"
/opt/module/hadoop-2.7.2/sbin/stop-dfs.sh
};;
esac
```

（2）增加脚本执行权限。

```
[atguigu@hadoop102 bin]$ chmod 777 cluster.sh
```

（3）cluster 集群启动脚本。

```
[atguigu@hadoop102 module]$ cluster.sh start
```

（4）cluster 集群停止脚本。

```
[atguigu@hadoop102 module]$ cluster.sh stop
```

4.6 本章总结

本章主要对用户行为数据采集模块的搭建进行了讲解，包括采集框架 Flume 的安装配置、Kafka 的安装部署和 Zookeeper 的安装部署，并对整个采集系统的整体框架进行了详细讲解。在本章中，读者除了需要学会搭建完整的大数据采集系统，还需要掌握数据采集框架 Flume 的基本用法。例如，如何编辑 Flume 的 Agent 配置文件，以及如何设置 Flume 的各项属性，此外，还应具备一定的 Shell 脚本编写能力，学会编写基本的程序启动、停止脚本。

第 5 章 业务数据采集模块

在数据仓库技术出现之前，人们对业务数据的分析、提取处于直接访问查询阶段，直接访问查询业务数据虽然速度快，但也存在很多问题。例如，业务数据的表结构为事务处理的性能而优化，有时并不适合查询与分析，事务处理的优先级通常高于分析系统，若二者运行于同一硬件上，分析系统的性能往往很差，而且很有可能影响业务系统的性能。所以将业务数据采集进数据仓库系统是非常有必要的。

本章主要讲解如何将业务数据采集进数据仓库系统，以及在业务数据的采集过程中需要注意的问题。

5.1 电商业务概述

在进行需求的实现之前，本节先对业务数据仓库的基础理论进行讲解，包含本项目主要涉及的电商业务流程、电商常识及电商表结构等。

5.1.1 电商业务流程

如图 5-1 所示，下面以一个普通用户的浏览足迹为例对电商的业务流程进行说明。用户打开电商网站首页开始浏览，可能通过分类查询或全文搜索寻找自己中意的商品，这些商品都存储在后台的管理系统中。

当用户找到自己中意的商品并想要购买时，可能将商品添加到购物车，此时发现需要登录。登录后对商品进行结算，这时候购物车的管理和商品订单信息的生成都会对业务数据仓库产生影响，会生成相应的订单数据和支付数据。

订单正式生成之后，系统还会对订单进行跟踪处理，直到订单全部完成。

电商的业务流程主要包括用户在前台浏览商品时的商品详情管理、用户将商品加入购物车进行支付时的用户个人中心和支付服务管理，以及用户支付完成后的订单后台服务管理，这些流程涉及十几张或几十张业务数据表，甚至更多。

数据仓库用于辅助管理者决策，与业务流程息息相关，建设数据模型的首要前提是了解业务流程，只有了解了业务流程，才能为数据仓库的建立提供指导方向，从而反过来为业务提供更好的决策数据支撑，让数据仓库的价值最大化。

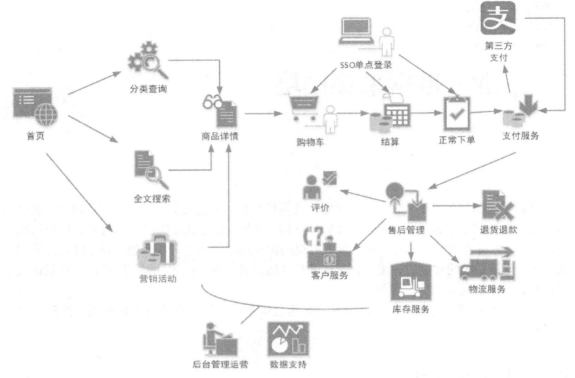

图 5-1　电商业务流程

5.1.2　电商常识

SKU 是 Stock Keeping Unit（库存量基本单位）的缩写，现在已经被引申为产品统一编号的简称，每种产品均对应唯一的 SKU。SPU 是 Standard Product Unit（标准产品单位）的缩写，是商品信息聚合的最小单位，是一组可复用、易检索的标准化信息集合。通过 SPU 表示一类商品的好处是可以共用商品的图片、海报、销售属性等。

例如，iPhone11 手机就是 SPU。一部白色、128GB 内存的 iPhone11，就是 SKU。在电商网站的商品详情页面，所有不同类型的 iPhone11 手机可以共用商品海报和商品图片等信息，避免了数据的冗余。

5.1.3　电商表结构

如图 5-2 所示为本电商数据仓库系统涉及的业务数据表结构，以订单表、用户表、SKU商品表、活动表和优惠券表为中心，延伸出优惠券领用表、支付流水表、活动订单关联表、订单详情表、订单状态表、商品评价表、编码字典表、退款表、SPU 商品表等。用户表提供用户的详细信息，支付流水表提供订单的支付详情，订单详情表提供订单的商品数量等信息，SKU 商品表为订单详情表提供商品的详细信息。本章只以图 5-2 中的 24 张表为例进行讲解，在实际项目中，业务数据库中的表格远远不止这些。

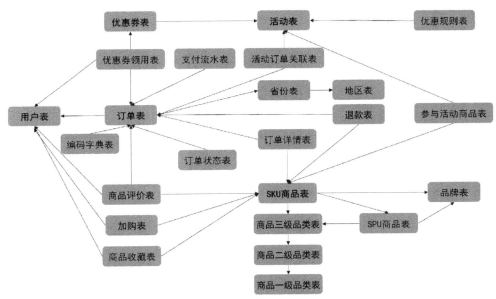

图 5-2 本电商数据仓库系统涉及的业务数据表结构

各张表的表结构如表 5-1～表 5-24 所示。

表 5-1 订单表

标　　签	含　　义
id	编号
consignee	收件人
consignee_tel	收件人电话
final_total_amount	总金额
order_status	订单状态
user_id	用户 id
delivery_address	送货地址
order_comment	订单备注
out_trade_no	订单交易编号（第三方支付用）
trade_body	订单描述（第三方支付用）
create_time	创建时间
operate_time	操作时间
expire_time	失效时间
tracking_no	物流订单编号
parent_order_id	父订单编号
img_url	图片路径
province_id	省份 id
benefit_reduce_amount	优惠金额
original_total_amount	原价金额
feight_fee	运费

表 5-2 订单详情表

标　　签	含　　义
id	编号
order_id	订单编号
sku_id	商品 id
sku_name	商品名称（冗余）
img_url	图片名称（冗余）
order_price	商品价格（下单时商品的价格）
sku_num	商品数量
create_time	创建时间
source_type	来源类型
source_id	来源编号

表 5-3 SKU 商品表

标　　签	含　　义
id	商品 id
spu_id	标准产品单位 id
price	价格
sku_name	商品名称
sku_desc	商品描述
weight	重量
tm_id	品牌 id
category3_id	三级品类 id
sku_default_img	默认显示图片（冗余）
create_time	创建时间

表 5-4 用户表

标　　签	含　　义
id	用户 id
login_name	用户名称
nick_name	用户昵称
passwd	用户密码
name	真实姓名
phone_num	手机号
email	邮箱
head_img	头像
user_level	用户级别
birthday	生日
gender	性别：男=M，女=F
create_time	创建时间
operate_time	操作时间

表 5-5 商品一级品类表

标签	含义
id	id
name	名称

表 5-6 商品二级品类表

标签	含义
id	id
name	名称
category1_id	一级品类 id

表 5-7 商品三级品类表

标签	含义
id	id
name	名称
category2_id	二级品类 id

表 5-8 支付流水表

标签	含义
id	编号
out_trade_no	对外业务编号
order_id	订单编号
user_id	用户 id
alipay_trade_no	支付宝交易流水编号
total_amount	支付金额
subject	交易内容
payment_type	支付类型
payment_time	支付时间

表 5-9 省份表

标签	含义
id	编号
name	省份名称
region_id	地区 id
area_code	地区编码
iso_code	国际编码

表 5-10 地区表

标签	含义
id	编号
name	地区名称

表 5-11 品牌表

标　　签	含　　义
id	id
tm_name	品牌名称

表 5-12 订单状态表

标　　签	含　　义
id	编号
order_id	订单编号
order_status	订单状态
operate_time	操作时间

表 5-13 SPU 商品表

标　　签	含　　义
id	商品 id
spu_name	标准产品单位名称
description	商品描述（后台简述）
category3_id	三级品类 id
tm_id	品牌 id

表 5-14 商品评价表

标　　签	含　　义
id	编号
user_id	用户 id
sku_id	商品 id
spu_id	标准产品单位 id
order_id	订单编号
appraise	评价：好评=1，中评=2，差评=3
comment_txt	评价内容
create_time	创建时间

表 5-15 退款表

标　　签	含　　义
id	编号
user_id	用户 id
order_id	订单编号
sku_id	商品 id
refund_type	退款类型
refund_amount	退款金额
refund_reason_type	退款原因类型
refund_reason_txt	退款原因内容
create_time	创建时间

表 5-16 加购表

标　签	含　义
id	编号
user_id	用户 id
sku_id	商品 id
cart_price	放入购物车时的价格
sku_num	数量
img_url	图片文件
sku_name	商品名称（冗余）
create_time	创建时间
operate_time	操作时间
is_ordered	是否已经下单
order_time	下单时间
source_type	来源类型
source_id	来源编号

表 5-17 商品收藏表

标　签	含　义
id	编号
user_id	用户 id
sku_id	商品 id
spu_id	标准产品单位 id
is_cancel	是否取消：正常=0，已取消=1
create_time	收藏时间
cancel_time	取消时间

表 5-18 优惠券领用表

标　签	含　义
id	编号
coupon_id	优惠券 id
user_id	用户 id
order_id	订单编号
coupon_status	优惠券状态
get_time	领券时间
using_time	使用（下单）时间
used_time	使用（支付）时间
expire_time	过期时间

表 5-19 优惠券表

标　签	含　义
id	优惠券编号
coupon_name	优惠券名称
coupon_type	优惠券类型：现金券=1，折扣券=2，满减券=3，满件打折券=4

87

续表

标签	含义
condition_amount	满减金额
condition_num	满减件数
activity_id	活动编号
benefit_amount	优惠金额
benefit_discount	优惠折扣
create_time	创建时间
range_type	范围类型：商品=1，品类=2，品牌=3
spu_id	标准产品单位 id
tm_id	品牌 id
category3_id	三级品类 id
limit_num	最多领用次数
operate_time	操作时间
expire_time	过期时间

表 5-20 活动表

标签	含义
id	活动 id
activity_name	活动名称
activity_type	活动类型
activity_desc	活动描述
start_time	开始时间
end_time	结束时间
create_time	创建时间

表 5-21 活动订单关联表

标签	含义
id	编号
activity_id	活动 id
order_id	订单编号
create_time	创建时间

表 5-22 优惠规则表

标签	含义
id	编号
activity_id	活动 id
condition_amount	满减金额
condition_num	满减件数
benefit_amount	优惠金额
benefit_discount	优惠折扣
benefit_level	优惠级别

表 5-23 编码字典表

标签	含义
dic_code	编号
dic_name	编码名称
parent_code	父编号
create_time	创建时间
operate_time	操作时间

表 5-24 参与活动商品表

标签	含义
id	编号
activity_id	活动 id
sku_id	商品 id
create_time	创建时间

5.1.4 数据同步策略

数据同步是指将数据从关系型数据库同步到大数据的存储系统中，针对不同类型的表应该有不同的同步策略。

表的类型包括每日全量表、每日增量表、每日新增及变化表和拉链表。
- 每日全量表：存储完整的数据。
- 每日增量表：存储新增加的数据。
- 每日新增及变化表：存储新增加的数据和变化的数据。
- 拉链表：对新增及变化表进行定期合并。

数据同步策略如下。

1. 每日全量同步策略

每日全量同步策略是指每天存储一份完整数据作为一个分区，适用于表数据量不大，且每天既会有新数据插入，又会有旧数据修改的场景。

维度表数据量通常比较小，可以进行每日全量同步，即每天存储一份完整数据。

2. 每日增量同步策略

每日增量同步策略是指每天存储一份增量数据作为一个分区，适用于表数据量大，且每天只会有新数据插入的场景。

3. 新增及变化策略

新增及变化策略只同步每日的新增及变化数据，适用于表的数据量大，有新增也有修改，但修改频率不高（缓慢变化维度）的场景。这类表从数据量的角度考虑，若采用每日全量同步策略，则数据量太大，冗余也太大；若采用每日增量同步策略，则无法反映数据的变化情况。利用每日新增及变化表，制作一张拉链表，可以方便地获取某个时间切片的快照数据。

拉链表是指在源表字段的基础上，增加一个开始时间和一个结束时间，该时间段用来表示一个状态的生命周期。制作拉链表，每天只需要同步新增和修改的数据即可，如表 5-25 所示，可以通过类似 "select * from user where start =<'2019-01-02' and end>='2019-01-02'" 的查询语句来获取 2019-01-02 当天的所有订单状态信息。

表 5-25　拉链表

name	start	end
张三	1990-01-01	2018-12-31
张小三	2019-01-01	2019-04-30
张大三	2019-05-01	9999-99-99
……	……	……

4．特殊维度同步策略

某些特殊的维度表可以不遵循上述数据同步策略。

1）客观世界维度

对于没有变化的客观世界的维度（比如，性别、地区、民族等），可以只存储一份固定值。

2）日期维度

对于日期维度，可以一次性导入一年或若干年的数据。

5.2　业务数据采集

业务数据通常存储在关系型数据库中，为了进行数据采集，我们首先要生成业务数据，然后选用合适的数据采集工具。在本项目中，我们选用 MySQL 作为业务数据的生成和存储数据库，选用 Sqoop 作为数据采集工具。本节主要讲解 MySQL 和 Sqoop 的安装部署、业务数据的生成和建模，以及业务数据导入数据仓库的相关内容。

5.2.1　MySQL 安装

1．安装包准备

（1）使用 rpm 命令配合管道符查看 MySQL 是否已经安装，其中，-q 选项为 query，-a 选项为 all，意思为查询全部安装，如果已经安装 MySQL，则将其卸载。

① 查看 MySQL 是否已经安装。

```
[root@hadoop102 桌面]# rpm -qa|grep MySQL
mysql-libs-5.1.73-7.el6.x86_64
```

② 卸载，-e 选项表示卸载，--nodeps 选项表示无视所有依赖强制卸载。

```
[root@hadoop102 桌面]# rpm -e --nodeps mysql-libs-5.1.73-7.el6.x86_64
```

（2）将 MySQL 安装包 mysql-libs.zip 上传至/opt/software 目录下并解压。

```
[root@hadoop102 software]# unzip mysql-libs.zip
[root@hadoop102 software]# ls
```

```
mysql-libs.zip
mysql-libs
```

2. 安装 MySQL 服务器端

（1）安装 MySQL 服务器端，使用 rpm 命令安装 MySQL，-i 选项为 install，-v 选项为 vision，-h 选项用于展示安装过程。

```
[root@hadoop102 mysql-libs]# rpm -ivh MySQL-server-5.6.24-1.el6.x86_64.rpm
```

（2）服务器端安装完成后会生成一个默认随机密码，存储在/root/.mysql_secret 文件中，root 用户可以直接使用 cat 命令或 sudo cat 命令查看产生的随机密码，登录 MySQL 后需要立即更改密码。

```
[root@hadoop102 mysql-libs]# cat /root/.mysql_secret
OEXaQuS8IWkG19Xs
```

（3）以 root 用户身份登录，或者使用 sudo 命令查看 MySQL 服务的运行状态，可以看到现在是"MySQL is not running…"状态。

```
[root@hadoop102 mysql-libs]# service mysql status
MySQL is not running...
```

（4）以 root 用户身份登录，或者使用 sudo 命令启动 MySQL。

```
[root@hadoop102 mysql-libs]# service mysql start
```

3. 安装 MySQL 客户端

（1）使用 rpm 命令安装 MySQL 客户端。

```
[root@hadoop102 mysql-libs]# rpm -ivh MySQL-client-5.6.24-1.el6.x86_64.rpm
```

（2）登录 MySQL，以 root 用户身份登录，密码为安装服务器端时自动生成的随机密码。

```
[root@hadoop102 mysql-libs]# mysql -uroot -pOEXaQuS8IWkG19Xs
```

（3）登录 MySQL 后，立即修改密码，修改密码后记住该密码。

```
mysql>SET PASSWORD=PASSWORD('000000');
```

（4）退出 MySQL。

```
mysql>exit
```

4. MySQL 中 user 表的主机配置

由于 MySQL 需要被 Hive 连接访问，为避免出现权限问题，我们可以配置在任何主机上只要使用 root 用户身份就可登录 MySQL。为此，我们需要修改 MySQL 的 user 表，具体步骤如下。

（1）以 root 用户身份输入设置的密码，登录 MySQL。

```
[root@hadoop102 mysql-libs]# mysql -uroot -p000000
```

（2）显示目前 MySQL 的所有数据库。

```
mysql>show databases;
```

（3）使用 MySQL 数据库。

```
mysql>use mysql;
```

（4）显示 MySQL 数据库中的所有表。

```
mysql>show tables;
```

（5）user 表中存储了允许登录 MySQL 的用户、密码等信息，展开 user 表，可以发现 user 表的字段非常多。

```
mysql>desc user;
```

（6）查询 user 表中的部分字段 host、user、password，显示关键信息，可以看到 4 条信息，host 分别是 localhost、hadoop102、127.0.0.1 和::1，这 4 个 host 都表明只有本机可以连接 MySQL。

```
mysql>select user, host, password from user;
```

（7）修改 user 表，把 host 字段值修改为%，使任何 host 都可以通过 root 用户+密码连接 MySQL。

```
mysql>update user set host='%' where host='localhost';
```

（8）只有删除 root 用户的其他 host，才能使"%"生效。

```
mysql>delete from user where Host='hadoop102 ';
mysql>delete from user where Host='127.0.0.1';
mysql>delete from user where Host='::1';
```

（9）刷新权限，使修改生效。

```
mysql>flush privileges;
```

（10）退出。

```
mysql> quit;
```

5.2.2 业务数据生成

业务数据的建库、建表和数据生成通过导入脚本完成，建议读者安装一个数据库可视化工具。本节以 SQLyog 为例进行讲解，SQLyog 的安装包可以在通过公众号获取的本书资料中获得，安装过程不再讲解，数据生成步骤如下。

1．建表语句的导入

（1）通过 SQLyog 创建数据库 gmall。

（2）设置数据库编码，如图 5-3 所示。

图 5-3　设置数据库编码

2. 数据生成

（1）在 hadoop102 的 /opt/module 目录下创建 db_log 文件夹。

```
[atguigu@hadoop102 module]$ mkdir db_log
```

（2）在通过公众号获取的本书资料中找到 gmall-mock-db-SNAPSHOT.jar 和 application.properties 文件，把 gmall-mock-db-SNAPSHOT.jar 和 application.properties 上传到 hadoop102 的 /opt/module/db_log 目录下。

（3）根据需求修改 application.properties 的相关配置，通过修改业务日期的配置可以生成不同日期的业务数据。

```
logging.level.root=info

spring.datasource.driver-class-name=com.mysql.jdbc.Driver
spring.datasource.url=jdbc:mysql://hadoop102:3306/gmall?characterEncoding=utf-8&useSSL=false&serverTimezone=GMT%2B8
spring.datasource.username=root
spring.datasource.password=000000

logging.pattern.console=%m%n
mybatis-plus.global-config.db-config.field-strategy=not_null

#业务日期
mock.date=2020-03-10
#是否重置
mock.clear=1

#生成新用户个数
mock.user.count=50
#生成新用户中男性的比例
mock.user.male-rate=20

#收藏和取消收藏的操作比例
mock.favor.cancel-rate=10
#收藏操作次数
mock.favor.count=100

#加入购物车操作次数
mock.cart.count=10
#每件商品最多加入购物车的数量
mock.cart.sku-maxcount-per-cart=3

#用户下单比例
mock.order.user-rate=80
#用户从购物车中结算商品的比例
mock.order.sku-rate=70
#是否参加活动
mock.order.join-activity=1
```

```
#是否使用优惠券
mock.order.use-coupon=1
#优惠券领取人数
mock.coupon.user-count=10

#用户提交订单后的支付比例
mock.payment.rate=70
#不同支付方式比例,支付宝:微信:银联
mock.payment.payment-type=30:60:10

#不同评价比例,好:中:差:自动
mock.comment.appraise-rate=30:10:10:50

#不同退款原因比例,质量问题:商品描述与实际描述不一致:缺货:号码不合适:拍错:不想买了:其他
mock.refund.reason-rate=30:10:20:5:15:5:5
```

(4) 在/opt/module/db_log 目录下执行如下命令,生成 2020-03-10 的数据。

```
[atguigu@hadoop102 db_log]$ java -jar gmall-mock-db-SNAPSHOT.jar
```

(5) 在配置文件 application.properties 中修改如下配置。

```
mock.date=2020-03-11
mock.clear=0
```

(6) 再次执行命令,生成 2020-03-11 的数据。

```
[atguigu@hadoop102 db_log]$ java -jar gmall-mock-db-SNAPSHOT.jar
```

5.2.3 业务数据建模

有时候,我们需要处理的业务数据并没有表与表之间的关系图,所以需要自己在 SQLyog 上建模。其他数据库可视化工具也有类似功能,此处不再赘述。建模具体过程如下。

(1) 在"文件"下拉菜单中选择"新架构设计器"命令,如图 5-4 所示。

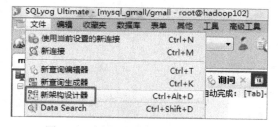

图 5-4 选择"新架构设计器"命令

(2) 打开架构设计器页面,如图 5-5 所示①。

---

① 图 5-5 中"拖拽"的正确写法为"拖曳"。

图 5-5　架构设计器页面

（3）依次将左侧 gmall 数据库中的表拖曳到右侧的架构设计器页面，如图 5-6 所示。

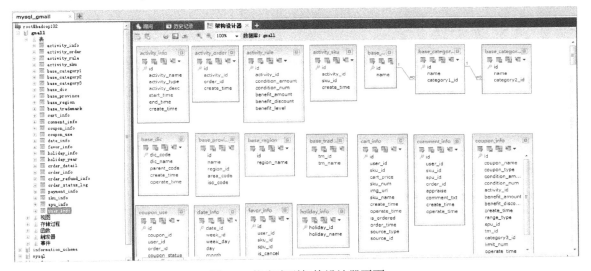

图 5-6　拖曳表到架构设计器页面

（4）根据外键关系，将所有表连接在一起。

例如，商品三级品类表关联商品二级品类表，如图 5-7 所示。

图 5-7　设置商品三级品类表外键

又如，商品二级品类表关联商品一级品类表，如图 5-8 所示。

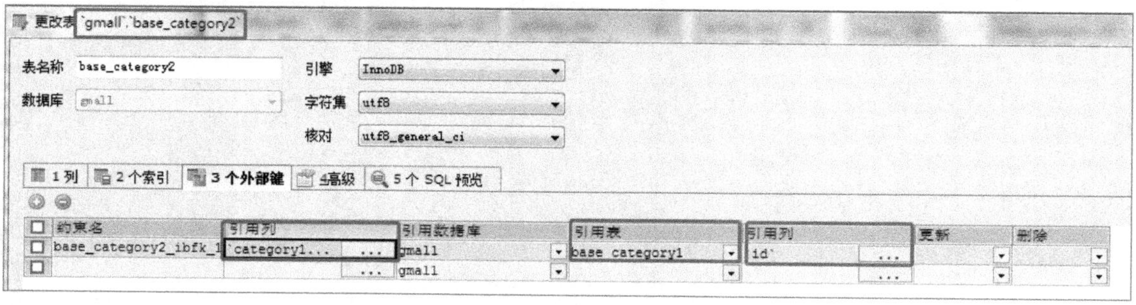

图 5-8　设置商品二级品类表外键

采用以上方法，将所有表关联起来，就可以得到与图 5-2 相似的电商业务数据表结构，方便用户清晰明了地看出所需处理的业务数据表之间的建模关系。

5.2.4　Sqoop 安装

Sqoop 是一个用于将关系型数据库和 Hadoop 中的数据进行相互转移的工具，可以将一个关系型数据库（如 MySQL、Oracle）中的数据导入 Hadoop（如 HDFS、Hive、HBase）中，也可以将 Hadoop（如 HDFS、Hive、HBase）中的数据导入关系型数据库（如 MySQL、Oracle）中。Sqoop 的安装步骤如下。

1．下载并解压

（1）下载安装包。
（2）上传安装包 sqoop-1.4.6.bin_hadoop-2.0.4-alpha.tar.gz 到虚拟机中。
（3）解压 Sqoop 安装包到指定目录。

```
[atguigu@hadoop102 software]$ tar -zxf sqoop-1.4.6.bin_hadoop-2.0.4-alpha.tar.gz -C /opt/module/
```

2．修改配置文件

Sqoop 的配置文件与大多数大数据框架类似，存储在 Sqoop 根目录的 conf 目录下。
（1）重命名配置文件。

```
[atguigu@hadoop102 conf]$ mv sqoop-env-template.sh sqoop-env.sh
```

（2）修改配置文件 sqoop-env.sh。

```
[atguigu@hadoop102 conf]$ vim sqoop-env.sh
```

增加如下内容。

```
export HADOOP_COMMON_HOME=/opt/module/hadoop-2.7.2
export HADOOP_MAPRED_HOME=/opt/module/hadoop-2.7.2
export ZOOKEEPER_HOME=/opt/module/zookeeper-3.4.10
export ZOOCFGDIR=/opt/module/zookeeper-3.4.10
```

3. 复制 JDBC 驱动

复制 JDBC 驱动到 Sqoop 的 lib 目录下。

```
[atguigu@hadoop102 mysql]$ cp mysql-connector-java-5.1.27-bin.jar /opt/module/
sqoop-1.4.6.bin__hadoop-2.0.4-alpha/lib/
```

4. 验证 Sqoop

我们可以通过执行如下命令来验证 Sqoop 的配置是否正确。

```
[atguigu@hadoop102 sqoop]$ bin/sqoop help
```

然后会出现 Warning 警告（警告信息已省略），并伴随着帮助命令的输出。

```
Available commands:
  codegen            Generate code to interact with database records
  create-hive-table  Import a table definition into Hive
  eval               Evaluate a SQL statement and display the results
  export             Export an HDFS directory to a database table
  help               List available commands
  import             Import a table from a database to HDFS
  import-all-tables  Import tables from a database to HDFS
  import-mainframe   Import datasets from a mainframe server to HDFS
  job                Work with saved jobs
  list-databases     List available databases on a server
  list-tables        List available tables in a database
  merge              Merge results of incremental imports
  metastore          Run a standalone Sqoop metastore
  version            Display version information
```

5. 测试 Sqoop 是否能够成功连接数据库

```
[atguigu@hadoop102 sqoop]$ bin/sqoop list-databases --connect jdbc:mysql://
hadoop102:3306/ --username root --password 000000
```

若出现如下输出，则表示连接成功。

```
information_schema
metastore
mysql
oozie
performance_schema
```

5.2.5 业务数据导入数据仓库

业务数据表同步策略分析如图 5-9 所示，根据表格性质的不同制定不同的数据同步策略。在后续的数据导入过程中，将根据不同的数据同步策略执行不同的数据导入命令。

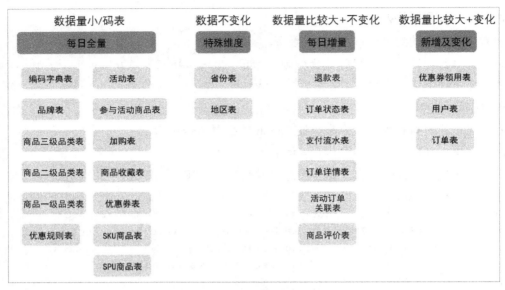

图 5-9　业务数据表同步策略分析

业务数据生成完毕之后，即可通过 Sqoop 数据导入命令将数据导入 HDFS 中，供后续数据仓库的搭建使用。在 Sqoop 中，"导入"的概念是指使用 import 关键字从非大数据集群（RDBMS）向大数据集群（HDFS、Hive、HBase）中传输数据，数据的导入指令示例如下。

```
/opt/module/sqoop/bin/sqoop import \
--connect \
--username \
--password \
--target-dir \
--delete-target-dir \
--num-mappers \
--fields-terminated-by \
--query "$2" "and $CONDITIONS;"
```

指令参数说明如下。

- --connect：指定 JDBC 的 url，需精确到数据库；
- --username：指定连接数据库使用的用户名；
- --password：指定连接数据库使用的密码；
- --target-dir：指定 HDFS 导入表的存储目录；
- --delete-target-dir：是否删除存在的 import 目标目录；
- --num-mappers：指定并发执行的数据导入作业个数；
- --fields-terminated-by：指定每个字段以什么符号结束；
- --query：指定查询语句，其中，"and $CONDITIONS;"必须存在。

1. Sqoop 定时导入脚本

（1）在/home/atguigu/bin 目录下创建脚本 mysql_to_hdfs.sh。

```
[atguigu@hadoop102 bin]$ vim mysql_to_hdfs.sh
```

在脚本中编写如下内容。

```bash
#!/bin/bash

sqoop=/opt/module/sqoop/bin/sqoop

if [ -n "$2" ] ;then
    do_date=$2
else
    do_date=`date -d '-1 day' +%F`
fi

#编写通用数据导入指令，通过第一个参数传入表名，通过第二个参数传入查询语句，对导入数据使用LZO
#压缩格式，并对LZO压缩文件创建索引
import_data(){
$sqoop import \
--connect jdbc:mysql://hadoop102:3306/gmall \
--username root \
--password 000000 \
--target-dir /origin_data/gmall/db/$1/$do_date \
--delete-target-dir \
--query "$2 and  \$CONDITIONS" \
--num-mappers 1 \
--fields-terminated-by '\t' \
--compress \
--compression-codec lzop \
--null-string '\\N' \
--null-non-string '\\N'

hadoop jar /opt/module/hadoop-2.7.2/share/hadoop/common/hadoop-lzo-0.4.20.jar com.hadoop.compression.lzo.DistributedLzoIndexer/origin_data/gmall/db/$1/$do_date
}

#针对不同表格分别调用不同的通用数据导入指令，全量导入数据的表格的查询条件为 where 1=1 ，增量
#导入数据的表格的查询条件为当天日期
import_order_info(){
  import_data order_info "select
                    id,
                    final_total_amount,
                    order_status,
                    user_id,
                    out_trade_no,
                    create_time,
                    operate_time,
                    province_id,
                    benefit_reduce_amount,
                    original_total_amount,
```

```
                    feight_fee
                from order_info
                where (date_format(create_time,'%Y-%m-%d')='$do_date'
                or date_format(operate_time,'%Y-%m-%d')='$do_date')"
}

import_coupon_use(){
  import_data coupon_use "select
                    id,
                    coupon_id,
                    user_id,
                    order_id,
                    coupon_status,
                    get_time,
                    using_time,
                    used_time
                from coupon_use
                where (date_format(get_time,'%Y-%m-%d')='$do_date'
                or date_format(using_time,'%Y-%m-%d')='$do_date'
                or date_format(used_time,'%Y-%m-%d')='$do_date')"
}

import_order_status_log(){
  import_data order_status_log "select
                        id,
                        order_id,
                        order_status,
                        operate_time
                    from order_status_log
                    where
date_format(operate_time,'%Y-%m-%d')='$do_date'"
}

import_activity_order(){
  import_data activity_order "select
                    id,
                    activity_id,
                    order_id,
                    create_time
                from activity_order
                where date_format(create_time,'%Y-%m-%d')='$do_date'"
}

import_user_info(){
  import_data "user_info" "select
                    id,
                    name,
```

```
                    birthday,
                    gender,
                    email,
                    user_level,
                    create_time,
                    operate_time
                from user_info
                where (DATE_FORMAT(create_time,'%Y-%m-%d')='$do_date'
                or DATE_FORMAT(operate_time,'%Y-%m-%d')='$do_date')"
}

import_order_detail(){
  import_data order_detail "select
                    od.id,
                    order_id,
                    user_id,
                    sku_id,
                    sku_name,
                    order_price,
                    sku_num,
                    od.create_time
                from order_detail od
                join order_info oi
                on od.order_id=oi.id
                where
DATE_FORMAT(od.create_time,'%Y-%m-%d')='$do_date'"
}

import_payment_info(){
  import_data "payment_info" "select
                      id,
                      out_trade_no,
                      order_id,
                      user_id,
                      alipay_trade_no,
                      total_amount,
                      subject,
                      payment_type,
                      payment_time
                  from payment_info
                  where
DATE_FORMAT(payment_time,'%Y-%m-%d')='$do_date'"
}

import_comment_info(){
  import_data comment_info "select
                    id,
```

```
                    user_id,
                    sku_id,
                    spu_id,
                    order_id,
                    appraise,
                    comment_txt,
                    create_time
                from comment_info
                where date_format(create_time,'%Y-%m-%d')='$do_date'"
}
import_order_refund_info(){
  import_data order_refund_info "select
                    id,
                    user_id,
                    order_id,
                    sku_id,
                    refund_type,
                    refund_num,
                    refund_amount,
                    refund_reason_type,
                    create_time
                from order_refund_info
                where date_format(create_time,'%Y-%m-%d')='$do_date'"
}
import_sku_info(){
  import_data sku_info "select
                    id,
                    spu_id,
                    price,
                    sku_name,
                    sku_desc,
                    weight,
                    tm_id,
                    category3_id,
                    create_time
                from sku_info where 1=1"
}
import_base_category1(){
  import_data "base_category1" "select
                    id,
                    name
                from base_category1 where 1=1"
}
```

```
import_base_category2(){
  import_data "base_category2" "select
                          id,
                          name,
                          category1_id
                    from base_category2 where 1=1"
}

import_base_category3(){
  import_data "base_category3" "select
                          id,
                          name,
                          category2_id
                    from base_category3 where 1=1"
}

import_base_province(){
  import_data base_province "select
                      id,
                      name,
                      region_id,
                      area_code,
                      iso_code
                    from base_province
                    where 1=1"
}

import_base_region(){
  import_data base_region "select
                      id,
                      region_name
                    from base_region
                    where 1=1"
}

import_base_trademark(){
  import_data base_trademark "select
                        tm_id,
                        tm_name
                    from base_trademark
                    where 1=1"
}

import_spu_info(){
  import_data spu_info "select
                    id,
                    spu_name,
```

```
                        category3_id,
                        tm_id
                    from spu_info
                    where 1=1"
}

import_favor_info(){
  import_data favor_info "select
                        id,
                        user_id,
                        sku_id,
                        spu_id,
                        is_cancel,
                        create_time,
                        cancel_time
                    from favor_info
                    where 1=1"
}

import_cart_info(){
  import_data cart_info "select
                        id,
                        user_id,
                        sku_id,
                        cart_price,
                        sku_num,
                        sku_name,
                        create_time,
                        operate_time,
                        is_ordered,
                        order_time
                    from cart_info
                    where 1=1"
}

import_coupon_info(){
  import_data coupon_info "select
                        id,
                        coupon_name,
                        coupon_type,
                        condition_amount,
                        condition_num,
                        activity_id,
                        benefit_amount,
                        benefit_discount,
                        create_time,
                        range_type,
```

```
                        spu_id,
                        tm_id,
                        category3_id,
                        limit_num,
                        operate_time,
                        expire_time
                    from coupon_info
                    where 1=1"
}

import_activity_info(){
  import_data activity_info "select
                        id,
                        activity_name,
                        activity_type,
                        start_time,
                        end_time,
                        create_time
                    from activity_info
                    where 1=1"
}

import_activity_rule(){
    import_data activity_rule "select
                                id,
                                activity_id,
                                condition_amount,
                                condition_num,
                                benefit_amount,
                                benefit_discount,
                                benefit_level
                        from activity_rule
                        where 1=1"
}

import_base_dic(){
    import_data base_dic "select
                        dic_code,
                        dic_name,
                        parent_code,
                        create_time,
                        operate_time
                    from base_dic
                    where 1=1"
}
import_activity_sku(){
    import_data activity_sku "select
```

```
                            id,
                            activity_id,
                            sku_id,
                            create_time
                        from activity_sku
                        where 1=1"
}
#对传入的第一个参数进行判断,根据传入参数的不同决定导入哪种表数据,若传入first,则表示初次执
#行脚本,导入所有表数据;若传入all,则导入除地区外的所有表数据
case $1 in
  "order_info")
     import_order_info
;;
  "base_category1")
     import_base_category1
;;
  "base_category2")
     import_base_category2
;;
  "base_category3")
     import_base_category3
;;
  "order_detail")
     import_order_detail
;;
  "sku_info")
     import_sku_info
;;
  "user_info")
     import_user_info
;;
  "payment_info")
     import_payment_info
;;
  "base_province")
     import_base_province
;;
  "base_region")
     import_base_region
;;
  "base_trademark")
     import_base_trademark
;;
  "activity_info")
     import_activity_info
;;
  "activity_order")
```

```
        import_activity_order
;;
  "cart_info")
        import_cart_info
;;
  "comment_info")
        import_comment_info
;;
  "coupon_info")
        import_coupon_info
;;
  "coupon_use")
        import_coupon_use
;;
  "favor_info")
        import_favor_info
;;
  "order_refund_info")
        import_order_refund_info
;;
  "order_status_log")
        import_order_status_log
;;
  "spu_info")
        import_spu_info
;;
  "activity_rule")
        import_activity_rule
;;
  "base_dic")
        import_base_dic
;;
"activity_sku")
        import_activity_sku
;;
"first")
    import_base_category1
    import_base_category2
    import_base_category3
    import_order_info
    import_order_detail
    import_sku_info
    import_user_info
    import_payment_info
    import_base_province
    import_base_region
    import_base_trademark
```

```
    import_activity_info
    import_activity_order
    import_cart_info
    import_comment_info
    import_coupon_use
    import_coupon_info
    import_favor_info
    import_order_refund_info
    import_order_status_log
    import_spu_info
    import_activity_rule
    import_base_dic
    import_activity_sku
;;
"all")
    import_base_category1
    import_base_category2
    import_base_category3
    import_order_info
    import_order_detail
    import_sku_info
    import_user_info
    import_payment_info
    import_base_trademark
    import_activity_info
    import_activity_order
    import_cart_info
    import_comment_info
    import_coupon_use
    import_coupon_info
    import_favor_info
    import_order_refund_info
    import_order_status_log
    import_spu_info
    import_activity_rule
    import_base_dic
    import_activity_sku
;;
esac
```

（2）增加脚本执行权限。

```
[atguigu@hadoop102 bin]$ chmod 777 mysql_to_hdfs.sh
```

（3）初次执行脚本，传入 first，导入 2020-03-10 的数据。

```
[atguigu@hadoop102 bin]$ mysql_to_hdfs.sh first 2020-03-10
```

（4）再次执行脚本，传入 all，导入 2020-03-11 的数据。

```
[atguigu@hadoop102 bin]$ mysql_to_hdfs.sh all 2020-03-11
```

2. Sqoop 导入数据异常处理

（1）问题描述：执行 Sqoop 导入数据脚本时，发生如下异常。

```
java.sql.SQLException: Streaming result set com.mysql.jdbc.RowDataDynamic@
65d6b83b is still active. No statements may be issued when any streaming result
sets are open and in use on a given connection. Ensure that you have
called .close() on any active streaming result sets before attempting more
queries.
 at com.mysql.jdbc.SQLError.createSQLException(SQLError.java:930)
 at com.mysql.jdbc.MysqlIO.checkForOutstandingStreamingData(MysqlIO.java:2646)
 at com.mysql.jdbc.MysqlIO.sendCommand(MysqlIO.java:1861)
 at com.mysql.jdbc.MysqlIO.sqlQueryDirect(MysqlIO.java:2101)
 at com.mysql.jdbc.ConnectionImpl.execSQL(ConnectionImpl.java:2548)
 at com.mysql.jdbc.ConnectionImpl.execSQL(ConnectionImpl.java:2477)
 at com.mysql.jdbc.StatementImpl.executeQuery(StatementImpl.java:1422)
 at com.mysql.jdbc.ConnectionImpl.getMaxBytesPerChar(ConnectionImpl.java:2945)
 at com.mysql.jdbc.Field.getMaxBytesPerCharacter(Field.java:582)
```

（2）问题解决方案：增加如下导入参数。

```
--driver com.mysql.jdbc.Driver \
```

5.3 本章总结

本章主要对业务数据采集模块进行了搭建，在搭建过程中，读者可以发现，业务数据具有数量众多且多种多样的数据表，所以需要针对不同类型的数据表制定不同的数据同步策略，然后在制定好策略的前提下选用合适的数据采集工具。通过本章的学习，希望读者对电商业务数据的采集工作有更多的了解。

第6章

数据仓库搭建模块

经过对数据采集模块的搭建，现在已经有了可以分析的数据源，但这些数据仅停留在大数据存储系统中是发挥不了任何作用的，我们只有对其进行分析重构才能发现其中的价值，也就是要搭建数据仓库。

数据仓库（Data Warehouse）是一个面向主题（Subject Oriented）的、集成（Integrate）的、相对稳定（Non-Volatile）的、反映历史变化（Time Variant）的数据集合，用于支持管理决策。我们可以从两个层次理解数据仓库的概念：首先，数据仓库用于支持管理决策，面向分析型数据处理，它不同于企业现有的操作型数据库；其次，数据仓库是对多个异构的数据源的有效集成，集成后按照主题进行重组，并包含历史数据，而且存放在数据仓库中的数据一般不再修改。

只有对数据仓库和数据仓库搭建工具有一定的了解，才能实现最终需求，本章从数据仓库的基础知识、环境搭建、需求实现几个方面展开介绍。

6.1 数据仓库理论准备

在搭建数据仓库之前，先对数据仓库的基础理论知识进行介绍。数据仓库包含的内容很多，包括架构、建模、方法论等，对应到具体工作中，可以概括为如下内容。

- 以 Hadoop、Spark、Hive 等组件为中心的数据架构体系。
- 各种数据建模方法，如维度建模。
- 数据同步策略。
- 数据仓库分层的相关理论。

无论数据仓库的规模有多大，在数据仓库搭建之初，只有对基础理论知识有一定的掌握，对数据仓库的整体架构有所规划，才能搭建出合理高效的数据仓库体系。

本节将围绕数据建模、数据同步策略、数据仓库分层理论几个方面，为读者介绍数据仓库的深层内核知识。

6.1.1 范式理论

关系型数据库在设计时，要遵守一定的规范要求，目前业界的范式包括第一范式（1NF）、第二范式（2NF）、第三范式（3NF）、巴斯-科德范式（BCNF）、第四范式（4NF）和第五范式（5NF）。范式可以理解为一张数据表的表结构符合的设计标准的级别。使用范式的根本目的包括如下两点。

（1）减少数据冗余，尽量让每个数据只出现一次。

（2）保证数据的一致性。

其缺点是在获取数据时，需要通过 join 连接出最后的数据。

1．什么是函数依赖

函数依赖示例如表 6-1 所示。

表 6-1 函数依赖示例：学生成绩表

学号	姓名	系名	系主任	课名	分数
1	李小明	经济系	王强	高等数学	95
1	李小明	经济系	王强	大学英语	87
1	李小明	经济系	王强	普通化学	76
2	张莉莉	经济系	王强	高等数学	72
2	张莉莉	经济系	王强	大学英语	98
2	张莉莉	经济系	王强	计算机基础	82
3	高芳芳	法律系	刘玲	高等数学	88
3	高芳芳	法律系	刘玲	法学基础	84

函数依赖分为完全函数依赖、部分函数依赖和传递函数依赖。

1）完全函数依赖

设 (X, Y) 是关系 R 的两个属性集合，X' 是 X 的真子集，存在 $X \rightarrow Y$，但对每一个 X' 都有 $X'! \rightarrow Y$，则称 Y 完全依赖于 X。

比如，通过（学号，课名）可推出分数，但单独用学号推不出分数，那么就可以说，分数完全依赖于（学号，课名）。

通过 (A, B) 能得出 C，但单独通过 A 或 B 得不出 C，那么就可以说，C 完全依赖于 (A, B)。

2）部分函数依赖

假如 Y 依赖于 X，但同时 Y 并不完全依赖于 X，那么就可以说，Y 部分依赖于 X。

比如，通过（学号，课名）可推出姓名，但直接通过学号也可以推出姓名，所以姓名部分依赖于（学号，课名）。

通过 (A, B) 能得出 C，通过 A 也能得出 C，或者通过 B 也能得出 C，那么就可以说，C 部分依赖于 (A, B)。

3）传递函数依赖

设 (X, Y, Z) 是关系 R 中互不相同的属性集合，存在 $X \rightarrow Y(Y! \rightarrow X), Y \rightarrow Z$，则称 Z 传递依赖于 X。

比如，通过学号可推出系名，通过系名可推出系主任，但通过系主任推不出学号，系主任主要依赖于系名。这种情况可以说，系主任传递依赖于学号。

通过 A 可得到 B，通过 B 可得到 C，但通过 C 得不到 A，那么就可以说，C 传递依赖于 A。

2. 第一范式

第一范式（1NF）的核心原则是属性不可分割。如表 6-2 所示，"商品"列中的数据不是原子数据项，是可以分割的，明显不符合第一范式。

表 6-2 不符合第一范式的表格设计

id	商 品	商家 id	用户 id
001	5 台计算机	×××旗舰店	00001

对表 6-2 进行修改，使表格符合第一范式的要求，如表 6-3 所示。

表 6-3 符合第一范式的表格设计

id	商 品	数量（台）	商家 id	用户 id
001	计算机	5	×××旗舰店	00001

实际上，第一范式是所有关系型数据库最基本的要求，在关系型数据库（RDBMS），如 SQL Server、Oracle、MySQL 中创建数据表时，如果数据表的设计不符合这个最基本的要求，那么操作一定是不能成功的。也就是说，只要在 RDBMS 中已经存在的数据表，一定是符合第一范式的。

3. 第二范式

第二范式（2NF）的核心原则是不能存在部分函数依赖。

如表 6-1 所示，该表格明显存在部分函数依赖。这张表的主键是（学号，课名），分数确实完全依赖于（学号，课名），但姓名并不完全依赖于（学号，课名）。

将表 6-1 进行调整，结果如表 6-4 和表 6-5 所示，即去掉部分函数依赖，使其符合第二范式。

表 6-4 学号-课名-分数表

学 号	课 名	分 数
1	高等数学	95
1	大学英语	87
1	普通化学	76
2	高等数学	72
2	大学英语	98
2	计算机基础	82
3	高等数学	88
3	法学基础	84

表 6-5 学号-姓名-系明细表

学 号	姓 名	系 名	系 主 任
1	李小明	经济系	王强
2	张莉莉	经济系	王强
3	高芳芳	法律系	刘玲

4. 第三范式

第三范式（3NF）的核心原则是不能存在传递函数依赖。

表 6-5 中存在传递函数依赖，通过系主任不能推出学号，将表格进一步拆分，使其符合第三范式，如表 6-6 和表 6-7 所示。

表 6-6 学号-姓名-系名表

学 号	姓 名	系 名
1	李小明	经济系
2	张莉莉	经济系
3	高芳芳	法律系

表 6-7 系名-系主任表

系 名	系 主 任
经济系	王强
法律系	刘玲

6.1.2 关系模型与维度模型

当今的数据处理大致可以分成两大类：联机事务处理（On-Line Transaction Processing，OLTP）、联机分析处理（On-Line Analytical Processing，OLAP）。OLTP 是传统关系型数据库的主要应用，主要是基本的、日常的事务处理，如银行交易。OLAP 是数据仓库系统的主要应用，支持复杂的分析操作，侧重决策支持，并且可提供直观、易懂的查询结果。二者的主要区别如表 6-8 所示。

表 6-8 OLTP 与 OLAP 的主要区别

对 比 属 性	OLTP	OLAP
读特性	每次查询只返回少量记录	对大量记录进行汇总
写特性	随机、低延时写入用户的输入	批量导入
使用场景	用户，Java EE 项目	内部分析师，为决策提供支持
数据表征	最新数据状态	随时间变化的历史状态
数据规模	GB	TP 到 PB

关系模型示意如图 6-1 所示，严格遵循第三范式（3NF）。从图 6-1 中可以看出，模型较为松散、零碎，物理表数量多，但数据冗余程度低。由于数据分布于众多的表中，因此这些数据可以更为灵活地被应用，功能性较强。关系模型主要应用于 OLTP 中，为了保证数据的一致性及避免冗余，大部分业务系统的表都是遵循第三范式的。

维度模型示意如图 6-2 所示，其主要应用于 OLAP 中，通常以某一张事实表为中心进行表的组织，主要面向业务，其特征是可能存在数据的冗余，但用户能方便地得到数据。

关系模型虽然数据冗余程度低，但在大规模数据中进行跨表分析、统计、查询时，会造成多表关联，这会大大降低执行效率。所以通常我们采用维度模型建模，把各种相关表整理成事实表和维度表两种。所有的维度表围绕事实表进行解释。

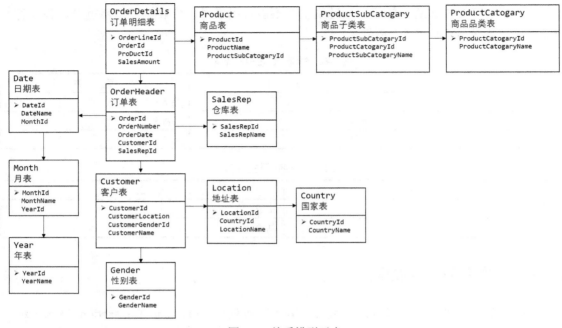

图 6-1 关系模型示意

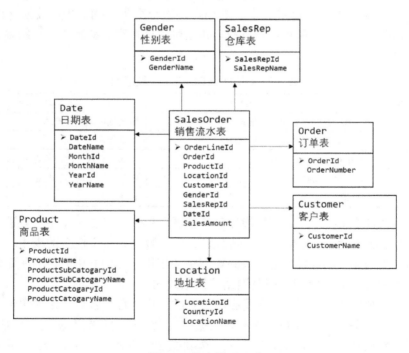

图 6-2 维度模型示意

6.1.3 星形模型、雪花模型与星座模型

在维度建模的基础上，有三种模型：星形模型、雪花模型与星座模型。

当所有维度表都直接连接到事实表上时，整个图解就像星星一样，故该模型称为星形模型，如图 6-3 所示。星形模型是一种非正规化的结构，多维数据集的每个维度都直接与事实表相连接，不存在渐变维度，所以数据有一定的冗余。例如，在地域维度表中，存在国家 A 省 B 的城市 C 及国家 A 省 B 的城市 D 两条记录，那么国家 A 和省 B 的信息分别存储了两次，即存在冗余。

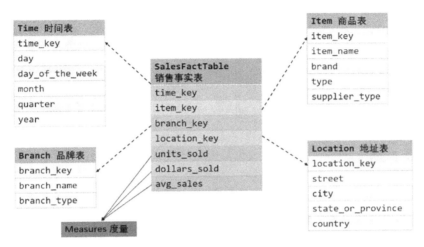

图 6-3　星形模型建模示意

当有一张或多张维度表没有直接连接到事实表上，而通过其他维度表连接到事实表上时，其图解就像多个雪花连接在一起，故该模型称为雪花模型。雪花模型是对星形模型的扩展。它对星形模型的维度表进行进一步层次化，原有的各维度表可能被扩展为小的事实表，形成一些局部的"层次"区域，这些被分解的表都连接到主维度表而不是事实表上，如图 6-4 所示。雪花模型的优点是可通过最大限度地减少数据存储量及联合较小的维度表来改善查询性能。雪花模型去除了数据冗余，比较靠近第三范式，但无法完全遵守，因为遵守第三范式的成本太高。

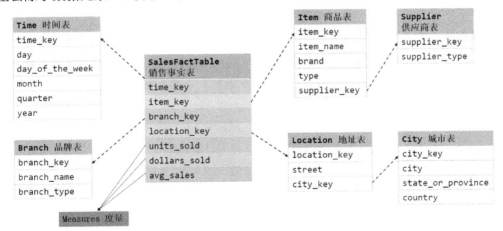

图 6-4　雪花模型建模示意

星座模型与前两种模型的区别是事实表的数量，星座模型是基于多张事实表的，且事实表之间共享一些维度表。星座模型与前两种模型并不冲突。如图 6-5 所示为星座模型建模示意。

星座模型基本上是很多数据仓库的常态,因为很多数据仓库都有多张事实表。

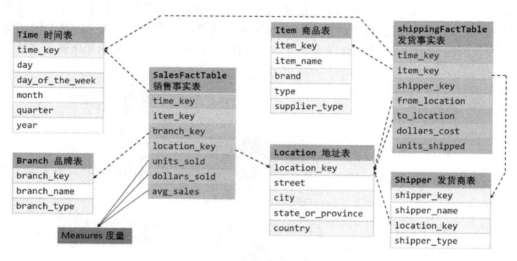

图 6-5 星座模型建模示意

目前在企业实际开发中,其不会只选择一种模型,而会根据情况灵活组合,甚至使各模型并存(一层维度和多层维度都保存)。但是,从整体来看,企业更倾向于选择维度更少的星形模型。尤其是对于 Hadoop 体系,减少 join 就是减少中间数据的传输和计算,可明显改善性能。

6.1.4 表的分类

在数据仓库建模理论中,通常将表分为事实表和维度表两类。事实表加维度表,能够描述一个完整的业务事件。例如,昨天早上张三在某电商平台花费 200 元购买了一个皮包,描述该业务事件需要三个维度,分别是时间维度(昨天早上)、商家维度(电商平台)和商品维度(皮包)。具体分类原则如下。

1. 事实表

事实表中的每行数据代表一个业务事件。"事实"这个术语表示的是业务事件的度量值。例如,订单事件中的下单金额。并且事实表会不断扩大,数据量一般较大。

1)事务性事实表

事务性事实表以每个事务或事件为单位,例如,以一笔支付记录作为事实表中的一行数据。

2)周期型快照事实表

周期型快照事实表中不会保留所有数据,只保留固定时间间隔的数据,例如,每天或每月的销售额,以及每月的账户余额等。

3)累积型快照事实表

累积型快照事实表(见表 6-9)用于跟踪业务事实的变化。例如,数据仓库中可能需要累计或存储从下订单开始到订单商品被打包、运输和签收的各业务阶段的时间点数据来跟踪订单生命周期的进展情况。当这个业务过程进行时,事实表的记录也要不断更新。

表 6-9 累积型快照事实表

订单 id	用户 id	下单时间	打包时间	发货时间	签收时间
001	000001	2019-02-12 10:10	2019-02-12 11:10	2019-02-12 12:10	2019-02-12 13:10

2. 维度表

维度表一般是指对应业务状态编号的解释表，也可以称为码表。例如，订单状态表、商品品类表等，如表 6-10 和表 6-11 所示。

表 6-10 订单状态表

订单状态编号	订单状态名称
1	未支付
2	支付
3	发货中
4	已发货
5	已完成

表 6-11 商品品类表

商品品类编号	品类名称
1	服装
2	保健
3	电器
4	图书

6.1.5 为什么要分层

数据仓库中的数据要想真正发挥最大的作用，必须对数据仓库进行分层，数据仓库分层的优点如下。

- 把复杂问题简单化。可以将一个复杂的任务分解成多个步骤来完成，每层只处理单一的步骤。
- 减少重复开发。规范数据分层，通过使用中间层数据，可以大大减少重复计算量，增加计算结果的复用性。
- 隔离原始数据。使真实数据与最终统计数据解耦。

数据仓库具体如何分层取决于设计者对数据仓库的整体规划，不过大部分的思路是相似的。本书将数据仓库分为五层，如下所述。

- ODS 层：原始数据层，存放原始数据，直接加载原始日志、数据，数据保持原貌不做处理。
- DWD 层：明细数据层，对 ODS 层数据进行清洗（去除空值、脏数据、超过极限范围的数据）、维度退化、脱敏等。
- DWS 层：服务数据层，以 DWD 层的数据为基础，按天进行轻度汇总。
- DWT 层：主题数据层，以 DWS 层的数据为基础，按主题进行汇总，获得每个主题的全量数据表。
- ADS 层：数据应用层，面向实际的数据需求，为各种统计报表提供数据。

数据仓库分层后要遵守一定的数据仓库命名规范，本项目中的规范如下。

1. 表命名

ODS 层命名为 ods_表名。
DWD 层命名为 dwd_dim/fact_表名。
DWS 层命名为 dws_表名。
DWT 层命名为 dwt_表名。
ADS 层命名为 ads_表名。
临时表命名为 tmp_×××。
用户行为表以.log 为后缀。

2. 脚本命名

脚本命名格式为数据源_to_目标_db/log.sh。
用户行为需求相关脚本以.log 为后缀；业务数据需求相关脚本以.db 为后缀。

6.1.6 数据仓库建模

在本项目中，数据仓库分为五层，分别为 ODS 层（原始数据层）、DWD 层（明细数据层）、DWS 层（服务数据层）、DWT 层（主题数据层）和 ADS 层（数据应用层）。本节主要探讨五层架构中的数据仓库建模思想。

1. ODS 层

ODS 层主要进行如下处理。
（1）数据保持原貌不进行任何修改，起到备份数据的作用。
（2）数据采用压缩格式，以减少磁盘存储空间（例如，100GB 的原始数据，可以被压缩到 10GB 左右）。
（3）创建分区表，防止后续进行全表扫描。

2. DWD 层

DWD 层需要构建维度模型，一般采用星形模型，而呈现的状态一般为星座模型。
维度建模一般按照以下四个步骤进行。

1）选择业务过程

在业务系统中，若业务表过多，则可挑选我们感兴趣的业务线，如下单业务、支付业务、退款业务及物流业务，一条业务线对应一张事实表。小型公司的业务表可能比较少，建议选择所有业务线。

2）声明粒度

数据粒度是指数据仓库中保存数据的细化程度或综合程度的级别。
声明粒度意味着精确定义事实表中的一行数据所表示的内容，应该尽可能选择最细粒度，以此来满足各种各样的需求。

典型的粒度声明如下。

- 将订单中的每个商品项作为下单事实表中的一行数据，粒度为每次。
- 将每周的订单次数作为一行数据，粒度为每周。
- 将每月的订单次数作为一行数据，粒度为每月。

如果在 DWD 层的粒度就是每周或者每月，那么后续就没有办法统计更细粒度（如每天）的指标了。所以建议采用最细粒度。

3）确定维度

维度的主要作用是描述业务事实，主要表示的是"谁、何处、何时"等信息，如时间维度、用户维度、地区维度等常见维度。

4）确定事实

此处的"事实"一词指的是业务中的度量值。例如，订单表的度量值就是订单件数、订单金额等。

在 DWD 层中，以业务过程为建模驱动，基于每个具体业务过程的特点，构建最细粒度的明细数据层事实表。事实表可进行适当的宽表化处理。

通过以上步骤，并结合本数据仓库项目的业务事实，得出业务总线矩阵表，如表 6-12 所示。业务总线矩阵的原则主要是判断维度表和事实表之间的关系，若两者有关联，则使用 √ 标记。

表 6-12 业务总线矩阵表

	时间	用户	地区	商品	优惠券	活动	编码	度量值
订单	√	√	√			√		件数/金额
订单详情	√	√	√	√				件数/金额
支付	√	√	√					次数/金额
加购	√	√		√				件数/金额
收藏	√	√		√				个数
评价	√	√		√				条数
退款	√	√		√				件数/金额
优惠券领用	√	√			√			个数

根据维度建模中的星形模型思想，将维度进行退化。如图 6-6 所示，地区表和省份表退化为地区维度表，SKU 商品表、品牌表、SPU 商品表、商品三级品类表、商品二级品类表、商品一级品类表退化为商品维度表，活动订单关联表和优惠规则表退化为活动维度表。

至此，数据仓库的维度建模已经完毕，DWS 层、DWT 层、ADS 层和维度建模已经没有关系了。

DWS 层和 DWT 层都是按照主题来创建宽表的，而主题相当于观察问题的角度，不同的维度表意味着不同的角度。

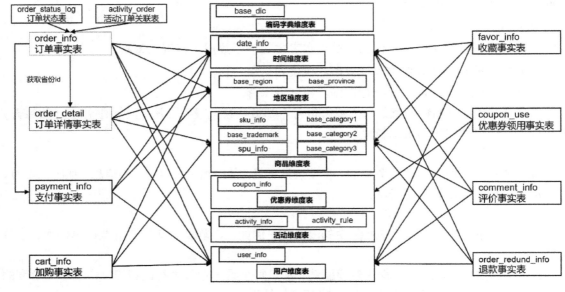

图 6-6　维度退化

3. DWS 层

DWS 层用于统计各主题对象的当天行为，服务于 DWT 层的主题宽表。如图 6-7 所示，DWS 层的宽表字段是站在不同维度的视角去看事实表的，重点关注事实表的度量值，通过与之关联的事实表，获得不同事实表的度量值。

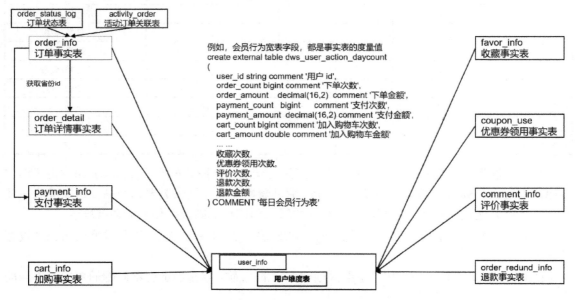

图 6-7　DWS 层宽表字段获取思路

4. DWT 层

DWT 层以分析的主题对象为建模驱动，基于上层的应用和产品的指标需求，构建主题对象的全量宽表。

DWT 层主题宽表都记录什么字段？

DWT 层的宽表字段站在维度表的角度去看事实表，重点关注事实表度量值的累计值、事实表行为的首次时间和末次时间，如图 6-8 所示。例如，订单事实表的度量值是下单次数、下单金额。订单事实表的行为是下单。我们站在用户维度表的角度去看订单事实表，重点关注订单事实表至今的累计下单次数、累计下单金额和某时间段内的累计下单次数、累计下单金额，以及关注下单行为的首次时间和末次时间。

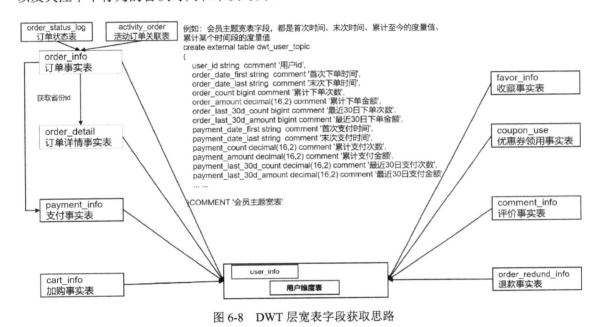

图 6-8　DWT 层宽表字段获取思路

5．ADS 层

ADS 层分别对设备主题、会员主题、商品主题和营销主题进行指标分析，其中，营销主题是会员主题和商品主题的跨主题分析案例。

6.1.7　业务术语

在进行需求的实现之前，我们先对相关的业务术语进行介绍，只有了解了这些业务术语的含义，才能找到实现需求的思路。

1）用户

用户以设备为判断标准，在移动统计中，每台独立设备被认为是一个独立用户。Android 系统根据 IMEI 号，iOS 系统根据 OpenUDID 来标识一个独立用户，每部手机是一个用户。

2）新增设备

新增设备是指首次联网使用应用的设备。如果在一台设备上首次打开某应用，那么这台设备被定义为新增设备；卸载再安装设备，不会算作一次新增。新增设备包括日新增设备、周新增设备及月新增设备。

3）活跃设备

打开应用的设备即活跃设备，不考虑设备的使用情况。一台设备一天内多次打开某应用会被记为一台活跃设备。

4）周（月）活跃设备

周（月）活跃设备是指某个自然周（月）内启动过应用的设备，该周（月）内多次启动某设备只记为一台活跃设备。

5）月活跃率

月活跃率是指月活跃设备占截至该月累计的设备总和的比例。

6）沉默设备

沉默设备是指设备仅在安装当天（次日）启动过一次应用，后续没有再次启动的行为。该指标可以反映新增设备质量和设备与应用的匹配程度。

7）版本分布

版本分布是指不同版本的应用周内各天的新增设备数、活跃设备数和启动次数。其有利于判断应用的各版本之间的优劣和用户行为习惯。

8）本周回流设备

本周回流设备是指上周未启动过应用而本周启动了应用的设备。

9）连续 n 周活跃设备

连续 n 周活跃设备是指连续 n 周，每周至少启动一次应用的设备。

10）忠诚用户

忠诚用户是指连续活跃 5 周以上的用户。

11）连续活跃设备

连续活跃设备是指连续 2 周及 2 周以上活跃的设备。

12）留存设备

某段时间内的新增设备，经过一段时间后，仍然使用应用的被认为是留存设备；这部分设备占当时新增设备的比例即留存率。

例如，5 月份新增设备 200 台，这 200 台设备在 6 月份启动过应用的有 100 台，7 月份启动过应用的有 80 台，8 月份启动过应用的有 50 台；则 5 月份新增设备一个月后的留存率是 50%，两个月后的留存率是 40%，三个月后的留存率是 25%。

13）用户新鲜度

用户新鲜度是指每天启动应用的新老用户比例，即新增用户数占活跃用户数的比例。

14）单次使用时长

单次使用时长是指设备每次启动使用的时间长度。

15）日使用时长

日使用时长是指设备累计一天内的使用时间长度。

16）启动次数计算标准

iOS 平台应用退到后台就算一次独立的启动；Android 平台规定，若两次启动之间的时间间隔小于 30 秒，则计算一次启动。例如，用户在使用应用过程中，若因收发短信或接电话等退出应用且 30 秒内又返回应用中，这两次行为应该是延续而非独立的，可以算作一次使用行为，即一次启动。业内大多使用 30 秒这个标准，但用户是可以自定义此时间间隔的。

6.2 数据仓库搭建环境准备

在第 2 章中,我们选取配置了 Tez 引擎的 Hive 作为数据仓库搭建工具,本节我们来进行 Hive 的安装部署。

Hive 是一款用类 SQL 语句来协助读/写、管理那些存储在分布式存储系统上的大数据集的数据仓库软件。Hive 可以将类 SQL 语句解析成 MapReduce 程序,从而避免编写繁杂的 MapReduce 程序,使用户分析数据变得容易。Hive 要分析的数据存储在 HDFS 上,所以它本身不提供数据存储功能。Hive 将数据映射成一张张的表,而将表的结构信息存储在关系型数据库(如 MySQL)中,所以在安装 Hive 之前,我们需要先安装 MySQL。

安装 MySQL 的具体操作在第 5 章中已经介绍过,读者需要在 hadoop103 和 hadoop104 两台节点服务器上分别操作一遍,以便为下一节内容做准备。

6.2.1 MySQL HA

MySQL 中存储了 Hive 所有表格的元数据信息,一旦 MySQL 中的数据丢失或损坏,会对整个数据仓库系统造成不可挽回的损失,为避免这种情况的发生,我们可以选择每天对元数据进行备份,进而实现 MySQL HA(High Availability,高可用)。

MySQL 的 HA 方案不止一种,本章介绍较为常用的一种——基于 Keepalived 的 MySQL HA。

MySQL 的 HA 离不开其主从复制技术。主从复制是指一台服务器充当主数据库服务器(master),另一台或多台服务器充当从数据库服务器(slave),从数据库服务器自动向主数据库服务器同步数据。实现 MySQL 的 HA,需要使两台服务器互为主从关系。

Keepalived 是基于 VRRP(Virtual Router Redundancy Protocol,虚拟路由器冗余协议)的一款高可用软件。Keepalived 有一台主数据库服务器和多台从数据库服务器,在主数据库服务器和从数据库服务器上面部署相同的服务配置,使用一个虚拟 IP 地址对外提供服务,当主数据库服务器出现故障时,虚拟 IP 地址会自动漂移到从数据库服务器上。

具体操作步骤如下。

1. MySQL 主从配置

1)主从集群规划

主从集群规划如表 6-13 所示。

表 6-13 主从集群规划

hadoop102	hadoop103	hadoop104
	MySQL(master)	MySQL(slave)

注意:MySQL 的安装步骤参考 5.2.1 节。

2)配置 master

修改 hadoop103 中 MySQL 的/usr/my.cnf 配置文件。

```
[mysqld]
```

```
#开启binlog
log_bin = mysql-bin
#binlog 日志类型
binlog_format = row
#MySQL 服务器的唯一 id
server_id = 1
```

重启 hadoop103 的 MySQL 服务。

```
[atguigu@hadoop103 ~]$ sudo service mysql restart
```

进入 MySQL 客户端，执行以下命令，查看 master 状态，如图 6-9 所示。

```
mysql> show master status;
+------------------+----------+--------------+------------------+-------------------+
| File             | Position | Binlog_Do_DB | Binlog_Ignore_DB | Executed_Gtid_Set |
+------------------+----------+--------------+------------------+-------------------+
| mysql-bin.000001 |      120 |              |                  |                   |
+------------------+----------+--------------+------------------+-------------------+
1 row in set (0.00 sec)
```

图 6-9　查看 MySQL 客户端的 master 状态（1）

3）配置 slave

修改 hadoop104 中 MySQL 的 /usr/my.cnf 配置文件。

```
[mysqld]
#MySQL 服务器的唯一 id
server_id = 2
#开启 slave 中继日志
relay_log=mysql-relay
```

重启 hadoop104 的 MySQL 服务。

```
[atguigu@hadoop104 ~]$ sudo service mysql restart
```

进入 hadoop104 的 MySQL 客户端，执行以下命令。

```
mysql>
CHANGE MASTER TO
MASTER_HOST='hadoop103',
MASTER_USER='root',
MASTER_PASSWORD='000000',
MASTER_LOG_FILE='mysql-bin.000001',
MASTER_LOG_POS=120;
```

启动 slave。

```
mysql> start slave;
```

查看 slave 状态，如图 6-10 所示。

```
mysql> show slave status\G;
*************************** 1. row ***************************
               Slave_IO_State: Waiting for master to send event
                  Master_Host: hadoop103
                  Master_User: root
                  Master_Port: 3306
                Connect_Retry: 60
              Master_Log_File: mysql-bin.000001
          Read_Master_Log_Pos: 120
               Relay_Log_File: mysql-relay.000002
                Relay_Log_Pos: 283
        Relay_Master_Log_File: mysql-bin.000001
             Slave_IO_Running: Yes         此处均为Yes，则表示主从复制搭建成功
            Slave_SQL_Running: Yes
```

图 6-10　查看 MySQL 客户端的 slave 状态（1）

2．MySQL 双主配置

1）双主集群规划

双主集群规划如表 6-14 所示。

表 6-14　双主集群规划

hadoop102	hadoop103	hadoop104
	MySQL（master，slave）	MySQL（slave，master）

2）配置 master

修改 hadoop104 中 MySQL 的/usr/my.cnf 配置文件。

```
[mysqld]

#开启binlog
log_bin = mysql-bin
#binlog日志类型
binlog_format = row
#MySQL服务器的唯一id
server_id = 2
#开启slave中继日志
relay_log=mysql-relay
```

重启 hadoop104 的 MySQL 服务。

```
[atguigu@hadoop104 ~]$ sudo service mysql restart
```

进入 hadoop104 的 MySQL 客户端，执行以下命令，查看 master 状态，如图 6-11 所示。

```
mysql> show master status;
+------------------+----------+--------------+------------------+-------------------+
| File             | Position | Binlog_Do_DB | Binlog_Ignore_DB | Executed_Gtid_Set |
+------------------+----------+--------------+------------------+-------------------+
| mysql-bin.000001 |      120 |              |                  |                   |
+------------------+----------+--------------+------------------+-------------------+
1 row in set (0.00 sec)
```

图 6-11　查看 MySQL 客户端的 master 状态（2）

3）配置 slave

修改 hadoop103 中 MySQL 的 /usr/my.cnf 配置文件。

```
[mysqld]
#MySQL 服务器的唯一 id
server_id = 1

#开启 binlog
log_bin = mysql-bin
#binlog 日志类型
binlog_format = row
#开启 slave 中继日志
relay_log=mysql-relay
```

重启 hadoop103 的 MySQL 服务。

```
[atguigu@hadoop103 ~]$ sudo service mysql restart
```

进入 hadoop103 的 MySQL 客户端，执行以下命令。

```
mysql>
CHANGE MASTER TO
MASTER_HOST='hadoop104',
MASTER_USER='root',
MASTER_PASSWORD='000000',
MASTER_LOG_FILE='mysql-bin.000001',
MASTER_LOG_POS=120;
```

启动 slave。

```
mysql> start slave;
```

查看 slave 状态，如图 6-12 所示。

```
mysql> show slave status\G;
*************************** 1. row ***************************
               Slave_IO_State: Waiting for master to send event
                  Master_Host: hadoop104
                  Master_User: root
                  Master_Port: 3306
                Connect_Retry: 60
              Master_Log_File: mysql-bin.000001
          Read_Master_Log_Pos: 120
               Relay_Log_File: mysql-relay.000002
                Relay_Log_Pos: 283
        Relay_Master_Log_File: mysql-bin.000001
             Slave_IO_Running: Yes         此处均为Yes，则表示主从复制搭建成功
            Slave_SQL_Running: Yes
```

图 6-12　查看 MySQL 客户端的 slave 状态（2）

3．Keepalived 安装部署

Keepalived 需要分别安装部署在 hadoop103 和 hadoop104 两台节点服务器上，具体操作步骤如下。

1）在hadoop103上安装部署

（1）执行yum命令，安装Keepalived。

```
[atguigu@hadoop103 ~]$ sudo yum install -y keepalived
```

（2）修改Keepalived的配置文件/etc/keepalived/keepalived.conf。

```
! Configuration File for keepalived
global_defs {
    router_id MySQL-ha
}
vrrp_instance VI_1 {
    state master #初始状态
    interface eth0 #网卡
    virtual_router_id 51 #虚拟路由id
    priority 100 #优先级
    advert_int 1 #Keepalived心跳间隔
    nopreempt #只在高优先级配置,原master恢复之后不重新上位
    authentication {
        auth_type PASS #认证方式
        auth_pass 1111
    }
    virtual_ipaddress {
        192.168.1.100 #虚拟IP地址
    }
}

#声明虚拟服务器
virtual_server 192.168.1.100 3306 {
    delay_loop 6
    persistence_timeout 30
    protocol TCP
    #声明真实服务器
    real_server 192.168.1.103 3306 {
        notify_down /var/lib/mysql/killkeepalived.sh #真实服务器出现故障后调用脚本
        TCP_CHECK {
            connect_timeout 3 #超时时间
            nb_get_retry 3 #重试次数
            delay_before_retry 2 #重试时间间隔
        }
    }
}
```

（3）编辑脚本文件/var/lib/mysql/killkeepalived.sh。

```
#! /bin/bash
sudo service keepalived stop
```

（4）增加脚本执行权限。

```
[atguigu@hadoop103 ~]$ sudo chmod +x /var/lib/mysql/killkeepalived.sh
```

（5）启动 Keepalived 服务。

```
[atguigu@hadoop103 ~]$ sudo service keepalived start
```

（6）设置 Keepalived 服务开机自启。

```
[atguigu@hadoop103 ~]$ sudo chkconfig keepalived on
```

2）在 hadoop104 上安装部署

（1）执行 yum 命令，安装 Keepalived。

```
[atguigu@hadoop104 ~]$ sudo yum install -y keepalived
```

（2）修改 Keepalived 的配置文件/etc/keepalived/keepalived.conf。

```
! Configuration File for keepalived
global_defs {
    router_id MySQL-ha
}
vrrp_instance VI_1 {
    state master #初始状态
    interface eth0 #网卡
    virtual_router_id 51 #虚拟路由 id
    priority 100 #优先级
    advert_int 1 #Keepalived 心跳间隔
    authentication {
        auth_type PASS #认证方式
        auth_pass 1111
    }
    virtual_ipaddress {
        192.168.1.100 #虚拟 IP 地址
    }
}

#声明虚拟服务器
virtual_server 192.168.1.100 3306 {
    delay_loop 6
    persistence_timeout 30
    protocol TCP
    #声明真实服务器
    real_server 192.168.1.104 3306 {
        notify_down /var/lib/mysql/killkeepalived.sh #真实服务器出现故障后调用脚本
        TCP_CHECK {
            connect_timeout 3 #超时时间
            nb_get_retry 3 #重试次数
            delay_before_retry 2 #重试时间间隔
        }
    }
}
```

（3）编辑脚本文件/var/lib/mysql/killkeepalived.sh。

```
#! /bin/bash
```

```
sudo service keepalived stop
```

（4）增加脚本执行权限。

```
[atguigu@hadoop104 ~]$ sudo chmod +x /var/lib/mysql/killkeepalived.sh
```

（5）启动 Keepalived 服务。

```
[atguigu@hadoop104 ~]$ sudo service keepalived start
```

（6）设置 Keepalived 服务开机自启。

```
[atguigu@hadoop104 ~]$ sudo chkconfig keepalived on
```

4．确保开机时 MySQL 先于 Keepalived 启动

需要分别在 hadoop103、hadoop104 节点服务器上进行如下操作。

第一步：查看开机时 MySQL 的启动次序，结果如图 6-13 所示。

```
[atguigu@hadoop104 ~]$ sudo vim /etc/init.d/mysql
```

```
#!/bin/sh
# Copyright Abandoned 1996 TCX DataKonsult AB & Monty Program KB & Detron HB
# This file is public domain and comes with NO WARRANTY of any kind

# MySQL daemon start/stop script.

# Usually this is put in /etc/init.d (at least on machines SYSV R4 based
# systems) and linked to /etc/rc3.d/S99mysql and /etc/rc0.d/K01mysql.
# When this is done the mysql server will be started when the machine is
# started and shut down when the systems goes down.
                              开机时的启动次序，数值越小，越先启动
# Comments to support chkconfig on RedHat Linux
# chkconfig: 2345 86 36
# description: A very fast and reliable SQL database engine.
```

图 6-13　查看开机时 MySQL 的启动次序

第二步：查看开机时 Keepalived 的启动次序，结果如图 6-14 所示。

```
[atguigu@hadoop104 ~]$ sudo vim /etc/init.d/keepalived
```

```
#!/bin/sh
#
# keepalived   High Availability monitor built upon LVS and VRRP
#
# chkconfig:   - 64 14
# description: Robust keepalive facility to the Linux Virtual Server project \
#              with multilayer TCP/IP stack checks.
```

图 6-14　查看开机时 Keepalived 的启动次序

第三步：若 Keepalived 先于 MySQL 启动，则需要按照以下步骤设置二者的启动次序。

（1）修改开机时 MySQL 的启动次序，如图 6-15 所示。

```
[atguigu@hadoop104 ~]$ sudo vim /etc/init.d/mysql
```

```
#!/bin/sh
# Copyright Abandoned 1996 TCX DataKonsult AB & Monty Program KB & Detron HB
# This file is public domain and comes with NO WARRANTY of any kind

# MySQL daemon start/stop script.

# Usually this is put in /etc/init.d (at least on machines SYSV R4 based
# systems) and linked to /etc/rc3.d/S99mysql and /etc/rc0.d/K01mysql.
# When this is done the mysql server will be started when the machine is
# started and shut down when the systems goes down.

# Comments to support chkconfig on RedHat Linux
# chkconfig: 2345 64 36
# description: A very fast and reliable SQL database engine.
```

图 6-15　修改开机时 MySQL 的启动次序

（2）重新设置 MySQL 开机自启。

```
[atguigu@hadoop104 ~]$ sudo chkconfig --del mysql
[atguigu@hadoop104 ~]$ sudo chkconfig --add mysql
[atguigu@hadoop104 ~]$ sudo chkconfig mysql on
```

（3）修改开机时 Keepalived 的启动次序，如图 6-16 所示。

```
[atguigu@hadoop104 ~]$ sudo vim /etc/init.d/keepalived
```

```
#!/bin/sh
#
# keepalived   High Availability monitor built upon LVS and VRRP
#
# chkconfig:   - 86 14
# description: Robust keepalive facility to the Linux Virtual Server project \
#              with multilayer TCP/IP stack checks.
```

图 6-16　修改开机时 Keepalived 的启动次序

（4）重新设置 Keepalived 开机自启。

```
[atguigu@hadoop104 ~]$ sudo chkconfig --del keepalived
[atguigu@hadoop104 ~]$ sudo chkconfig --add keepalived
[atguigu@hadoop104 ~]$ sudo chkconfig keepalived on
```

6.2.2　Hive 安装

在安装了 MySQL 后，接下来可以着手对 Hive 进行正式的安装部署。

1．安装及配置 Hive

（1）把 Hive 的安装包 apache-hive-3.1.2-bin.tar.gz 上传到 Linux 的/opt/software 目录下，然后解压 apache-hive-3.1.2-bin.tar.gz 到/opt/module 目录下。

```
[atguigu@hadoop102 software]$ tar -zxvf apache-hive-3.1.2-bin.tar.gz -C /opt/module
```

（2）修改 apache-hive-3.1.2-bin 的名称为 hive。

```
[atguigu@hadoop102 module]$ mv apache-hive-3.1.2-bin/ hive
```

（3）修改/etc/profile 文件，添加环境变量。

```
[atguigu@hadoop102 software]$ sudo vim /etc/profile
```

添加以下内容。

```
#HIVE_HOME
export HIVE_HOME=/opt/module/hive
export PATH=$PATH:$HIVE_HOME/bin
```

执行以下命令，使环境变量生效。

```
[atguigu@hadoop102 software]$ source /etc/profile
```

（4）进入/opt/module/hive/lib 目录下执行以下命令，解决日志 jar 包冲突。

```
[atguigu@hadoop102 lib]$ mv log4j-slf4j-impl-2.10.0.jar log4j-slf4j-impl-2.10.0.jar.bak
```

2. 复制驱动

(1) 在/opt/software/mysql-libs 目录下解压 mysql-connector-java-5.1.27.tar.gz 驱动包。

```
[root@hadoop102 mysql-libs]# tar -zxvf mysql-connector-java-5.1.27.tar.gz
```

(2) 将/opt/software/mysql-libs/mysql-connector-java-5.1.27 目录下的 mysql-connector-java-5.1.27-bin.jar 复制到/opt/module/hive/lib 目录下,用于稍后启动 Hive 时连接 MySQL。

```
[root@hadoop102 mysql-connector-java-5.1.27]# cp mysql-connector-java-5.1.27-bin.jar
 /opt/module/hive/lib
```

3. 配置 Metastore 到 MySQL

(1) 在/opt/module/hive/conf 目录下创建一个 hive-site.xml 文件。

```
[atguigu@hadoop102 conf]$ touch hive-site.xml
[atguigu@hadoop102 conf]$ vim hive-site.xml
```

(2) 根据官方文档配置参数,并复制数据到 hive-site.xml 文件中。

```xml
<?xml version="1.0"?>
<?xml-stylesheet type="text/xsl" href="configuration.xsl"?>
<configuration>
<!--配置 Hive 保存元数据信息所需的 MySQL URL 地址,此处使用 Keepalived 服务对外提供的虚拟 IP
地址-->
<property>
  <name>javax.jdo.option.ConnectionURL</name>
  <value>jdbc:mysql://hadoop100:3306/metastore?createDatabaseIfNotExist=true
</value>
  <description>JDBC connect string for a JDBC metastore</description>
</property>
<!--配置 Hive 连接 MySQL 的驱动全类名-->
<property>
  <name>javax.jdo.option.ConnectionDriverName</name>
  <value>com.mysql.jdbc.Driver</value>
  <description>Driver class name for a JDBC metastore</description>
</property>
<!--配置 Hive 连接 MySQL 的用户名 -->
<property>
  <name>javax.jdo.option.ConnectionUserName</name>
  <value>root</value>
  <description>username to use against metastore database</description>
</property>
<!--配置 Hive 连接 MySQL 的密码 -->
<property>
  <name>javax.jdo.option.ConnectionPassword</name>
  <value>000000</value>
  <description>password to use against metastore database</description>
</property>
<property>
```

```xml
        <name>hive.metastore.warehouse.dir</name>
        <value>/user/hive/warehouse</value>
    </property>

    <property>
        <name>hive.metastore.schema.verification</name>
        <value>false</value>
    </property>

    <property>
        <name>hive.metastore.uris</name>
        <value>thrift://hadoop102:9083</value>
    </property>

    <property>
    <name>hive.server2.thrift.port</name>
    <value>10000</value>
    </property>

    <property>
        <name>hive.server2.thrift.bind.host</name>
        <value>hadoop102</value>
    </property>

    <property>
        <name>hive.metastore.event.db.notification.api.auth</name>
        <value>false</value>
    </property>

    <property>
        <name>hive.cli.print.header</name>
        <value>true</value>
    </property>

    <property>
        <name>hive.cli.print.current.db</name>
        <value>true</value>
    </property>
</configuration>
```

4．初始化元数据库

（1）启动 MySQL。

```
[atguigu@hadoop103 mysql-libs]$ mysql -uroot -p000000
```

（2）新建 Hive 元数据库。

```
mysql> create database metastore;
mysql> quit;
```

（3）初始化 Hive 元数据库。

`[atguigu@hadoop102 conf]$ schematool -initSchema -dbType mysql -verbose`

5. 启动 Hive

（1）Hive 2.x 以上版本，需要先启动 Metastore 和 HiveServer2 服务，否则会报错。

```
FAILED: HiveException java.lang.RuntimeException: Unable to instantiate org.apache.hadoop.hive.ql.metadata.SessionHiveMetaStoreClient
```

（2）在/opt/module/hive/bin 目录下编写 Hive 服务启动脚本，在脚本中启动 Metastore 和 HiveServer2 服务。

`[atguigu@hadoop102 bin]$ vim hiveservices.sh`

脚本内容如下。

```bash
#!/bin/bash
HIVE_LOG_DIR=$HIVE_HOME/logs

mkdir -p $HIVE_LOG_DIR

#检查进程是否运行正常，参数1为进程名，参数2为进程端口
function check_process()
{
    pid=$(ps -ef 2>/dev/null | grep -v grep | grep -i $1 | awk '{print $2}')
    ppid=$(netstat -nltp 2>/dev/null | grep $2 | awk '{print $7}' | cut -d '/' -f 1)
    echo $pid
    [[ "$pid" =~ "$ppid" ]] && [ "$ppid" ] && return 0 || return 1
}

function hive_start()
{
    metapid=$(check_process HiveMetastore 9083)
    cmd="nohup hive --service metastore >$HIVE_LOG_DIR/metastore.log 2>&1 &"
    cmd=$cmd" sleep 4; hdfs dfsadmin -safemode wait >/dev/null 2>&1"
    [ -z "$metapid" ] && eval $cmd || echo "Metastore 服务已启动"
    server2pid=$(check_process HiveServer2 10000)
    cmd="nohup hive --service hiveserver2 >$HIVE_LOG_DIR/hiveServer2.log 2>&1 &"
    [ -z "$server2pid" ] && eval $cmd || echo "HiveServer2 服务已启动"
}

function hive_stop()
{
    metapid=$(check_process HiveMetastore 9083)
    [ "$metapid" ] && kill $metapid || echo "Metastore 服务未启动"
    server2pid=$(check_process HiveServer2 10000)
    [ "$server2pid" ] && kill $server2pid || echo "HiveServer2 服务未启动"
}
```

```
case $1 in
"start")
    hive_start
    ;;
"stop")
    hive_stop
    ;;
"restart")
    hive_stop
    sleep 2
    hive_start
    ;;
"status")
    check_process HiveMetastore 9083 >/dev/null && echo "Metastore 服务运行正常"
|| echo "Metastore 服务运行异常"
    check_process HiveServer2 10000 >/dev/null && echo "HiveServer2 服务运行正常"
|| echo "HiveServer2 服务运行异常"
    ;;
*)
    echo Invalid Args!
    echo 'Usage: '$(basename $0)' start|stop|restart|status'
    ;;
esac
```

(3) 增加脚本执行权限。

```
[atguigu@hadoop102 bin]$ chmod +x hiveservices.sh
```

(4) 启动 Hive 后台服务。

```
[atguigu@hadoop102 bin]$ hiveservices.sh start
```

(5) 查看 Hive 后台服务运行情况。

```
[atguigu@hadoop102 bin]$ hiveservices.sh status
Metastore 服务运行正常
HiveServer2 服务运行正常
```

(6) 启动 Hive 客户端。

```
[atguigu@hadoop102 hive]$ bin/hive
```

6.2.3 Tez 引擎安装

Tez 是 Hive 的一个运行引擎，性能优于 MR。为什么 Tez 的性能优于 MR 呢？从如图 6-17 所示的 Tez 引擎原理中可得出结论。

图 6-17 左图表示用 Hive 直接编写 MR 程序，假设形成了四个有依赖关系的 MR 作业，其中每一个椭圆形代表一个 MR 作业，浅色方块代表 Map Task，深色方块代表 Reduce Task，云状表示需要将上一步产生的结果数据持久化到 HDFS 中才能供下游作业使用，这会产生大量

的磁盘 IO。

  Tez 可以将多个有依赖关系的作业转换为一个作业，这样只需写一次 HDFS 即可，且中间节点较少，从而大大提升了作业的计算性能。

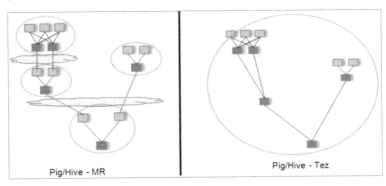

图 6-17 Tez 引擎原理示意

1. 安装包准备

（1）下载 Tez 的依赖包。

（2）将 apache-tez-0.9.1-bin.tar.gz 复制到 hadoop102 的/opt/software 目录下。

```
[atguigu@hadoop102 software]$ ls
apache-tez-0.9.1-bin.tar.gz
```

（3）将 apache-tez-0.9.1-bin.tar.gz 解压至/opt/module 目录下。

```
[atguigu@hadoop102 module]$ tar -zxvf apache-tez-0.9.1-bin.tar.gz -C /opt/module
```

（4）修改名称。

```
[atguigu@hadoop102 module]$ mv apache-tez-0.9.1-bin/ tez-0.9.1
```

2. 在 Hive 中配置 Tez

（1）进入 Hive 的配置目录/opt/module/hive/conf。

```
[atguigu@hadoop102 conf]$ pwd
/opt/module/hive/conf
```

（2）在 hive-env.sh 文件中添加 Tez 环境变量配置和依赖包环境变量配置。

```
[atguigu@hadoop102 conf]$ vim hive-env.sh
```

  添加的配置如下。

```
# 设置 Hadoop 集群的安装目录
export HADOOP_HOME=/opt/module/hadoop-2.7.2

# 设置 Hive 的配置文件目录
export HIVE_CONF_DIR=/opt/module/hive/conf

# 设置运行 Tez 环境所需的 jar 包路径
export TEZ_HOME=/opt/module/tez-0.9.1      #此处是读者自己的 Tez 的解压目录
export TEZ_JARS=""
```

```
for jar in `ls $TEZ_HOME |grep jar`; do
    export TEZ_JARS=$TEZ_JARS:$TEZ_HOME/$jar
done
for jar in `ls $TEZ_HOME/lib`; do
    export TEZ_JARS=$TEZ_JARS:$TEZ_HOME/lib/$jar
done

export HIVE_AUX_JARS_PATH=/opt/module/hadoop-2.7.2/share/hadoop/common/hadoop-lzo-0.4.20.jar$TEZ_JARS
```

（3）在 hive-site.xml 文件中添加如下配置，将 Hive 的计算引擎更改为 Tez。

```
<property>
    <name>hive.execution.engine</name>
    <value>tez</value>
</property>
```

3. 配置 Tez

在 Hive 的/opt/module/hive/conf 目录下创建一个 tez-site.xml 文件。

```
[atguigu@hadoop102 conf]$ pwd
/opt/module/hive/conf
[atguigu@hadoop102 conf]$ vim tez-site.xml
```

在文件中添加如下内容。

```
<?xml version="1.0" encoding="UTF-8"?>
<?xml-stylesheet type="text/xsl" href="configuration.xsl"?>
<configuration>
#配置在 Tez 中使用的 uris 的 jar 包路径
<property>
 <name>tez.lib.uris</name>     <value>${fs.defaultFS}/tez/tez-0.9.1,${fs.defaultFS}/tez/tez-0.9.1/lib</value>
</property>
#配置在 Tez 中使用的 uris 的类路径
<property>
 <name>tez.lib.uris.classpath</name>     <value>${fs.defaultFS}/tez/tez-0.9.1,${fs.defaultFS}/tez/tez-0.9.1/lib</value>
</property>
#是否使用 Hadoop 依赖
<property>
    <name>tez.use.cluster.hadoop-libs</name>
    <value>true</value>
</property>
#配置 Tez 自己的历史服务器
<property>
    <name>tez.history.logging.service.class</name>
<value>org.apache.tez.dag.history.logging.ats.ATSHistoryLoggingService</value>
</property>
</configuration>
```

4．上传 Tez 到集群

将/opt/module/tez-0.9.1 上传到 HDFS 的/tez 目录下。

```
[atguigu@hadoop102 conf]$ hadoop fs -mkdir /tez
[atguigu@hadoop102 conf]$ hadoop fs -put /opt/module/tez-0.9.1/ /tez
[atguigu@hadoop102 conf]$ hadoop fs -ls /tez
/tez/tez-0.9.1
```

5．测试

（1）启动 Hive。

```
[atguigu@hadoop102 hive]$ bin/hive
```

（2）创建 LZO 表。

```
hive (default)> create table student(
id int,
name string);
```

（3）向表中插入数据。

```
hive (default)> insert into student values(1,"zhangsan");
```

（4）如果没有报错，就表示配置成功了。

```
hive (default)> select * from student;
1       zhangsan
```

6．小结

运行 Tez 时，有可能遇到因为 Container 使用过多内存而被 NodeManager 杀死进程的问题，如下所示。

```
Caused by: org.apache.tez.dag.api.SessionNotRunning: TezSession has already shutdown. Application application_1546781144082_0005 failed 2 times due to AM Container for appattempt_1546781144082_0005_000002 exited with exitCode: -103
For more detailed output, check application tracking page:http://hadoop103:8088/cluster/app/application_1546781144082_0005Then, click on links to logs of each attempt.
Diagnostics: Container [pid=11116,containerID=container_1546781144082_0005_02_000001] is running beyond virtual memory limits. Current usage: 216.3 MB of 1 GB physical memory used; 2.6 GB of 2.1 GB virtual memory used. Killing container.
```

产生的原因是 NodeManager 上运行的 Container 试图使用过多的内存，而被 NodeManager 杀掉了，如下所示。

```
[摘录] The NodeManager is killing your container. It sounds like you are trying to use hadoop streaming which is running as a child process of the map-reduce task. The NodeManager monitors the entire process tree of the task and if it eats up more memory than the maximum set in mapreduce.map.memory.mb or mapreduce.reduce.memory.mb respectively, we would expect the Nodemanager to kill the task, otherwise your task is stealing memory belonging to other containers, which you don't want.
```

解决方法：修改 Hadoop 的配置文件 yarn-site.xml，增加如下配置，关掉虚拟内存检查，修改后，分发配置文件，并重启集群。

```
<property>
    <name>yarn.nodemanager.vmem-check-enabled</name>
    <value>false</value>
</property>
```

6.3 数据仓库搭建——ODS 层

ODS 层为原始数据层，其保持数据原貌不进行任何修改，起到备份数据的作用；数据采用 LZO 压缩格式，以减少磁盘存储空间；创建分区表，可以避免后续对表查询时进行全表扫描。在进行 ODS 层数据的导入之前，先要创建数据库，用于存储整个电商数据仓库项目的所有数据信息。

6.3.1 创建数据库

（1）启动 Hive。

```
[atguigu@hadoop102 hive]$ bin/hiveservices.sh start
[atguigu@hadoop102 hive]$ bin/hive
```

（2）显示数据库。

```
hive (default)> show databases;
```

（3）创建数据库。

```
hive (default)> create database gmall;
```

说明：当数据库已存在且有数据，需要强制删除时，执行如下命令。

```
drop database gmall cascade;
```

（4）使用 gmall 数据库，以下所有操作均在该数据库下进行。

```
hive (default)> use gmall;
```

6.3.2 用户行为数据

ODS 层主要加载采集日志系统落盘在 HDFS 的日志数据文件，在采集日志系统的部署过程中，我们根据日志的不同类型，将日志分为启动日志和事件日志，在搭建 ODS 层时，我们也将根据两种数据类型分别进行建表导入处理。

分析思路如下。

（1）在进行建表之前，若要创建的表已经存在，则先删除该表，以防止脚本出错，影响后续执行。

（2）创建外部表。外部表即只建立表与原始数据之间的映射关系，而不改变数据的位置，在对表执行删除操作时,只会删除表的元数据,而不会删除表的数据,相对来说更加安全,这种方式在实际工作环境中应用十分广泛。

(3)表格的字段为 JSON 格式的 String 类型字符串。

(4)表格按照日期进行分区,方便用户按照日期进行查询。

(5)文件输入格式为 LZO 压缩格式,由于 LZO 压缩格式的文件不支持 HDFS 对其进行分片,因此需要对 LZO 压缩格式的文件创建索引。

(6)存储路径根据存储数据的不同分别指定,启动日志表存储路径为/user/hive/warehouse/gmall/ods/ods_start_log;事件日志表存储路径为/user/hive/warehouse/gmall/ods/ods_event_log。

(7)外部表创建成功后,将 Flume 采集落盘的文件加载到 Hive 表中,ODS 层即搭建成功。具体操作如下。

(1)按照上述思路创建输入数据是 LZO 压缩格式、输出数据是 TEXT 存储格式、支持 JSON 解析的分区启动日志表。

```
hive (gmall)>
drop table if exists ods_start_log;
CREATE EXTERNAL TABLE ods_start_log (`line` string)
PARTITIONED BY (`dt` string)
STORED AS
  INPUTFORMAT 'com.hadoop.mapred.DeprecatedLzoTextInputFormat'
  OUTPUTFORMAT 'org.apache.hadoop.hive.ql.io.HiveIgnoreKeyTextOutputFormat'
LOCATION '/warehouse/gmall/ods/ods_start_log';
```

(2)加载数据,指定每天数据的分区信息为具体到日的日期。

```
hive (gmall)>
load data inpath '/origin_data/gmall/log/topic_start/2020-03-10' into table gmall.ods_start_log partition(dt='2020-03-10');
```

注意:日期格式都配置成 YYYY-MM-DD 格式,这是 Hive 默认支持的日期格式。

(3)查看分区启动日志表是否加载成功。

```
hive (gmall)> select * from ods_start_log limit 2;
```

(4)由于 LZO 压缩格式的文件不支持 HDFS 对其进行分片,因此对 LZO 压缩格式的文件创建索引。

```
hadoop jar /opt/module/hadoop-2.7.2/share/hadoop/common/hadoop-lzo-0.4.20.jar com.hadoop.compression.lzo.DistributedLzoIndexer /warehouse/gmall/ods/ods_start_log/dt=2020-03-10
```

(5)创建输入数据是 LZO 压缩格式、支持 JSON 解析的分区事件日志表。

```
hive (gmall)>
drop table if exists ods_event_log;
CREATE EXTERNAL TABLE ods_event_log(`line` string)
PARTITIONED BY (`dt` string)
STORED AS
  INPUTFORMAT 'com.hadoop.mapred.DeprecatedLzoTextInputFormat'
```

```
  OUTPUTFORMAT 'org.apache.hadoop.hive.ql.io.HiveIgnoreKeyTextOutputFormat'
LOCATION '/warehouse/gmall/ods/ods_event_log';
```

（6）加载数据。

```
hive (gmall)>
load data inpath '/origin_data/gmall/log/topic_event/2020-03-10' into table
gmall.ods_event_log partition(dt='2020-03-10');
```

注意：日期格式都配置成 YYYY-MM-DD 格式，这是 Hive 默认支持的日期格式。

（7）查看分区事件日志表是否加载成功。

```
hive (gmall)> select * from ods_event_log limit 2;
```

（8）为 LZO 压缩格式的文件创建索引。

```
hadoop jar /opt/module/hadoop-2.7.2/share/hadoop/common/hadoop-lzo-0.4.20.jar
com.hadoop.compression.lzo.DistributedLzoIndexer
/warehouse/gmall/ods/ods_event_log/dt=2020-03-10
```

Shell 中单引号和双引号的区别如下。

（1）在/home/atguigu/bin 目录下创建一个 test.sh 文件。

```
[atguigu@hadoop102 bin]$ vim test.sh
```

在文件中添加如下内容。

```
#!/bin/bash
do_date=$1

echo '$do_date'
echo "$do_date"
echo "'$do_date'"
echo '"$do_date"'
echo `date`
```

（2）查看执行结果。

```
[atguigu@hadoop102 bin]$ test.sh 2020-03-10
$do_date
2020-03-10
'2020-03-10'
"$do_date"
2020 年 05 月 02 日 星期四 21:02:08 CST
```

（3）总结如下。

- 单引号表示不取出变量值。
- 双引号表示取出变量值。
- 反引号表示执行引号中的命令。
- 双引号内部嵌套单引号表示取出变量值。
- 单引号内部嵌套双引号表示不取出变量值。

6.3.3 ODS 层用户行为数据导入脚本

将 ODS 层用户行为数据的加载过程编写成脚本，方便每日调用执行。
（1）在 hadoop102 的 /home/atguigu/bin 目录下创建脚本 ods_log.sh。

```
[atguigu@hadoop102 bin]$ vim ods_log.sh
```

在脚本中编写如下内容。

```
#!/bin/bash

# 定义变量，方便后续修改
APP=gmall
hive=/opt/module/hive/bin/hive
hadoop=/opt/module/hadoop-2.7.2/bin/hadoop

# 若输入了日期参数，则取输入参数作为日期值；若没有输入日期参数，则取当前时间的前一天作为日期值
if [ -n "$1" ] ;then
   do_date=$1
else
   do_date=`date -d "-1 day" +%F`
fi

echo "===日志日期为 $do_date==="
sql="
load data inpath '/origin_data/gmall/log/topic_start/$do_date' into table ${db}.ods_start_log partition(dt='$do_date');
load data inpath '/origin_data/gmall/log/topic_event/$do_date' into table ${db}.ods_event_log partition(dt='$do_date');
"

$hive -e "$sql"
$hadoop jar /opt/module/hadoop-2.7.2/share/hadoop/common/hadoop-lzo-0.4.20.jar com.hadoop.compression.lzo.DistributedLzoIndexer /warehouse/gmall/ods/ods_start_log/dt=$do_date
$hadoop jar /opt/module/hadoop-2.7.2/share/hadoop/common/hadoop-lzo-0.4.20.jar com.hadoop.compression.lzo.DistributedLzoIndexer /warehouse/gmall/ods/ods_event_log/dt=$do_date
```

说明：
[-n 变量值] 用于判断变量的值是否为空。
- 如果变量的值非空，则返回 true。
- 如果变量的值为空，则返回 false。

（2）增加脚本执行权限。

```
[atguigu@hadoop102 bin]$ chmod 777 ods_log.sh
```

(3) 执行脚本,导入数据。

```
[atguigu@hadoop102 module]$ ods_log.sh 2020-03-11
```

(4) 查询结果数据。

```
hive (gmall)>
select * from ods_log where dt='2020-03-11' limit 2;
```

6.3.4 业务数据

业务数据的 ODS 层搭建与用户行为数据的 ODS 层搭建相同,都是保留原始数据,不对数据进行任何转换处理,根据需求分析选取业务数据库中表的必需字段进行建表,然后将 Sqoop 导入的原始数据加载(Load)至所建表格中。

1. 创建订单表

```
hive (gmall)>
drop table if exists ods_order_info;
create external table ods_order_info (
    `id` string COMMENT '编号',
    `final_total_amount` decimal(10,2) COMMENT '总金额',
    `order_status` string COMMENT '订单状态',
    `user_id` string COMMENT '用户id',
    `out_trade_no` string COMMENT '订单交易编号',
    `create_time` string COMMENT '创建时间',
    `operate_time` string COMMENT '操作时间',
    `province_id` string COMMENT '省份id',
    `benefit_reduce_amount` decimal(10,2) COMMENT '优惠金额',
    `original_total_amount` decimal(10,2) COMMENT '原价金额',
    `feight_fee` decimal(10,2) COMMENT '运费'
) COMMENT '订单表'
PARTITIONED BY (`dt` string)
row format delimited fields terminated by '\t'
STORED AS
  INPUTFORMAT 'com.hadoop.mapred.DeprecatedLzoTextInputFormat'
  OUTPUTFORMAT 'org.apache.hadoop.hive.ql.io.HiveIgnoreKeyTextOutputFormat'
location '/warehouse/gmall/ods/ods_order_info/';
```

2. 创建订单详情表

```
hive (gmall)>
drop table if exists ods_order_detail;
create external table ods_order_detail(
    `id` string COMMENT '编号',
    `order_id` string COMMENT '订单编号',
    `user_id` string COMMENT '用户id',
    `sku_id` string COMMENT '商品id',
    `sku_name` string COMMENT '商品名称',
    `order_price` decimal(10,2) COMMENT '商品价格',
```

```
    `sku_num` bigint COMMENT '商品数量',
    `create_time` string COMMENT '创建时间'
) COMMENT '订单详情表'
PARTITIONED BY (`dt` string)
row format delimited fields terminated by '\t'
STORED AS
  INPUTFORMAT 'com.hadoop.mapred.DeprecatedLzoTextInputFormat'
  OUTPUTFORMAT 'org.apache.hadoop.hive.ql.io.HiveIgnoreKeyTextOutputFormat'
location '/warehouse/gmall/ods/ods_order_detail/';
```

3. 创建 SKU 商品表

```
hive (gmall)>
drop table if exists ods_sku_info;
create external table ods_sku_info(
    `id` string COMMENT '商品id',
    `spu_id` string  COMMENT '标准产品单位id',
    `price` decimal(10,2) COMMENT '价格',
    `sku_name` string COMMENT '商品名称',
    `sku_desc` string COMMENT '商品描述',
    `weight` string COMMENT '重量',
    `tm_id` string COMMENT '品牌id',
    `category3_id` string COMMENT '三级品类id',
    `create_time` string COMMENT '创建时间'
) COMMENT 'SKU商品表'
PARTITIONED BY (`dt` string)
row format delimited fields terminated by '\t'
STORED AS
  INPUTFORMAT 'com.hadoop.mapred.DeprecatedLzoTextInputFormat'
  OUTPUTFORMAT 'org.apache.hadoop.hive.ql.io.HiveIgnoreKeyTextOutputFormat'
location '/warehouse/gmall/ods/ods_sku_info/';
```

4. 创建用户表

```
hive (gmall)>
drop table if exists ods_user_info;
create external table ods_user_info(
    `id` string COMMENT '用户id',
    `name` string COMMENT '真实姓名',
    `birthday` string COMMENT '生日',
    `gender` string COMMENT '性别',
    `email` string COMMENT '邮箱',
    `user_level` string COMMENT '用户级别',
    `create_time` string COMMENT '创建时间',
    `operate_time` string COMMENT '操作时间'
) COMMENT '用户表'
PARTITIONED BY (`dt` string)
row format delimited fields terminated by '\t'
STORED AS
```

```
    INPUTFORMAT 'com.hadoop.mapred.DeprecatedLzoTextInputFormat'
    OUTPUTFORMAT 'org.apache.hadoop.hive.ql.io.HiveIgnoreKeyTextOutputFormat'
location '/warehouse/gmall/ods/ods_user_info/';
```

5. 创建商品一级品类表

```
hive (gmall)>
drop table if exists ods_base_category1;
create external table ods_base_category1(
    `id` string COMMENT 'id',
    `name` string COMMENT '名称'
) COMMENT '商品一级品类表'
PARTITIONED BY (`dt` string)
row format delimited fields terminated by '\t'
STORED AS
    INPUTFORMAT 'com.hadoop.mapred.DeprecatedLzoTextInputFormat'
    OUTPUTFORMAT 'org.apache.hadoop.hive.ql.io.HiveIgnoreKeyTextOutputFormat'
location '/warehouse/gmall/ods/ods_base_category1/';
```

6. 创建商品二级品类表

```
hive (gmall)>
drop table if exists ods_base_category2;
create external table ods_base_category2(
    `id` string COMMENT 'id',
    `name` string COMMENT '名称',
    category1_id string COMMENT '一级品类id'
) COMMENT '商品二级品类表'
PARTITIONED BY (`dt` string)
row format delimited fields terminated by '\t'
STORED AS
    INPUTFORMAT 'com.hadoop.mapred.DeprecatedLzoTextInputFormat'
    OUTPUTFORMAT 'org.apache.hadoop.hive.ql.io.HiveIgnoreKeyTextOutputFormat'
location '/warehouse/gmall/ods/ods_base_category2/';
```

7. 创建商品三级品类表

```
hive (gmall)>
drop table if exists ods_base_category3;
create external table ods_base_category3(
    `id` string COMMENT 'id',
    `name` string COMMENT '名称',
    category2_id string COMMENT '二级品类id'
) COMMENT '商品三级品类表'
PARTITIONED BY (`dt` string)
row format delimited fields terminated by '\t'
STORED AS
    INPUTFORMAT 'com.hadoop.mapred.DeprecatedLzoTextInputFormat'
    OUTPUTFORMAT 'org.apache.hadoop.hive.ql.io.HiveIgnoreKeyTextOutputFormat'
location '/warehouse/gmall/ods/ods_base_category3/';
```

8. 创建支付流水表

```
hive (gmall)>
drop table if exists ods_payment_info;
create external table ods_payment_info(
    `id`             bigint COMMENT '编号',
    `out_trade_no`   string COMMENT '对外业务编号',
    `order_id`       string COMMENT '订单编号',
    `user_id`        string COMMENT '用户id',
    `alipay_trade_no` string COMMENT '支付宝交易流水编号',
    `total_amount`   decimal(16,2) COMMENT '支付金额',
    `subject`        string COMMENT '交易内容',
    `payment_type`   string COMMENT '支付类型',
    `payment_time`   string COMMENT '支付时间'
) COMMENT '支付流水表'
PARTITIONED BY (`dt` string)
row format delimited fields terminated by '\t'
STORED AS
  INPUTFORMAT 'com.hadoop.mapred.DeprecatedLzoTextInputFormat'
  OUTPUTFORMAT 'org.apache.hadoop.hive.ql.io.HiveIgnoreKeyTextOutputFormat'
location '/warehouse/gmall/ods/ods_payment_info/';
```

9. 创建省份表

```
hive (gmall)>
drop table if exists ods_base_province;
create external table ods_base_province (
    `id`         bigint COMMENT '编号',
    `name`       string COMMENT '省份名称',
    `region_id`  string COMMENT '地区id',
    `area_code`  string COMMENT '地区编码',
    `iso_code`   string COMMENT '国际编码'
) COMMENT '省份表'
row format delimited fields terminated by '\t'
STORED AS
  INPUTFORMAT 'com.hadoop.mapred.DeprecatedLzoTextInputFormat'
  OUTPUTFORMAT 'org.apache.hadoop.hive.ql.io.HiveIgnoreKeyTextOutputFormat'
location '/warehouse/gmall/ods/ods_base_province/';
```

10. 创建地区表

```
hive (gmall)>
drop table if exists ods_base_region;
create external table ods_base_region (
    `id`   bigint COMMENT '编号',
    `name` string COMMENT '地区名称'
) COMMENT '地区表'
row format delimited fields terminated by '\t'
STORED AS
  INPUTFORMAT 'com.hadoop.mapred.DeprecatedLzoTextInputFormat'
```

```
    OUTPUTFORMAT 'org.apache.hadoop.hive.ql.io.HiveIgnoreKeyTextOutputFormat'
location '/warehouse/gmall/ods/ods_base_region/';
```

11. 创建品牌表

```
hive (gmall)>
drop table if exists ods_base_trademark;
create external table ods_base_trademark (
    `id`   bigint COMMENT 'id',
    `tm_name` string COMMENT '品牌名称'
) COMMENT '品牌表'
PARTITIONED BY (`dt` string)
row format delimited fields terminated by '\t'
STORED AS
  INPUTFORMAT 'com.hadoop.mapred.DeprecatedLzoTextInputFormat'
  OUTPUTFORMAT 'org.apache.hadoop.hive.ql.io.HiveIgnoreKeyTextOutputFormat'
location '/warehouse/gmall/ods/ods_base_trademark/';
```

12. 创建订单状态表

```
hive (gmall)>
drop table if exists ods_order_status_log;
create external table ods_order_status_log (
    `id`  bigint COMMENT '编号',
    `order_id` string COMMENT '订单编号',
    `order_status` string COMMENT '订单状态',
    `operate_time` string COMMENT '操作时间'
) COMMENT '订单状态表'
PARTITIONED BY (`dt` string)
row format delimited fields terminated by '\t'
STORED AS
  INPUTFORMAT 'com.hadoop.mapred.DeprecatedLzoTextInputFormat'
  OUTPUTFORMAT 'org.apache.hadoop.hive.ql.io.HiveIgnoreKeyTextOutputFormat'
location '/warehouse/gmall/ods/ods_order_status_log/';
```

13. 创建 SPU 商品表

```
hive (gmall)>
drop table if exists ods_spu_info;
create external table ods_spu_info(
    `id` string COMMENT '商品id',
    `spu_name` string COMMENT '标准产品单位名称',
    `category3_id` string COMMENT '三级品类id',
    `tm_id` string COMMENT '品牌id'
) COMMENT 'SPU 商品表'
PARTITIONED BY (`dt` string)
row format delimited fields terminated by '\t'
STORED AS
  INPUTFORMAT 'com.hadoop.mapred.DeprecatedLzoTextInputFormat'
  OUTPUTFORMAT 'org.apache.hadoop.hive.ql.io.HiveIgnoreKeyTextOutputFormat'
location '/warehouse/gmall/ods/ods_spu_info/';
```

14. 创建商品评价表

```
hive (gmall)>
drop table if exists ods_comment_info;
create external table ods_comment_info(
    `id` string COMMENT '编号',
    `user_id` string COMMENT '用户id',
    `sku_id` string COMMENT '商品id',
    `spu_id` string COMMENT '标准产品单位id',
    `order_id` string COMMENT '订单编号',
    `appraise` string COMMENT '评价',
    `create_time` string COMMENT '创建时间'
) COMMENT '商品评价表'
PARTITIONED BY (`dt` string)
row format delimited fields terminated by '\t'
STORED AS
  INPUTFORMAT 'com.hadoop.mapred.DeprecatedLzoTextInputFormat'
  OUTPUTFORMAT 'org.apache.hadoop.hive.ql.io.HiveIgnoreKeyTextOutputFormat'
location '/warehouse/gmall/ods/ods_comment_info/';
```

15. 创建退款表

```
hive (gmall)>
drop table if exists ods_order_refund_info;
create external table ods_order_refund_info(
    `id` string COMMENT '编号',
    `user_id` string COMMENT '用户id',
    `order_id` string COMMENT '订单编号',
    `sku_id` string COMMENT '商品id',
    `refund_type` string COMMENT '退款类型',
    `refund_num` bigint COMMENT '退款件数',
    `refund_amount` decimal(16,2) COMMENT '退款金额',
    `refund_reason_type` string COMMENT '退款原因类型',
    `create_time` string COMMENT '创建时间'
) COMMENT '退款表'
PARTITIONED BY (`dt` string)
row format delimited fields terminated by '\t'
STORED AS
  INPUTFORMAT 'com.hadoop.mapred.DeprecatedLzoTextInputFormat'
  OUTPUTFORMAT 'org.apache.hadoop.hive.ql.io.HiveIgnoreKeyTextOutputFormat'
location '/warehouse/gmall/ods/ods_order_refund_info/';
```

16. 创建加购表

```
hive (gmall)>
drop table if exists ods_cart_info;
create external table ods_cart_info(
    `id` string COMMENT '编号',
    `user_id` string COMMENT '用户id',
    `sku_id` string COMMENT '商品id',
```

```
    `cart_price` string COMMENT '放入购物车时的价格',
    `sku_num` string COMMENT '数量',
    `sku_name` string COMMENT '商品名称 (冗余)',
    `create_time` string COMMENT '创建时间',
    `operate_time` string COMMENT '操作时间',
    `is_ordered` string COMMENT '是否已经下单',
    `order_time` string COMMENT '下单时间'
) COMMENT '加购表'
PARTITIONED BY (`dt` string)
row format delimited fields terminated by '\t'
STORED AS
  INPUTFORMAT 'com.hadoop.mapred.DeprecatedLzoTextInputFormat'
  OUTPUTFORMAT 'org.apache.hadoop.hive.ql.io.HiveIgnoreKeyTextOutputFormat'
location '/warehouse/gmall/ods/ods_cart_info/';
```

17. 创建商品收藏表

```
hive (gmall)>
drop table if exists ods_favor_info;
create external table ods_favor_info(
    `id` string COMMENT '编号',
    `user_id` string COMMENT '用户id',
    `sku_id` string COMMENT '商品id',
    `spu_id` string COMMENT '标准产品单位id',
    `is_cancel` string COMMENT '是否取消',
    `create_time` string COMMENT '收藏时间',
    `cancel_time` string COMMENT '取消时间'
) COMMENT '商品收藏表'
PARTITIONED BY (`dt` string)
row format delimited fields terminated by '\t'
STORED AS
  INPUTFORMAT 'com.hadoop.mapred.DeprecatedLzoTextInputFormat'
  OUTPUTFORMAT 'org.apache.hadoop.hive.ql.io.HiveIgnoreKeyTextOutputFormat'
location '/warehouse/gmall/ods/ods_favor_info/';
```

18. 创建优惠券领用表

```
hive (gmall)>
drop table if exists ods_coupon_use;
create external table ods_coupon_use(
    `id` string COMMENT '编号',
    `coupon_id` string COMMENT '优惠券id',
    `user_id` string COMMENT '用户id',
    `order_id` string COMMENT '订单编号',
    `coupon_status` string COMMENT '优惠券状态',
    `get_time` string COMMENT '领取时间',
    `using_time` string COMMENT '使用(下单)时间',
    `used_time` string COMMENT '使用(支付)时间'
) COMMENT '优惠券领用表'
```

```sql
PARTITIONED BY (`dt` string)
row format delimited fields terminated by '\t'
STORED AS
  INPUTFORMAT 'com.hadoop.mapred.DeprecatedLzoTextInputFormat'
  OUTPUTFORMAT 'org.apache.hadoop.hive.ql.io.HiveIgnoreKeyTextOutputFormat'
location '/warehouse/gmall/ods/ods_coupon_use/';
```

19. 创建优惠券表

```sql
hive (gmall)>
drop table if exists ods_coupon_info;
create external table ods_coupon_info(
    `id` string COMMENT '优惠券编号',
    `coupon_name` string COMMENT '优惠券名称',
    `coupon_type` string COMMENT '优惠券类型 1 现金券 2 折扣券 3 满减券 4 满件打折券',
    `condition_amount` string COMMENT '满减金额',
    `condition_num` string COMMENT '满减件数',
    `activity_id` string COMMENT '活动编号',
    `benefit_amount` string COMMENT '优惠金额',
    `benefit_discount` string COMMENT '优惠折扣',
    `create_time` string COMMENT '创建时间',
    `range_type` string COMMENT '范围类型 1 商品 2 品类 3 品牌',
    `spu_id` string COMMENT '标准产品单位 id',
    `tm_id` string COMMENT '品牌 id',
    `category3_id` string COMMENT '三级品类 id',
    `limit_num` string COMMENT '最多领用次数',
    `operate_time` string COMMENT '操作时间',
    `expire_time` string COMMENT '过期时间'
) COMMENT '优惠券表'
PARTITIONED BY (`dt` string)
row format delimited fields terminated by '\t'
STORED AS
  INPUTFORMAT 'com.hadoop.mapred.DeprecatedLzoTextInputFormat'
  OUTPUTFORMAT 'org.apache.hadoop.hive.ql.io.HiveIgnoreKeyTextOutputFormat'
location '/warehouse/gmall/ods/ods_coupon_info/';
```

20. 创建活动表

```sql
hive (gmall)>
drop table if exists ods_activity_info;
create external table ods_activity_info(
    `id` string COMMENT '活动 id',
    `activity_name` string COMMENT '活动名称',
    `activity_type` string COMMENT '活动类型',
    `start_time` string COMMENT '开始时间',
    `end_time` string COMMENT '结束时间',
    `create_time` string COMMENT '创建时间'
) COMMENT '活动表'
PARTITIONED BY (`dt` string)
```

```
row format delimited fields terminated by '\t'
STORED AS
  INPUTFORMAT 'com.hadoop.mapred.DeprecatedLzoTextInputFormat'
  OUTPUTFORMAT 'org.apache.hadoop.hive.ql.io.HiveIgnoreKeyTextOutputFormat'
location '/warehouse/gmall/ods/ods_activity_info/';
```

21. 创建活动订单关联表

```
hive (gmall)>
drop table if exists ods_activity_order;
create external table ods_activity_order(
    `id` string COMMENT '编号',
    `activity_id` string COMMENT '活动id',
    `order_id` string COMMENT '订单编号',
    `create_time` string COMMENT '创建时间'
) COMMENT '活动订单关联表'
PARTITIONED BY (`dt` string)
row format delimited fields terminated by '\t'
STORED AS
  INPUTFORMAT 'com.hadoop.mapred.DeprecatedLzoTextInputFormat'
  OUTPUTFORMAT 'org.apache.hadoop.hive.ql.io.HiveIgnoreKeyTextOutputFormat'
location '/warehouse/gmall/ods/ods_activity_order/';
```

22. 创建优惠规则表

```
hive (gmall)>
drop table if exists ods_activity_rule;
create external table ods_activity_rule(
    `id` string COMMENT '编号',
    `activity_id` string COMMENT '活动id',
    `condition_amount` string COMMENT '满减金额',
    `condition_num` string COMMENT '满减件数',
    `benefit_amount` string COMMENT '优惠金额',
    `benefit_discount` string COMMENT '优惠折扣',
    `benefit_level` string COMMENT '优惠级别'
) COMMENT '优惠规则表'
PARTITIONED BY (`dt` string)
row format delimited fields terminated by '\t'
STORED AS
  INPUTFORMAT 'com.hadoop.mapred.DeprecatedLzoTextInputFormat'
  OUTPUTFORMAT 'org.apache.hadoop.hive.ql.io.HiveIgnoreKeyTextOutputFormat'
location '/warehouse/gmall/ods/ods_activity_rule/';
```

23. 创建编码字典表

```
hive (gmall)>
drop table if exists ods_base_dic;
create external table ods_base_dic(
    `dic_code` string COMMENT '编号',
    `dic_name` string COMMENT '编码名称',
    `parent_code` string COMMENT '父编码',
```

```
    `create_time` string COMMENT '创建时间',
    `operate_time` string COMMENT '修改时间'
) COMMENT '编码字典表'
PARTITIONED BY (`dt` string)
row format delimited fields terminated by '\t'
STORED AS
  INPUTFORMAT 'com.hadoop.mapred.DeprecatedLzoTextInputFormat'
  OUTPUTFORMAT 'org.apache.hadoop.hive.ql.io.HiveIgnoreKeyTextOutputFormat'
location '/warehouse/gmall/ods/ods_base_dic/';
```

24. 创建参与活动商品表

```
hive (gmall)>
drop table if exists ods_activity_sku;
create external table ods_activity_sku (
    `id` string COMMENT '编号',
    `activity_id` string COMMENT '活动id',
    `sku_id` string COMMENT '商品id',
    `create_time` string COMMENT '创建时间'
) COMMENT '参与活动商品表'
PARTITIONED BY (`dt` string)
row format delimited fields terminated by '\t'
STORED AS
  INPUTFORMAT 'com.hadoop.mapred.DeprecatedLzoTextInputFormat'
  OUTPUTFORMAT 'org.apache.hadoop.hive.ql.io.HiveIgnoreKeyTextOutputFormat'
location '/warehouse/gmall/ods/ods_activity_sku/';
```

6.3.5 ODS 层业务数据导入脚本

将 ODS 层业务数据的加载过程编写成脚本，方便每日调用执行。
（1）在/home/atguigu/bin 目录下创建脚本 hdfs_to_ods_db.sh。

```
[atguigu@hadoop102 bin]$ vim hdfs_to_ods_db.sh
```

在脚本中编写如下内容。

```bash
#!/bin/bash

APP=gmall
hive=/opt/module/hive/bin/hive

#若输入了日期参数，则取输入参数作为日期值；若没有输入日期参数，则取当前时间的前一天作为日期值
if [ -n "$2" ] ;then
    do_date=$2
else
    do_date=`date -d "-1 day" +%F`
fi

sql1="
```

```
load data inpath '/origin_data/$APP/db/order_info/$do_date' OVERWRITE into
table ${APP}.ods_order_info partition(dt='$do_date');

load data inpath '/origin_data/$APP/db/order_detail/$do_date' OVERWRITE into
table ${APP}.ods_order_detail partition(dt='$do_date');

load data inpath '/origin_data/$APP/db/sku_info/$do_date' OVERWRITE into table
${APP}.ods_sku_info partition(dt='$do_date');

load data inpath '/origin_data/$APP/db/user_info/$do_date' OVERWRITE into table
${APP}.ods_user_info partition(dt='$do_date');

load data inpath '/origin_data/$APP/db/payment_info/$do_date' OVERWRITE into
table ${APP}.ods_payment_info partition(dt='$do_date');

load data inpath '/origin_data/$APP/db/base_category1/$do_date' OVERWRITE into
table ${APP}.ods_base_category1 partition(dt='$do_date');

load data inpath '/origin_data/$APP/db/base_category2/$do_date' OVERWRITE into
table ${APP}.ods_base_category2 partition(dt='$do_date');

load data inpath '/origin_data/$APP/db/base_category3/$do_date' OVERWRITE into
table ${APP}.ods_base_category3 partition(dt='$do_date');

load data inpath '/origin_data/$APP/db/base_trademark/$do_date' OVERWRITE into
table ${APP}.ods_base_trademark partition(dt='$do_date');

load data inpath '/origin_data/$APP/db/activity_info/$do_date' OVERWRITE into
table ${APP}.ods_activity_info partition(dt='$do_date');

load data inpath '/origin_data/$APP/db/activity_order/$do_date' OVERWRITE into
table ${APP}.ods_activity_order partition(dt='$do_date');

load data inpath '/origin_data/$APP/db/cart_info/$do_date' OVERWRITE into table
${APP}.ods_cart_info partition(dt='$do_date');

load data inpath '/origin_data/$APP/db/comment_info/$do_date' OVERWRITE into
table ${APP}.ods_comment_info partition(dt='$do_date');

load data inpath '/origin_data/$APP/db/coupon_info/$do_date' OVERWRITE into
table ${APP}.ods_coupon_info partition(dt='$do_date');

load data inpath '/origin_data/$APP/db/coupon_use/$do_date' OVERWRITE into
table ${APP}.ods_coupon_use partition(dt='$do_date');

load data inpath '/origin_data/$APP/db/favor_info/$do_date' OVERWRITE into
table ${APP}.ods_favor_info partition(dt='$do_date');
```

```
load data inpath '/origin_data/$APP/db/order_refund_info/$do_date' OVERWRITE
into table ${APP}.ods_order_refund_info partition(dt='$do_date');

load data inpath '/origin_data/$APP/db/order_status_log/$do_date' OVERWRITE
into table ${APP}.ods_order_status_log partition(dt='$do_date');

load data inpath '/origin_data/$APP/db/spu_info/$do_date' OVERWRITE into table
${APP}.ods_spu_info partition(dt='$do_date');

load data inpath '/origin_data/$APP/db/activity_rule/$do_date' OVERWRITE into
table ${APP}.ods_activity_rule partition(dt='$do_date');

load data inpath '/origin_data/$APP/db/base_dic/$do_date' OVERWRITE into table
${APP}.ods_base_dic partition(dt='$do_date');
load data inpath '/origin_data/$APP/db/activity_sku/$do_date' OVERWRITE into
table ${APP}.ods_activity_sku partition(dt='$do_date');
"

sql2="
load data inpath '/origin_data/$APP/db/base_province/$do_date' OVERWRITE into
table ${APP}.ods_base_province;

load data inpath '/origin_data/$APP/db/base_region/$do_date' OVERWRITE into
table ${APP}.ods_base_region;
"
case $1 in
"first"){
    $hive -e "$sql1"
    $hive -e "$sql2"
};;
"all"){
    $hive -e "$sql1"
};;
esac
```

（2）增加脚本执行权限。

`[atguigu@hadoop102 bin]$ chmod 777 hdfs_to_ods_db.sh`

（3）初次执行脚本，传入 first，导入 2020-03-10 的数据。

`[atguigu@hadoop102 bin]$ hdfs_to_ods_db.sh first 2020-03-10`

（4）再次执行脚本，传入 all，导入 2020-03-11 的数据。

`[atguigu@hadoop102 bin]$ hdfs_to_ods_db.sh all 2020-03-11`

（5）测试数据是否导入成功。

`hive (gmall)> select * from ods_order_detail where dt='2020-03-11' limit 10;`

6.4 数据仓库搭建——DWD 层

对 ODS 层的数据进行判空过滤,然后对商品品类表进行维度退化(降维),使用 Parquet 格式进行存储,并保存为 LZO 压缩格式,以减少存储空间的占用。

6.4.1 用户行为启动日志表解析

在第 4 章中,通过日志生成程序生成的启动日志数据示例如下。

```
{
    "mid":"995",
    "uid":"995",
    "vc":"10",
    "vn":"1.3.4",
    "l":"en",
    "sr":"B",
    "os":"8.1.2",
    "ar":"MX",
    "md":"HTC-2",
    "ba":"HTC",
    "sv":"V2.0.6",
    "g":"43R2SEQX@gmail.com",
    "hw":"640*960",
    "t":"1561472502444",
    "nw":"4G",
    "ln":"-99.3",
    "la":"20.4",
    "entry":"2",
    "open_ad_type":"2",
    "action":"1",
    "loading_time":"2",
    "detail":"",
    "extend1":"",
    "en":"start",
}
```

日志数据为 JSON 格式,Hive 内置了 JSON 字符串解析工具,从而可以得到字符串内字段的对应信息,根据示例数据中的字段信息,可确定启动日志表中所包含的字段。启动日志表为分区表,以日期为分区,方便用户进行分区查询。将数据存储为 Parquet 格式,并保存为 LZO 压缩格式,可大大减少存储空间的占用。

1. 创建启动日志表

在 gmall 数据库中执行如下建表语句。

```
hive (gmall)>
drop table if exists dwd_start_log;
```

```sql
CREATE EXTERNAL TABLE dwd_start_log( --字段为示例数据中出现的字段
    `mid_id` string,
    `user_id` string,
    `version_code` string,
    `version_name` string,
    `lang` string,
    `source` string,
    `os` string,
    `area` string,
    `model` string,
    `brand` string,
    `sdk_version` string,
    `gmail` string,
    `height_width` string,
    `app_time` string,
    `network` string,
    `lng` string,
    `lat` string,
    `entry` string,
    `open_ad_type` string,
    `action` string,
    `loading_time` string,
    `detail` string,
    `extend1` string
)
PARTITIONED BY (dt string) --以日期为分区
stored as parquet  --存储格式为Parquet
location '/warehouse/gmall/dwd/dwd_start_log/' --指定存储目录
tblproperties ("parquet.compression"="lzo") --指定压缩格式为LZO
;
```

2. 向启动日志表中导入数据

（1）使用 insert 语句向已经创建的启动日志表中导入数据，其中，字段信息由 Hive 内置的 JSON 字符串解析函数 get_json_object()解析得到。get_json_object()函数的第一个参数填写 JSON 对象变量，第二个参数使用$表示 JSON 变量标识。

```
hive (gmall)>
insert overwrite table dwd_start_log -PARTITION (dt='2020-03-10') --指定分区
select
    --使用Hive内置的JSON字符串解析函数解析JSON字符串，获取字段值
    get_json_object(line,'$.mid') mid_id,
    get_json_object(line,'$.uid') user_id,
    get_json_object(line,'$.vc') version_code,
    get_json_object(line,'$.vn') version_name,
    get_json_object(line,'$.l') lang,
    get_json_object(line,'$.sr') source,
    get_json_object(line,'$.os') os,
```

```
    get_json_object(line,'$.ar') area,
    get_json_object(line,'$.md') model,
    get_json_object(line,'$.ba') brand,
    get_json_object(line,'$.sv') sdk_version,
    get_json_object(line,'$.g') gmail,
    get_json_object(line,'$.hw') height_width,
    get_json_object(line,'$.t') app_time,
    get_json_object(line,'$.nw') network,
    get_json_object(line,'$.ln') lng,
    get_json_object(line,'$.la') lat,
    get_json_object(line,'$.entry') entry,
    get_json_object(line,'$.open_ad_type') open_ad_type,
    get_json_object(line,'$.action') action,
    get_json_object(line,'$.loading_time') loading_time,
    get_json_object(line,'$.detail') detail,
    get_json_object(line,'$.extend1') extend1
from ods_start_log  --数据来源为ODS层的启动日志表
where dt='2020-03-10'; --指定数据日期
```

（2）测试。

```
hive (gmall)> select * from dwd_start_log limit 2;
```

3. DWD层启动日志表加载数据脚本

（1）在hadoop102的/home/atguigu/bin目录下创建脚本dwd_start_log.sh。

```
[atguigu@hadoop102 bin]$ vim dwd_start_log.sh
```

在脚本中编写如下内容。

```
#!/bin/bash
# 定义变量, 方便后续修改
APP=gmall
hive=/opt/module/hive/bin/hive

#如果输入了日期参数, 则取输入参数作为日期值; 如果没有输入日期参数, 则取当前时间的前一天作为日期值
if [ -n "$1" ] ;then
 do_date=$1
else
 do_date=`date -d "-1 day" +%F`
fi

sql="
insert overwrite table "$APP".dwd_start_log
PARTITION (dt='$do_date')
select
    get_json_object(line,'$.mid') mid_id,
    get_json_object(line,'$.uid') user_id,
    get_json_object(line,'$.vc') version_code,
    get_json_object(line,'$.vn') version_name,
```

```
        get_json_object(line,'$.l') lang,
        get_json_object(line,'$.sr') source,
        get_json_object(line,'$.os') os,
        get_json_object(line,'$.ar') area,
        get_json_object(line,'$.md') model,
        get_json_object(line,'$.ba') brand,
        get_json_object(line,'$.sv') sdk_version,
        get_json_object(line,'$.g') gmail,
        get_json_object(line,'$.hw') height_width,
        get_json_object(line,'$.t') app_time,
        get_json_object(line,'$.nw') network,
        get_json_object(line,'$.ln') lng,
        get_json_object(line,'$.la') lat,
        get_json_object(line,'$.entry') entry,
        get_json_object(line,'$.open_ad_type') open_ad_type,
        get_json_object(line,'$.action') action,
        get_json_object(line,'$.loading_time') loading_time,
        get_json_object(line,'$.detail') detail,
        get_json_object(line,'$.extend1') extend1
from "$APP".ods_start_log
where dt='$do_date';
"

$hive -e "$sql"
```

（2）增加脚本执行权限。

```
[atguigu@hadoop102 bin]$ chmod 777 dwd_start_log.sh
```

（3）执行脚本，导入数据。

```
[atguigu@hadoop102 module]$ dwd_start_log.sh 2020-03-11
```

（4）查询结果数据。

```
hive (gmall)>
select * from dwd_start_log where dt='2020-03-11' limit 2;
```

6.4.2 用户行为事件表拆分

我们对 ODS 层的事件表进行解析的基本思路如图 6-18 和图 6-19 所示，通过自定义 UDF 函数和自定义 UDTF 函数将事件表解析成 dwd_base_event_log 表，再将 dwd_base_event_log 表拆解成分类事件表。

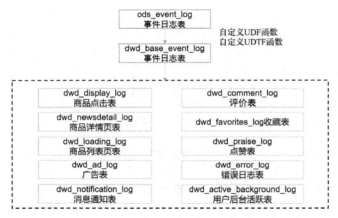

图 6-18 用户行为事件表拆分思路

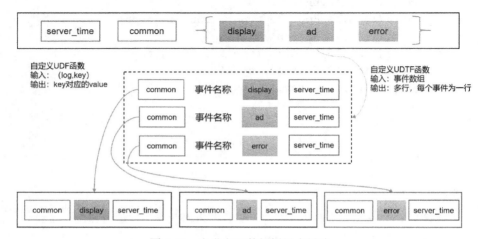

图 6-19 自定义函数解析日志思路

如图 6-20 所示为 DWD 层用户行为事件表数据解析思路。

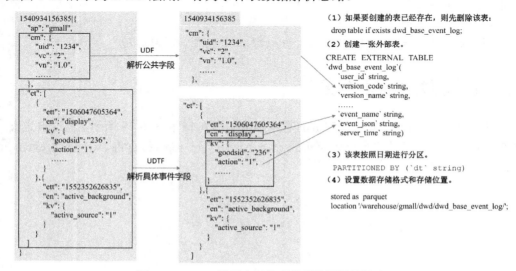

图 6-20 DWD 层用户行为事件表数据解析思路

1. 创建基础明细表

基础明细表用于存储从 ODS 层原始表中转换过来的明细数据。

创建事件日志基础明细表。

```
hive (gmall)>
drop table if exists dwd_base_event_log;
CREATE EXTERNAL TABLE dwd_base_event_log(
    `mid_id` string,
    `user_id` string,
    `version_code` string,
    `version_name` string,
    `lang` string,
    `source` string,
    `os` string,
    `area` string,
    `model` string,
    `brand` string,
    `sdk_version` string,
    `gmail` string,
    `height_width` string,
    `app_time` string,
    `network` string,
    `lng` string,
    `lat` string,
    `event_name` string,
    `event_json` string,
    `server_time` string
)
PARTITIONED BY (`dt` string)
stored as parquet
location '/warehouse/gmall/dwd/dwd_base_event_log/'
TBLPROPERTIES('parquet.compression'='lzo');
```

其中，event_name 和 event_json 分别对应事件名称和整个事件。这个地方将原始日志一对多的形式拆分出来了，操作的时候我们需要将原始日志进行解析，将用到 UDF 和 UDTF。

UDF 是什么？

UDF 的全称为 User-Defined Function，意思为用户定义函数，它的作用是什么呢？有的时候，用户的需求无法通过 Hive 提供的内置函数来实现，这时候通过编写 UDF，用户可以方便地定义自己需要的处理逻辑，并在查询中使用它们。

UDTF 是什么？

Hive 中有三种 UDF：（普通）UDF、用户定义聚集函数（User-Defined Aggregate Function，UDAF）和用户定义表生成函数（User-Defined Table-generating Function，UDTF）。UDTF 操作作用于单个数据行，并且将产生多个数据行作为输出。

2. 自定义 UDF 函数（解析公共字段）

自定义 UDF 函数，解析公共字段，思路如图 6-21 所示。

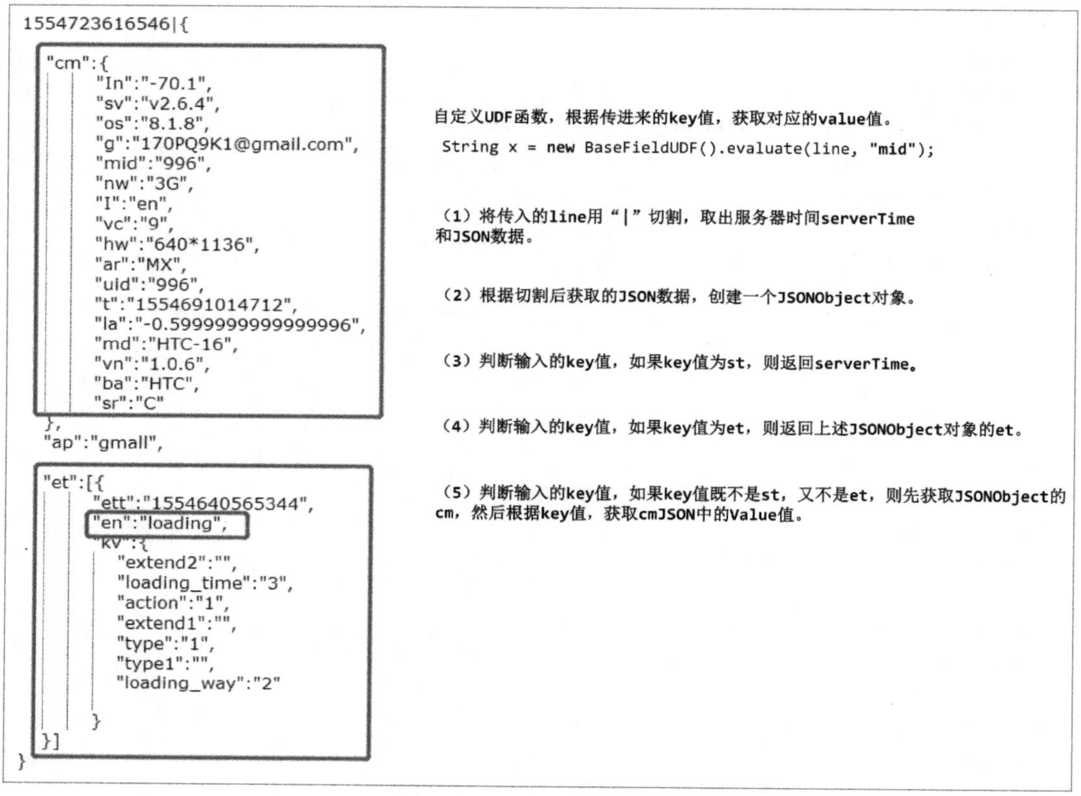

```
1554723616546|{
    "cm":{
        "ln":"-70.1",
        "sv":"V2.6.4",
        "os":"8.1.8",
        "g":"170PQ9K1@gmail.com",
        "mid":"996",
        "nw":"3G",
        "l":"en",
        "vc":"9",
        "hw":"640*1136",
        "ar":"MX",
        "uid":"996",
        "t":"1554691014712",
        "la":"-0.5999999999999996",
        "md":"HTC-16",
        "vn":"1.0.6",
        "ba":"HTC",
        "sr":"C"
    },
    "ap":"gmall",
    "et":[{
        "ett":"1554640565344",
        "en":"loading",
        "kv":{
            "extend2":"",
            "loading_time":"3",
            "action":"1",
            "extend1":"",
            "type":"1",
            "type1":"",
            "loading_way":"2"
        }
    }]
}
```

自定义 UDF 函数，根据传进来的 key 值，获取对应的 value 值。
String x = new BaseFieldUDF().evaluate(line, "mid");

（1）将传入的 line 用 "|" 切割，取出服务器时间 serverTime 和 JSON 数据。

（2）根据切割后获取的 JSON 数据，创建一个 JSONObject 对象。

（3）判断输入的 key 值，如果 key 值为 st，则返回 serverTime。

（4）判断输入的 key 值，如果 key 值为 et，则返回上述 JSONObject 对象的 et。

（5）判断输入的 key 值，如果 key 值既不是 st，又不是 et，则先获取 JSONObject 的 cm，然后根据 key 值，获取 cmJSON 中的 Value 值。

图 6-21　自定义 UDF 函数解析公共字段的思路

（1）创建一个 maven 工程：hivefunction。
（2）创建包名：com.atguigu.udf。
（3）在 pom.xml 文件中添加如下内容。

```xml
<properties>
    <project.build.sourceEncoding>UTF8</project.build.sourceEncoding>
    <hive.version>1.2.1</hive.version>
</properties>

<dependencies>
    <!--添加 Hive 依赖-->
    <dependency>
        <groupId>org.apache.hive</groupId>
        <artifactId>hive-exec</artifactId>
        <version>${hive.version}</version>
    </dependency>
</dependencies>

<build>
```

```xml
<plugins>
    <plugin>
        <artifactId>maven-compiler-plugin</artifactId>
        <version>2.3.2</version>
        <configuration>
            <source>1.8</source>
            <target>1.8</target>
        </configuration>
    </plugin>
    <plugin>
        <artifactId>maven-assembly-plugin</artifactId>
        <configuration>
            <descriptorRefs>
                <descriptorRef>jar-with-dependencies</descriptorRef>
            </descriptorRefs>
        </configuration>
        <executions>
            <execution>
                <id>make-assembly</id>
                <phase>package</phase>
                <goals>
                    <goal>single</goal>
                </goals>
            </execution>
        </executions>
    </plugin>
</plugins>
</build>
```

（4）自定义 UDF 函数，解析公共字段。

```java
package com.atguigu.udf;
import org.apache.commons.lang.StringUtils;
import org.apache.hadoop.hive.ql.exec.UDF;
import org.json.JSONException;
import org.json.JSONObject;

public class BaseFieldUDF extends UDF {

    public String evaluate(String line, String key) throws JSONException {

        // 1 按"\\|"对日志line进行切割
        String[] log = line.split("\\|");

        // 2 合法性校验
        if (log.length != 2 || StringUtils.isBlank(log[1])) {
            return "";
        }
```

```java
        // 3 开始处理JSON
        JSONObject baseJson = new JSONObject(log[1].trim());

        String result = "";

        // 4 根据传进来的key值，查找对应的value值
        if ("et".equals(key)) {
            if (baseJson.has("et")) {
                result = baseJson.getString("et");
            }
        } else if ("st".equals(key)) {
            result = log[0].trim();
        } else {
            JSONObject cm = baseJson.getJSONObject("cm");
            if (cm.has(key)) {
                result = cm.getString(key);
            }
        }

        return result;
}
 public static void main(String[] args) {

        String line =
"1541217850324|{\"cm\":{\"mid\":\"m7856\",\"uid\":\"u8739\",\"ln\":\"-74.8\",\"sv\":\"V2.2.2\",\"os\":\"8.1.3\",\"g\":\"P7XC9126@gmail.com\",\"nw\":\"3G\",\"l\":\"es\",\"vc\":\"6\",\"hw\":\"640*960\",\"ar\":\"MX\",\"t\":\"1541204134250\",\"la\":\"-31.7\",\"md\":\"huawei-17\",\"vn\":\"1.1.2\",\"sr\":\"O\",\"ba\":\"Huawei\"},\"ap\":\"weather\",\"et\":[{\"ett\":\"1541146624055\",\"en\":\"display\",\"kv\":{\"goodsid\":\"n4195\",\"copyright\":\"ESPN\",\"content_provider\":\"CNN\",\"extend2\":\"5\",\"action\":\"2\",\"extend1\":\"2\",\"place\":\"3\",\"showtype\":\"2\",\"category\":\"72\",\"newstype\":\"5\"}},{\"ett\":\"1541213331817\",\"en\":\"loading\",\"kv\":{\"extend2\":\"\",\"loading_time\":\"15\",\"action\":\"3\",\"extend1\":\"\",\"type1\":\"\",\"type\":\"3\",\"loading_way\":\"1\"}},{\"ett\":\"1541126195645\",\"en\":\"ad\",\"kv\":{\"entry\":\"3\",\"show_style\":\"0\",\"action\":\"2\",\"detail\":\"325\",\"source\":\"4\",\"behavior\":\"2\",\"content\":\"1\",\"newstype\":\"5\"}},{\"ett\":\"1541202678812\",\"en\":\"notification\",\"kv\":{\"ap_time\":\"1541184614380\",\"action\":\"3\",\"type\":\"4\",\"content\":\"\"}},{\"ett\":\"1541194686688\",\"en\":\"active_background\",\"kv\":{\"active_source\":\"3\"}}]}";
        String x = new BaseFieldUDF().evaluate(line, "mid,uid,vc,vn,l,sr,os,ar,md,ba,sv,g,hw,nw,ln,la,t");
        System.out.println(x);
    }
}
```

注意：main()函数主要用于模拟数据测试。

3. 自定义 UDTF 函数（解析具体事件字段）

自定义 UDTF 函数，解析具体事件字段，获取事件名称，思路如图 6-22 所示。

图 6-22　自定义 UDTF 函数获取事件名称的思路

（1）创建包名：com.atguigu.udtf。
（2）在 com.atguigu.udtf 包下创建类名：EventJsonUDTF。
（3）EventJsonUDTF 继承 GenericUDTF 类，并重写该抽象类的相关方法。

```java
package com.atguigu.udtf;

import org.apache.commons.lang.StringUtils;
import org.apache.hadoop.hive.ql.exec.UDFArgumentException;
import org.apache.hadoop.hive.ql.metadata.HiveException;
import org.apache.hadoop.hive.ql.udf.generic.GenericUDTF;
import org.apache.hadoop.hive.serde2.objectinspector.ObjectInspector;
import org.apache.hadoop.hive.serde2.objectinspector.ObjectInspectorFactory;
import org.apache.hadoop.hive.serde2.objectinspector.StructObjectInspector;
import org.apache.hadoop.hive.serde2.objectinspector.primitive.PrimitiveObjectInspectorFactory;
import org.json.JSONArray;
import org.json.JSONException;

import java.util.ArrayList;

public class EventJsonUDTF extends GenericUDTF {
```

```java
//在该方法中，我们将指定输出参数的名称和参数类型
@Override
public StructObjectInspector initialize(ObjectInspector[] argOIs) throws UDFArgumentException {

    ArrayList<String> fieldNames = new ArrayList<String>();
    ArrayList<ObjectInspector> fieldOIs = new ArrayList<ObjectInspector>();

    fieldNames.add("event_name");
    fieldOIs.add(PrimitiveObjectInspectorFactory.javaStringObjectInspector);
    fieldNames.add("event_json");
    fieldOIs.add(PrimitiveObjectInspectorFactory.javaStringObjectInspector);

    return ObjectInspectorFactory.getStandardStructObjectInspector(fieldNames, fieldOIs);
}

//输入1条记录，输出若干条结果
@Override
public void process(Object[] objects) throws HiveException {

    // 获取传入的 et
    String input = objects[0].toString();

    // 如果传进来的数据为空，则直接返回，过滤掉该数据
    if (StringUtils.isBlank(input)) {
        return;
    } else {

        try {
            // 获取事件的个数（ad/facoriters）
            JSONArray ja = new JSONArray(input);

            if (ja == null)
                return;

            // 循环遍历每个事件
            for (int i = 0; i < ja.length(); i++) {
                String[] result = new String[2];

                try {
                    // 取出每个事件的名称（ad/facoriters）
                    result[0] = ja.getJSONObject(i).getString("en");

                    // 取出每个事件整体
                    result[1] = ja.getString(i);
```

```
            } catch (JSONException e) {
                continue;
            }
            // 将结果返回
            forward(result);
        }
    } catch (JSONException e) {
        e.printStackTrace();
    }
    }
}

//当没有记录处理的时候，该方法会被调用，用来清理代码或者产生额外的输出
@Override
public void close() throws HiveException {

}
}
```

（4）打包自定义函数，如图 6-23 所示。

图 6-23　打包自定义函数

（5）将 hivefunction-1.0-SNAPSHOT 上传到 HDFS 的/user/hive/jars 目录下。

（6）创建永久函数，与创建好的自定义函数类进行关联。

```
hive (gmall)>
create function base_analizer as 'com.atguigu.udf.BaseFieldUDF' using jar
'hdfs://hadoop102:9000/user/hive/jars/hivefunction-1.0-SNAPSHOT.jar';

create function flat_analizer as 'com.atguigu.udtf.EventJsonUDTF' using jar
'hdfs://hadoop102:9000/user/hive/jars/hivefunction-1.0-SNAPSHOT.jar';
```

4．解析事件日志基础明细表

（1）解析事件日志基础明细表。

```
hive (gmall)>
insert overwrite table dwd_base_event_log partition(dt='2020-03-10')
select
    base_analizer(line,'mid') as mid_id,
    base_analizer(line,'uid') as user_id,
    base_analizer(line,'vc') as version_code,
    base_analizer(line,'vn') as version_name,
    base_analizer(line,'l') as lang,
    base_analizer(line,'sr') as source,
    base_analizer(line,'os') as os,
```

```sql
    base_analizer(line,'ar') as area,
    base_analizer(line,'md') as model,
    base_analizer(line,'ba') as brand,
    base_analizer(line,'sv') as sdk_version,
    base_analizer(line,'g') as gmail,
    base_analizer(line,'hw') as height_width,
    base_analizer(line,'t') as app_time,
    base_analizer(line,'nw') as network,
    base_analizer(line,'ln') as lng,
    base_analizer(line,'la') as lat,
    event_name,
    event_json,
    base_analizer(line,'st') as server_time
from ods_event_log lateral view flat_analizer(base_analizer(line,'et')) tmp_flat
as event_name,event_json
where dt='2020-03-10' and base_analizer(line,'et')<>'';
```

（2）测试。

```
hive (gmall)> select * from dwd_base_event_log limit 2;
```

5. DWD 层数据解析脚本

（1）在 hadoop102 的 /home/atguigu/bin 目录下创建脚本 dwd_base_log.sh。

```
[atguigu@hadoop102 bin]$ vim dwd_base_log.sh
```

在脚本中编写如下内容。

```bash
#!/bin/bash

# 定义变量，方便后续修改
APP=gmall
hive=/opt/module/hive/bin/hive

# 如果输入了日期参数，则取输入参数作为日期值；如果没有输入日期参数，则取当前时间的前一天作为日期值
if [ -n "$1" ] ;then
 do_date=$1
else
 do_date=`date -d "-1 day" +%F`
fi

sql="
insert overwrite table "$APP".dwd_base_event_log partition(dt='$do_date')
select
    "$APP".base_analizer(line,'mid') as mid_id,
    "$APP".base_analizer(line,'uid') as user_id,
    "$APP".base_analizer(line,'vc') as version_code,
    "$APP".base_analizer(line,'vn') as version_name,
    "$APP".base_analizer(line,'l') as lang,
    "$APP".base_analizer(line,'sr') as source,
```

```
    "$APP".base_analizer(line,'os') as os,
    "$APP".base_analizer(line,'ar') as area,
    "$APP".base_analizer(line,'md') as model,
    "$APP".base_analizer(line,'ba') as brand,
    "$APP".base_analizer(line,'sv') as sdk_version,
    "$APP".base_analizer(line,'g') as gmail,
    "$APP".base_analizer(line,'hw') as height_width,
    "$APP".base_analizer(line,'t') as app_time,
    "$APP".base_analizer(line,'nw') as network,
    "$APP".base_analizer(line,'ln') as lng,
    "$APP".base_analizer(line,'la') as lat,
    event_name,
    event_json,
    "$APP".base_analizer(line,'st') as server_time
from "$APP".ods_event_log lateral view "$APP".flat_analizer("$APP".base_
analizer(line,'et')) tem_flat as event_name,event_json
where dt='$do_date'  and "$APP".base_analizer(line,'et')<>'';
"

$hive -e "$sql"
```

(2) 增加脚本执行权限。

```
[atguigu@hadoop102 bin]$ chmod 777 dwd_base_log.sh
```

(3) 执行脚本，导入数据。

```
[atguigu@hadoop102 module]$ dwd_base_log.sh 2020-03-11
```

(4) 查询结果数据。

```
hive (gmall)>
select * from dwd_base_event_log where dt='2020-03-11' limit 2;
```

6.4.3 用户行为事件表解析

得到 DWD 层的事件日志基础明细表后，可将各具体事件表从事件日志基础明细表的 event_json 字段中解析出来。根据不同的事件名称字段 event_name，我们需要分别创建商品点击表、商品详情页表、商品列表页表、广告表、消息通知表、用户后台活跃表、评价表、收藏表、点赞表和错误日志表，并根据事件名称字段对事件日志基础明细表进行查询，然后将查询结果一一插入创建好的具体事件表中，各具体事件表的建表语句相似，只有 event_json 解析出的字段信息不同，所以不再进行单独讲解，具体操作步骤如下。

1. 商品点击表

1) 建表语句

```
hive (gmall)>
drop table if exists dwd_display_log;
CREATE EXTERNAL TABLE dwd_display_log(
    `mid_id` string,
```

```
    `user_id` string,
    `version_code` string,
    `version_name` string,
    `lang` string,
    `source` string,
    `os` string,
    `area` string,
    `model` string,
    `brand` string,
    `sdk_version` string,
    `gmail` string,
    `height_width` string,
    `app_time` string,
    `network` string,
    `lng` string,
    `lat` string,
    `action` string,
    `goodsid` string,
    `place` string,
    `extend1` string,
    `category` string,
    `server_time` string
)
PARTITIONED BY (dt string)
location '/warehouse/gmall/dwd/dwd_display_log/';
```

2）导入数据

```
hive (gmall)>
insert overwrite table dwd_display_log
PARTITION (dt='2020-03-10')
select
    mid_id,
    user_id,
    version_code,
    version_name,
    lang,
    source,
    os,
    area,
    model,
    brand,
    sdk_version,
    gmail,
    height_width,
    app_time,
    network,
    lng,
```

```sql
    lat,
    get_json_object(event_json,'$.kv.action') action,
    get_json_object(event_json,'$.kv.goodsid') goodsid,
    get_json_object(event_json,'$.kv.place') place,
    get_json_object(event_json,'$.kv.extend1') extend1,
    get_json_object(event_json,'$.kv.category') category,
    server_time
from dwd_base_event_log
where dt='2020-03-10' and event_name='display';
```

3）测试

```
hive (gmall)> select * from dwd_display_log limit 2;
```

2．商品详情页表

1）建表语句

```sql
hive (gmall)>
drop table if exists dwd_newsdetail_log;
CREATE EXTERNAL TABLE dwd_newsdetail_log(
    `mid_id` string,
    `user_id` string,
    `version_code` string,
    `version_name` string,
    `lang` string,
    `source` string,
    `os` string,
    `area` string,
    `model` string,
    `brand` string,
    `sdk_version` string,
    `gmail` string,
    `height_width` string,
    `app_time` string,
    `network` string,
    `lng` string,
    `lat` string,
    `entry` string,
    `action` string,
    `goodsid` string,
    `showtype` string,
    `news_staytime` string,
    `loading_time` string,
    `type1` string,
    `category` string,
    `server_time` string
)
PARTITIONED BY (dt string)
location '/warehouse/gmall/dwd/dwd_newsdetail_log/';
```

2）导入数据

```
hive (gmall)>
insert overwrite table dwd_newsdetail_log PARTITION (dt='2020-03-10')
select
    mid_id,
    user_id,
    version_code,
    version_name,
    lang,
    source,
    os,
    area,
    model,
    brand,
    sdk_version,
    gmail,
    height_width,
    app_time,
    network,
    lng,
    lat,
    get_json_object(event_json,'$.kv.entry') entry,
    get_json_object(event_json,'$.kv.action') action,
    get_json_object(event_json,'$.kv.goodsid') goodsid,
    get_json_object(event_json,'$.kv.showtype') showtype,
    get_json_object(event_json,'$.kv.news_staytime') news_staytime,
    get_json_object(event_json,'$.kv.loading_time') loading_time,
    get_json_object(event_json,'$.kv.type1') type1,
    get_json_object(event_json,'$.kv.category') category,
    server_time
from dwd_base_event_log
where dt='2020-03-10' and event_name='newsdetail';
```

3）测试

```
hive (gmall)> select * from dwd_newsdetail_log limit 2;
```

3. 商品列表页表

1）建表语句

```
hive (gmall)>
drop table if exists dwd_loading_log;
CREATE EXTERNAL TABLE dwd_loading_log(
    `mid_id` string,
    `user_id` string,
    `version_code` string,
    `version_name` string,
    `lang` string,
    `source` string,
```

```
    `os` string,
    `area` string,
    `model` string,
    `brand` string,
    `sdk_version` string,
    `gmail` string,
    `height_width` string,
    `app_time` string,
    `network` string,
    `lng` string,
    `lat` string,
    `action` string,
    `loading_time` string,
    `loading_way` string,
    `extend1` string,
    `extend2` string,
    `type` string,
    `type1` string,
    `server_time` string
)
PARTITIONED BY (dt string)
location '/warehouse/gmall/dwd/dwd_loading_log/';
```

2）导入数据

```
hive (gmall)>

insert overwrite table dwd_loading_log PARTITION (dt='2020-03-10')
select
    mid_id,
    user_id,
    version_code,
    version_name,
    lang,
    source,
    os,
    area,
    model,
    brand,
    sdk_version,
    gmail,
    height_width,
    app_time,
    network,
    lng,
    lat,
    get_json_object(event_json,'$.kv.action') action,
    get_json_object(event_json,'$.kv.loading_time') loading_time,
    get_json_object(event_json,'$.kv.loading_way') loading_way,
```

```
    get_json_object(event_json,'$.kv.extend1') extend1,
    get_json_object(event_json,'$.kv.extend2') extend2,
    get_json_object(event_json,'$.kv.type') type,
    get_json_object(event_json,'$.kv.type1') type1,
    server_time
from dwd_base_event_log
where dt='2020-03-10' and event_name='loading';
```

3）测试

```
hive (gmall)> select * from dwd_loading_log limit 2;
```

4．广告表

1）建表语句

```
hive (gmall)>

drop table if exists dwd_ad_log;
CREATE EXTERNAL TABLE dwd_ad_log(
    `mid_id` string,
    `user_id` string,
    `version_code` string,
    `version_name` string,
    `lang` string,
    `source` string,
    `os` string,
    `area` string,
    `model` string,
    `brand` string,
    `sdk_version` string,
    `gmail` string,
    `height_width` string,
    `app_time` string,
    `network` string,
    `lng` string,
    `lat` string,
    `entry` string,
    `action` string,
    `content` string,
    `detail` string,
    `ad_source` string,
    `behavior` string,
    `newstype` string,
    `show_style` string,
    `server_time` string
)
PARTITIONED BY (dt string)
location '/warehouse/gmall/dwd/dwd_ad_log/';
```

2）导入数据

```
hive (gmall)>
insert overwrite table dwd_ad_log PARTITION (dt='2020-03-10')
select
    mid_id,
    user_id,
    version_code,
    version_name,
    lang,
    source,
    os,
    area,
    model,
    brand,
    sdk_version,
    gmail,
    height_width,
    app_time,
    network,
    lng,
    lat,
    get_json_object(event_json,'$.kv.entry') entry,
    get_json_object(event_json,'$.kv.action') action,
    get_json_object(event_json,'$.kv.content') content,
    get_json_object(event_json,'$.kv.detail') detail,
    get_json_object(event_json,'$.kv.source') ad_source,
    get_json_object(event_json,'$.kv.behavior') behavior,
    get_json_object(event_json,'$.kv.newstype') newstype,
    get_json_object(event_json,'$.kv.show_style') show_style,
    server_time
from dwd_base_event_log
where dt='2020-03-10' and event_name='ad';
```

3）测试

```
hive (gmall)> select * from dwd_ad_log limit 2;
```

5. 消息通知表

1）建表语句

```
hive (gmall)>
drop table if exists dwd_notification_log;
CREATE EXTERNAL TABLE dwd_notification_log(
    `mid_id` string,
    `user_id` string,
    `version_code` string,
    `version_name` string,
    `lang` string,
```

```
    `source` string,
    `os` string,
    `area` string,
    `model` string,
    `brand` string,
    `sdk_version` string,
    `gmail` string,
    `height_width` string,
    `app_time` string,
    `network` string,
    `lng` string,
    `lat` string,
    `action` string,
    `noti_type` string,
    `ap_time` string,
    `content` string,
    `server_time` string
)
PARTITIONED BY (dt string)
location '/warehouse/gmall/dwd/dwd_notification_log/';
```

2）导入数据

```
hive (gmall)>

insert overwrite table dwd_notification_log
PARTITION (dt='2020-03-10')
select
    mid_id,
    user_id,
    version_code,
    version_name,
    lang,
    source,
    os,
    area,
    model,
    brand,
    sdk_version,
    gmail,
    height_width,
    app_time,
    network,
    lng,
    lat,
    get_json_object(event_json,'$.kv.action') action,
    get_json_object(event_json,'$.kv.noti_type') noti_type,
    get_json_object(event_json,'$.kv.ap_time') ap_time,
    get_json_object(event_json,'$.kv.content') content,
```

```
    server_time
from dwd_base_event_log
where dt='2020-03-10' and event_name='notification';
```

3）测试

```
hive (gmall)> select * from dwd_notification_log limit 2;
```

6．用户后台活跃表

1）建表语句

```
hive (gmall)>
drop table if exists dwd_active_background_log;
CREATE EXTERNAL TABLE dwd_active_background_log(
    `mid_id` string,
    `user_id` string,
    `version_code` string,
    `version_name` string,
    `lang` string,
    `source` string,
    `os` string,
    `area` string,
    `model` string,
    `brand` string,
    `sdk_version` string,
    `gmail` string,
    `height_width` string,
    `app_time` string,
    `network` string,
    `lng` string,
    `lat` string,
    `active_source` string,
    `server_time` string
)
PARTITIONED BY (dt string)
location '/warehouse/gmall/dwd/dwd_background_log/';
```

2）导入数据

```
hive (gmall)>

insert overwrite table dwd_active_background_log
PARTITION (dt='2020-03-10')
select
    mid_id,
    user_id,
    version_code,
    version_name,
    lang,
    source,
    os,
```

```
    area,
    model,
    brand,
    sdk_version,
    gmail,
    height_width,
    app_time,
    network,
    lng,
    lat,
    get_json_object(event_json,'$.kv.active_source') active_source,
    server_time
from dwd_base_event_log
where dt='2020-03-10' and event_name='active_background';
```

3)测试

```
hive (gmall)> select * from dwd_active_background_log limit 2;
```

7. 评价表

1)建表语句

```
hive (gmall)>
drop table if exists dwd_comment_log;
CREATE EXTERNAL TABLE dwd_comment_log(
    `mid_id` string,
    `user_id` string,
    `version_code` string,
    `version_name` string,
    `lang` string,
    `source` string,
    `os` string,
    `area` string,
    `model` string,
    `brand` string,
    `sdk_version` string,
    `gmail` string,
    `height_width` string,
    `app_time` string,
    `network` string,
    `lng` string,
    `lat` string,
    `comment_id` int,
    `userid` int,
    `p_comment_id` int,
    `content` string,
    `addtime` string,
    `other_id` int,
    `praise_count` int,
```

```
    `reply_count` int,
    `server_time` string
)
PARTITIONED BY (dt string)
location '/warehouse/gmall/dwd/dwd_comment_log/';
```

2）导入数据

```
hive (gmall)>

insert overwrite table dwd_comment_log
PARTITION (dt='2020-03-10')
select
    mid_id,
    user_id,
    version_code,
    version_name,
    lang,
    source,
    os,
    area,
    model,
    brand,
    sdk_version,
    gmail,
    height_width,
    app_time,
    network,
    lng,
    lat,
    get_json_object(event_json,'$.kv.comment_id') comment_id,
    get_json_object(event_json,'$.kv.userid') userid,
    get_json_object(event_json,'$.kv.p_comment_id') p_comment_id,
    get_json_object(event_json,'$.kv.content') content,
    get_json_object(event_json,'$.kv.addtime') addtime,
    get_json_object(event_json,'$.kv.other_id') other_id,
    get_json_object(event_json,'$.kv.praise_count') praise_count,
    get_json_object(event_json,'$.kv.reply_count') reply_count,
    server_time
from dwd_base_event_log
where dt='2020-03-10' and event_name='comment';
```

3）测试

```
hive (gmall)> select * from dwd_comment_log limit 2;
```

8．收藏表

1）建表语句

```
hive (gmall)>
```

```
drop table if exists dwd_favorites_log;
CREATE EXTERNAL TABLE dwd_favorites_log(
    `mid_id` string,
    `user_id` string,
    `version_code` string,
    `version_name` string,
    `lang` string,
    `source` string,
    `os` string,
    `area` string,
    `model` string,
    `brand` string,
    `sdk_version` string,
    `gmail` string,
    `height_width` string,
    `app_time` string,
    `network` string,
    `lng` string,
    `lat` string,
    `id` int,
    `course_id` int,
    `userid` int,
    `add_time` string,
    `server_time` string
)
PARTITIONED BY (dt string)
location '/warehouse/gmall/dwd/dwd_favorites_log/';
```

2）导入数据

```
hive (gmall)>

insert overwrite table dwd_favorites_log
PARTITION (dt='2020-03-10')
select
    mid_id,
    user_id,
    version_code,
    version_name,
    lang,
    source,
    os,
    area,
    model,
    brand,
    sdk_version,
    gmail,
    height_width,
```

```
    app_time,
    network,
    lng,
    lat,
    get_json_object(event_json,'$.kv.id') id,
    get_json_object(event_json,'$.kv.course_id') course_id,
    get_json_object(event_json,'$.kv.userid') userid,
    get_json_object(event_json,'$.kv.add_time') add_time,
    server_time
from dwd_base_event_log
where dt='2020-03-10' and event_name='favorites';
```

3）测试

```
hive (gmall)> select * from dwd_favorites_log limit 2;
```

9. 点赞表

1）建表语句

```
hive (gmall)>
drop table if exists dwd_praise_log;
CREATE EXTERNAL TABLE dwd_praise_log(
    `mid_id` string,
    `user_id` string,
    `version_code` string,
    `version_name` string,
    `lang` string,
    `source` string,
    `os` string,
    `area` string,
    `model` string,
    `brand` string,
    `sdk_version` string,
    `gmail` string,
    `height_width` string,
    `app_time` string,
    `network` string,
    `lng` string,
    `lat` string,
    `id` string,
    `userid` string,
    `target_id` string,
    `type` string,
    `add_time` string,
    `server_time` string
)
PARTITIONED BY (dt string)
location '/warehouse/gmall/dwd/dwd_praise_log/';
```

2）导入数据

```
hive (gmall)>

insert overwrite table dwd_praise_log
PARTITION (dt='2020-03-10')
select
    mid_id,
    user_id,
    version_code,
    version_name,
    lang,
    source,
    os,
    area,
    model,
    brand,
    sdk_version,
    gmail,
    height_width,
    app_time,
    network,
    lng,
    lat,
    get_json_object(event_json,'$.kv.id') id,
    get_json_object(event_json,'$.kv.userid') userid,
    get_json_object(event_json,'$.kv.target_id') target_id,
    get_json_object(event_json,'$.kv.type') type,
    get_json_object(event_json,'$.kv.add_time') add_time,
    server_time
from dwd_base_event_log
where dt='2020-03-10' and event_name='praise';
```

3）测试

```
hive (gmall)> select * from dwd_praise_log limit 2;
```

10. 错误日志表

1）建表语句

```
hive (gmall)>
drop table if exists dwd_error_log;
CREATE EXTERNAL TABLE dwd_error_log(
    `mid_id` string,
    `user_id` string,
    `version_code` string,
    `version_name` string,
    `lang` string,
    `source` string,
    `os` string,
```

```
    `area` string,
    `model` string,
    `brand` string,
    `sdk_version` string,
    `gmail` string,
    `height_width` string,
    `app_time` string,
    `network` string,
    `lng` string,
    `lat` string,
    `errorBrief` string,
    `errorDetail` string,
    `server_time` string
)
PARTITIONED BY (dt string)
location '/warehouse/gmall/dwd/dwd_error_log/';
```

2）导入数据

```
hive (gmall)>

insert overwrite table dwd_error_log
PARTITION (dt='2020-03-10')
select
    mid_id,
    user_id,
    version_code,
    version_name,
    lang,
    source,
    os,
    area,
    model,
    brand,
    sdk_version,
    gmail,
    height_width,
    app_time,
    network,
    lng,
    lat,
    get_json_object(event_json,'$.kv.errorBrief') errorBrief,
    get_json_object(event_json,'$.kv.errorDetail') errorDetail,
    server_time
from dwd_base_event_log
where dt='2020-03-10' and event_name='error';
```

3）测试

```
hive (gmall)> select * from dwd_error_log limit 2;
```

11. DWD 层事件表加载数据脚本

将加载及解析数据的过程编写成脚本,方便每日调用执行。

(1) 在 hadoop102 的 /home/atguigu/bin 目录下创建脚本 ods_to_dwd_event_log.sh。

```
[atguigu@hadoop102 bin]$ vim ods_to_dwd_event_log.sh
```

在脚本中编写如下内容。

```
#!/bin/bash

# 定义变量,方便后续修改
APP=gmall
hive=/opt/module/hive/bin/hive

# 如果输入了日期参数,则取输入参数作为日期值;如果没有输入日期参数,则取当前时间的前一天作为日期值
if [ -n "$1" ] ;then
    do_date=$1
else
    do_date=`date -d "-1 day" +%F`
fi

sql="
insert overwrite table "$APP".dwd_display_log
PARTITION (dt='$do_date')
select
    mid_id,
    user_id,
    version_code,
    version_name,
    lang,
    source,
    os,
    area,
    model,
    brand,
    sdk_version,
    gmail,
    height_width,
    app_time,
    network,
    lng,
    lat,
    get_json_object(event_json,'$.kv.action') action,
    get_json_object(event_json,'$.kv.goodsid') goodsid,
    get_json_object(event_json,'$.kv.place') place,
    get_json_object(event_json,'$.kv.extend1') extend1,
    get_json_object(event_json,'$.kv.category') category,
    server_time
```

```sql
from "$APP".dwd_base_event_log
where dt='$do_date' and event_name='display';


insert overwrite table "$APP".dwd_newsdetail_log
PARTITION (dt='$do_date')
select
    mid_id,
    user_id,
    version_code,
    version_name,
    lang,
    source,
    os,
    area,
    model,
    brand,
    sdk_version,
    gmail,
    height_width,
    app_time,
    network,
    lng,
    lat,
    get_json_object(event_json,'$.kv.entry') entry,
    get_json_object(event_json,'$.kv.action') action,
    get_json_object(event_json,'$.kv.goodsid') goodsid,
    get_json_object(event_json,'$.kv.showtype') showtype,
    get_json_object(event_json,'$.kv.news_staytime') news_staytime,
    get_json_object(event_json,'$.kv.loading_time') loading_time,
    get_json_object(event_json,'$.kv.type1') type1,
    get_json_object(event_json,'$.kv.category') category,
    server_time
from "$APP".dwd_base_event_log
where dt='$do_date' and event_name='newsdetail';


insert overwrite table "$APP".dwd_loading_log
PARTITION (dt='$do_date')
select
    mid_id,
    user_id,
    version_code,
    version_name,
    lang,
    source,
    os,
```

```sql
    area,
    model,
    brand,
    sdk_version,
    gmail,
    height_width,
    app_time,
    network,
    lng,
    lat,
    get_json_object(event_json,'$.kv.action') action,
    get_json_object(event_json,'$.kv.loading_time') loading_time,
    get_json_object(event_json,'$.kv.loading_way') loading_way,
    get_json_object(event_json,'$.kv.extend1') extend1,
    get_json_object(event_json,'$.kv.extend2') extend2,
    get_json_object(event_json,'$.kv.type') type,
    get_json_object(event_json,'$.kv.type1') type1,
    server_time
from "$APP".dwd_base_event_log
where dt='$do_date' and event_name='loading';


insert overwrite table "$APP".dwd_ad_log
PARTITION (dt='$do_date')
select
    mid_id,
    user_id,
    version_code,
    version_name,
    lang,
    source,
    os,
    area,
    model,
    brand,
    sdk_version,
    gmail,
    height_width,
    app_time,
    network,
    lng,
    lat,
    get_json_object(event_json,'$.kv.entry') entry,
    get_json_object(event_json,'$.kv.action') action,
    get_json_object(event_json,'$.kv.contentType') contentType,
    get_json_object(event_json,'$.kv.displayMills') displayMills,
    get_json_object(event_json,'$.kv.itemId') itemId,
```

```sql
        get_json_object(event_json,'$.kv.activityId') activityId,
    server_time
from "$APP".dwd_base_event_log
where dt='$do_date' and event_name='ad';


insert overwrite table "$APP".dwd_notification_log
PARTITION (dt='$do_date')
select
    mid_id,
    user_id,
    version_code,
    version_name,
    lang,
    source,
    os,
    area,
    model,
    brand,
    sdk_version,
    gmail,
    height_width,
    app_time,
    network,
    lng,
    lat,
    get_json_object(event_json,'$.kv.action') action,
    get_json_object(event_json,'$.kv.noti_type') noti_type,
    get_json_object(event_json,'$.kv.ap_time') ap_time,
    get_json_object(event_json,'$.kv.content') content,
    server_time
from "$APP".dwd_base_event_log
where dt='$do_date' and event_name='notification';


insert overwrite table "$APP".dwd_active_background_log
PARTITION (dt='$do_date')
select
    mid_id,
    user_id,
    version_code,
    version_name,
    lang,
    source,
    os,
    area,
    model,
```

```sql
    brand,
    sdk_version,
    gmail,
    height_width,
    app_time,
    network,
    lng,
    lat,
    get_json_object(event_json,'$.kv.active_source') active_source,
    server_time
from "$APP".dwd_base_event_log
where dt='$do_date' and event_name='active_background';


insert overwrite table "$APP".dwd_comment_log
PARTITION (dt='$do_date')
select
    mid_id,
    user_id,
    version_code,
    version_name,
    lang,
    source,
    os,
    area,
    model,
    brand,
    sdk_version,
    gmail,
    height_width,
    app_time,
    network,
    lng,
    lat,
    get_json_object(event_json,'$.kv.comment_id') comment_id,
    get_json_object(event_json,'$.kv.userid') userid,
    get_json_object(event_json,'$.kv.p_comment_id') p_comment_id,
    get_json_object(event_json,'$.kv.content') content,
    get_json_object(event_json,'$.kv.addtime') addtime,
    get_json_object(event_json,'$.kv.other_id') other_id,
    get_json_object(event_json,'$.kv.praise_count') praise_count,
    get_json_object(event_json,'$.kv.reply_count') reply_count,
    server_time
from "$APP".dwd_base_event_log
where dt='$do_date' and event_name='comment';
```

```sql
insert overwrite table "$APP".dwd_favorites_log
PARTITION (dt='$do_date')
select
    mid_id,
    user_id,
    version_code,
    version_name,
    lang,
    source,
    os,
    area,
    model,
    brand,
    sdk_version,
    gmail,
    height_width,
    app_time,
    network,
    lng,
    lat,
    get_json_object(event_json,'$.kv.id') id,
    get_json_object(event_json,'$.kv.course_id') course_id,
    get_json_object(event_json,'$.kv.userid') userid,
    get_json_object(event_json,'$.kv.add_time') add_time,
    server_time
from "$APP".dwd_base_event_log
where dt='$do_date' and event_name='favorites';


insert overwrite table "$APP".dwd_praise_log
PARTITION (dt='$do_date')
select
    mid_id,
    user_id,
    version_code,
    version_name,
    lang,
    source,
    os,
    area,
    model,
    brand,
    sdk_version,
    gmail,
    height_width,
    app_time,
    network,
```

```
    lng,
    lat,
    get_json_object(event_json,'$.kv.id') id,
    get_json_object(event_json,'$.kv.userid') userid,
    get_json_object(event_json,'$.kv.target_id') target_id,
    get_json_object(event_json,'$.kv.type') type,
    get_json_object(event_json,'$.kv.add_time') add_time,
    server_time
from "$APP".dwd_base_event_log
where dt='$do_date' and event_name='praise';


insert overwrite table "$APP".dwd_error_log
PARTITION (dt='$do_date')
select
    mid_id,
    user_id,
    version_code,
    version_name,
    lang,
    source,
    os,
    area,
    model,
    brand,
    sdk_version,
    gmail,
    height_width,
    app_time,
    network,
    lng,
    lat,
    get_json_object(event_json,'$.kv.errorBrief') errorBrief,
    get_json_object(event_json,'$.kv.errorDetail') errorDetail,
    server_time
from "$APP".dwd_base_event_log
where dt='$do_date' and event_name='error';
"

$hive -e "$sql"
```

（2）增加脚本执行权限。

```
[atguigu@hadoop102 bin]$ chmod 777 ods_to_dwd_event_log.sh
```

（3）执行脚本，导入数据。

```
[atguigu@hadoop102 module]$ ods_to_dwd_event_log.sh 2020-03-11
```

（4）查询结果数据。
```
hive (gmall)>
select * from dwd_comment_log where dt='2020-03-11' limit 2;
```

6.4.4 业务数据维度表解析

关于业务数据，DWD 层的搭建主要需要注意维度的退化，ODS 层的业务数据有二十多张表，形成了比较复杂的关系模型，这种情况下想要获取一些细节维度的信息，通常需要进行多表 join 才能得到，为了使查询更加方便，也为了避免进行大量的表 join 计算，将关系模型进行适度的维度退化很有必要。数据仓库建模基本思路如图 6-24 所示，从图 6-24 中可以看出，构建业务数据 DWD 层的思路是：将众多维度按照类型和关系退化为 7 张主要的维度表。

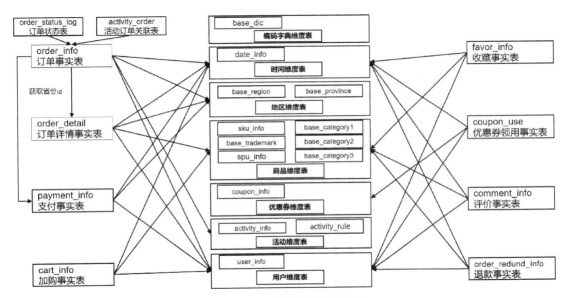

图 6-24 数据仓库建模基本思路

接下来对进行维度退化后的几张主要的维度表进行讲解。

1. 商品维度表（全量表）

1）建表语句

```
hive (gmall)>
DROP TABLE IF EXISTS `dwd_dim_sku_info`;
CREATE EXTERNAL TABLE `dwd_dim_sku_info` (
    `id` string COMMENT '商品id',
    `spu_id` string COMMENT '标准产品单位id',
    `price` double COMMENT '商品价格',
    `sku_name` string COMMENT '商品名称',
    `sku_desc` string COMMENT '商品描述',
    `weight` double COMMENT '重量',
    `tm_id` string COMMENT '品牌id',
```

```
    `tm_name` string COMMENT '品牌名称',
    `category3_id` string COMMENT '三级品类id',
    `category2_id` string COMMENT '二级品类id',
    `category1_id` string COMMENT '一级品类id',
    `category3_name` string COMMENT '三级品类名称',
    `category2_name` string COMMENT '二级品类名称',
    `category1_name` string COMMENT '一级品类名称',
    `spu_name` string COMMENT '标准产品单位名称',
    `create_time` string COMMENT '创建时间'
) COMMENT '商品维度表'
PARTITIONED BY (`dt` string)
stored as parquet
location '/warehouse/gmall/dwd/dwd_dim_sku_info/'
tblproperties ("parquet.compression"="lzo");
```

2）导入数据

```
hive (gmall)>
insert overwrite table dwd_dim_sku_info partition(dt='2020-03-10')
select
    sku.id,
    sku.spu_id,
    sku.price,
    sku.sku_name,
    sku.sku_desc,
    sku.weight,
    sku.tm_id,
    ob.tm_name,
    sku.category3_id,
    c2.id category2_id,
    c1.id category1_id,
    c3.name category3_name,
    c2.name category2_name,
    c1.name category1_name,
    spu.spu_name,
    sku.create_time
from
(
    select * from ods_sku_info where dt='2020-03-10'
)sku
join
(
    select * from ods_base_trademark where dt='2020-03-10'
)ob on sku.tm_id=ob.tm_id
join
(
    select * from ods_spu_info where dt='2020-03-10'
)spu on spu.id = sku.spu_id
```

```
join
(
    select * from ods_base_category3 where dt='2020-03-10'
)c3 on sku.category3_id=c3.id
join
(
    select * from ods_base_category2 where dt='2020-03-10'
)c2 on c3.category2_id=c2.id
join
(
    select * from ods_base_category1 where dt='2020-03-10'
)c1 on c2.category1_id=c1.id;
```

3）查询结果数据

```
hive (gmall)> select * from dwd_dim_sku_info where dt='2020-03-10';
```

2. 优惠券维度表（全量）

将 ODS 层 ods_coupon_info 表中的数据导入 DWD 层优惠券维度表中，在导入过程中可以进行适当的清洗。

1）建表语句

```
hive (gmall)>
drop table if exists dwd_dim_coupon_info;
create external table dwd_dim_coupon_info(
    `id` string COMMENT '优惠券编号',
    `coupon_name` string COMMENT '优惠券名称',
    `coupon_type` string COMMENT '优惠券类型 1 现金券 2 折扣券 3 满减券 4 满件打折券',
    `condition_amount` string COMMENT '满额数',
    `condition_num` string COMMENT '满件数',
    `activity_id` string COMMENT '活动编号',
    `benefit_amount` string COMMENT '满减金额',
    `benefit_discount` string COMMENT '折扣',
    `create_time` string COMMENT '创建时间',
    `range_type` string COMMENT '范围类型 1 商品 2 品类 3 品牌',
    `spu_id` string COMMENT '标准产品单位 id',
    `tm_id` string COMMENT '品牌 id',
    `category3_id` string COMMENT '品类 id',
    `limit_num` string COMMENT '最多领用次数',
    `operate_time` string COMMENT '操作时间',
    `expire_time` string COMMENT '过期时间'
) COMMENT '优惠券维度表'
PARTITIONED BY (`dt` string)
row format delimited fields terminated by '\t'
stored as parquet
location '/warehouse/gmall/dwd/dwd_dim_coupon_info/';
tblproperties ("parquet.compression"="lzo");
```

2)导入数据

```
hive (gmall)>
insert overwrite table dwd_dim_coupon_info partition(dt='2020-03-10')
select
    id,
    coupon_name,
    coupon_type,
    condition_amount,
    condition_num,
    activity_id,
    benefit_amount,
    benefit_discount,
    create_time,
    range_type,
    spu_id,
    tm_id,
    category3_id,
    limit_num,
    operate_time,
    expire_time
from ods_coupon_info
where dt='2020-03-10';
```

3)查询结果数据

```
hive (gmall)> select * from dwd_dim_coupon_info where dt='2020-03-10';
```

3. 活动维度表（全量）

1)建表语句

```
hive (gmall)>
drop table if exists dwd_dim_activity_info;
create external table dwd_dim_activity_info(
    `id` string COMMENT '编号',
    `activity_name` string COMMENT '活动名称',
    `activity_type` string COMMENT '活动类型',
    `condition_amount` string COMMENT '满减金额',
    `condition_num` string COMMENT '满减件数',
    `benefit_amount` string COMMENT '优惠金额',
    `benefit_discount` string COMMENT '优惠折扣',
    `benefit_level` string COMMENT '优惠级别',
    `start_time` string COMMENT '开始时间',
    `end_time` string COMMENT '结束时间',
    `create_time` string COMMENT '创建时间'
) COMMENT '活动维度表'
PARTITIONED BY (`dt` string)
row format delimited fields terminated by '\t'
stored as parquet
location '/warehouse/gmall/dwd/dwd_dim_activity_info/';
```

```
tblproperties ("parquet.compression"="lzo");
```

2）导入数据

```
hive (gmall)>
insert overwrite table dwd_dim_activity_info partition(dt='2020-03-10')
select
    info.id,
    info.activity_name,
    info.activity_type,
    rule.condition_amount,
    rule.condition_num,
    rule.benefit_amount,
    rule.benefit_discount,
    rule.benefit_level,
    info.start_time,
    info.end_time,
    info.create_time
from
(
    select * from ods_activity_info where dt='2020-03-10'
)info
left join
(
    select * from ods_activity_rule where dt='2020-03-10'
)rule on info.id = rule.activity_id;
```

3）查询结果数据

```
hive (gmall)> select * from dwd_dim_activity_info where dt='2020-03-10';
```

4．地区维度表（特殊）

1）建表语句

```
hive (gmall)>
DROP TABLE IF EXISTS `dwd_dim_base_province`;
CREATE EXTERNAL TABLE `dwd_dim_base_province` (
    `id` string COMMENT 'id',
    `province_name` string COMMENT '省市名称',
    `area_code` string COMMENT '地区编码',
    `iso_code` string COMMENT 'ISO 编码',
    `region_id` string COMMENT '地区id',
    `region_name` string COMMENT '地区名称'
) COMMENT '地区维度表'
stored as parquet
location '/warehouse/gmall/dwd/dwd_dim_base_province/';
tblproperties ("parquet.compression"="lzo");
```

2）导入数据

```
hive (gmall)>
insert overwrite table dwd_dim_base_province
```

```
select
    bp.id,
    bp.name,
    bp.area_code,
    bp.iso_code,
    bp.region_id,
    br.region_name
from ods_base_province bp
join ods_base_region br
on bp.region_id=br.id;
```

3）查询结果数据

```
hive (gmall)> select * from dwd_dim_base_province;
```

5．时间维度表（特殊）

时间维度表比较特殊，时间的维度是不会发生改变的，所以只需导入一份固定的数据即可。

（1）建表语句。

```
hive (gmall)>
DROP TABLE IF EXISTS `dwd_dim_date_info`;
CREATE EXTERNAL TABLE `dwd_dim_date_info`(
    `date_id` string COMMENT '日',
    `week_id` int COMMENT '周',
    `week_day` int COMMENT '周的第几天',
    `day` int COMMENT '每月的第几天',
    `month` int COMMENT '第几月',
    `quarter` int COMMENT '第几季度',
    `year` int COMMENT '年',
    `is_workday` int COMMENT '是否是周末',
    `holiday_id` int COMMENT '是否是节假日'
) COMMENT '时间维度表'
row format delimited fields terminated by '\t'
stored as parquet
location '/warehouse/gmall/dwd/dwd_dim_date_info/';
tblproperties ("parquet.compression"="lzo");
```

（2）将 date_info.txt 文件上传到 hadoop102 的 /opt/module/db_log/ 目录下。

（3）导入数据。

```
hive (gmall)>
load data local inpath '/opt/module/db_log/date_info.txt' into table dwd_dim_date_info;
```

（4）查询结果数据。

```
hive (gmall)> select * from dwd_dim_date_info;
```

6.4.5 业务数据事实表解析

DWD 层中事实表的创建，则需要根据各张表的特点进行不同的处理。本节主要对 DWD 层业务数据事实表的创建和数据导入进行讲解。

1. 订单详情事实表（事务型事实表）

如表 6-15 所示，订单详情事实表与时间、用户、地区、商品四个维度有关，其中，与时间、用户、商品的关联分别通过 dt、user_id、sku_id 字段建立，与地区的关联则需要通过与 ODS 层的订单详情表进行 join 获得。

表 6-15 订单详情事实表相关维度

	时间	用户	地区	商品	优惠券	活动	编码	度量值
订单详情	√	√	√	√				件数/金额

1）建表语句

```
hive (gmall)>
drop table if exists dwd_fact_order_detail;
create external table dwd_fact_order_detail (
    `id` string COMMENT 'id',
    `order_id` string COMMENT '订单编号',
    `province_id` string COMMENT '省份id',
    `user_id` string COMMENT '用户id',
    `sku_id` string COMMENT '商品id',
    `create_time` string COMMENT '创建时间',
    `total_amount` decimal(16,2) COMMENT '总金额',
    `sku_num` bigint COMMENT '商品数量'
) COMMENT '订单详情事实表'
PARTITIONED BY (`dt` string)
stored as parquet
location '/warehouse/gmall/dwd/dwd_fact_order_detail/'
tblproperties ("parquet.compression"="lzo")
;
```

2）导入数据

```
insert overwrite table dwd_fact_order_detail partition(dt='2020-03-10')
select
    od.id,
    od.order_id,
    oi.province_id,
    od.user_id,
    od.sku_id,
    od.create_time,
    od.order_price*od.sku_num,
    od.sku_num
from
(
```

```
    select * from ods_order_detail where dt='2020-03-10'
) od
join
(
    select * from ods_order_info where dt='2020-03-10'
) oi
on od.order_id=oi.id;
```

3）查询结果数据

```
hive (gmall)> select * from dwd_fact_order_detail where dt='2020-03-10';
```

2. 支付事实表（事务型事实表）

如表 6-16 所示，支付事实表与时间、用户、地区三个维度有关，其中，与时间、用户的关联分别通过 dt、user_id 字段建立，与地区的关联则需要通过与 ODS 层的订单详情表进行 join 获得。

表 6-16 支付事实表相关维度

	时间	用户	地区	商品	优惠券	活动	编码	度量值
支付	√	√	√					次数/金额

1）建表语句

```
hive (gmall)>
drop table if exists dwd_fact_payment_info;
create external table dwd_fact_payment_info (
    `id` string COMMENT '',
    `out_trade_no` string COMMENT '对外业务编号',
    `order_id` string COMMENT '订单编号',
    `user_id` string COMMENT '用户编号',
    `alipay_trade_no` string COMMENT '支付宝交易流水编号',
    `payment_amount`    decimal(16,2) COMMENT '支付金额',
    `subject`           string COMMENT '交易内容',
    `payment_type` string COMMENT '支付类型',
    `payment_time` string COMMENT '支付时间',
    `province_id` string COMMENT '省份id'
) COMMENT '支付事实表'
PARTITIONED BY (`dt` string)
stored as parquet
location '/warehouse/gmall/dwd/dwd_fact_payment_info/'
tblproperties ("parquet.compression"="lzo");
```

2）导入数据

```
hive (gmall)>
insert overwrite table dwd_fact_payment_info partition(dt='2020-03-10')
select
    pi.id,
    pi.out_trade_no,
    pi.order_id,
```

```
    pi.user_id,
    pi.alipay_trade_no,
    pi.total_amount,
    pi.subject,
    pi.payment_type,
    pi.payment_time,
    oi.province_id
from
(
    select * from ods_payment_info where dt='2020-03-10'
)pi
join
(
    select id, province_id from ods_order_info where dt='2020-03-10'
)oi
on pi.order_id = oi.id;
```

3）查询结果数据

```
hive (gmall)> select * from dwd_fact_payment_info where dt='2020-03-10';
```

3. 退款事实表（事务型事实表）

如表 6-17 所示，退款事实表与时间、用户、商品三个维度有关，ODS 层的退单表已经具有所有关联字段，所以无须从其他表格获得关联，直接将原 ODS 层退单表中的数据导入即可。

表 6-17 退款事实表相关维度

	时间	用户	地区	商品	优惠券	活动	编码	度量值
退款	√	√		√				件数/金额

1）建表语句

```
hive (gmall)>
drop table if exists dwd_fact_order_refund_info;
create external table dwd_fact_order_refund_info(
    `id` string COMMENT '编号',
    `user_id` string COMMENT '用户id',
    `order_id` string COMMENT '订单编号',
    `sku_id` string COMMENT '商品id',
    `refund_type` string COMMENT '退款类型',
    `refund_num` bigint COMMENT '退款件数',
    `refund_amount` decimal(16,2) COMMENT '退款金额',
    `refund_reason_type` string COMMENT '退款原因类型',
    `create_time` string COMMENT '退款时间'
) COMMENT '退款事实表'
PARTITIONED BY (`dt` string)
stored as parquet
row format delimited fields terminated by '\t'
location '/warehouse/gmall/dwd/dwd_fact_order_refund_info/'
tblproperties ("parquet.compression"="lzo");
```

2）导入数据

```
hive (gmall)>
insert overwrite table dwd_fact_order_refund_info partition(dt='2020-03-10')
select
    id,
    user_id,
    order_id,
    sku_id,
    refund_type,
    refund_num,
    refund_amount,
    refund_reason_type,
    create_time
from ods_order_refund_info
where dt='2020-03-10';
```

3）查询结果数据

```
hive (gmall)> select * from dwd_fact_order_refund_info where dt='2020-03-10';
```

4．评价事实表（事务型事实表）

如表 6-18 所示，评价事实表与时间、用户、商品三个维度有关，ODS 层的商品评价表已经具有所有关联字段，所以无须从其他表格中获得关联，直接将 ODS 层商品评价表中的数据导入即可。

表 6-18　评价事实表相关维度

	时间	用户	地区	商品	优惠券	活动	编码	度量值
评价	√	√		√				条数

1）建表语句

```
hive (gmall)>
drop table if exists dwd_fact_comment_info;
create external table dwd_fact_comment_info(
    `id` string COMMENT '编号',
    `user_id` string COMMENT '用户id',
    `sku_id` string COMMENT '商品id',
    `spu_id` string COMMENT '标准产品单位id',
    `order_id` string COMMENT '订单编号',
    `appraise` string COMMENT '评价',
    `create_time` string COMMENT '评价时间'
) COMMENT '评价事实表'
PARTITIONED BY (`dt` string)
stored as parquet
row format delimited fields terminated by '\t'
location '/warehouse/gmall/dwd/dwd_fact_comment_info/'
tblproperties ("parquet.compression"="lzo");
```

2）导入数据

```
hive (gmall)>
insert overwrite table dwd_fact_comment_info partition(dt='2020-03-10')
select
    id,
    user_id,
    sku_id,
    spu_id,
    order_id,
    appraise,
    create_time
from ods_comment_info
where dt='2020-03-10';
```

3）查询结果数据

```
hive (gmall)> select * from dwd_fact_comment_info where dt='2020-03-10';
```

5. 加购事实表（周期型快照事实表，每日快照）

由于购物车中的数据经常会发生变化，所以不适合采用每日增量同步策略导入数据。我们采用的策略是每天做一次快照，进行全量数据导入。这样做的劣势是存储的数据量会比较大。

由于周期型快照事实表存储的数据比较注重时效性，存储时间过于久远的数据存在的意义不大，所以可以定时删除以前的数据来释放内存。

如表 6-19 所示，加购事实表与时间、用户、商品三个维度有关。

表 6-19 加购事实表相关维度

	时间	用户	地区	商品	优惠券	活动	编码	度量值
加购	√	√		√				件数/金额

1）建表语句

```
hive (gmall)>
drop table if exists dwd_fact_cart_info;
create external table dwd_fact_cart_info(
    `id` string COMMENT '编号',
    `user_id` string COMMENT '用户id',
    `sku_id` string COMMENT '商品id',
    `cart_price` string COMMENT '放入购物车时的价格',
    `sku_num` string COMMENT '数量',
    `sku_name` string COMMENT '商品名称（冗余）',
    `create_time` string COMMENT '创建时间',
    `operate_time` string COMMENT '操作时间',
    `is_ordered` string COMMENT '是否已经下单,1为已下单;0为未下单',
    `order_time` string COMMENT '下单时间'
) COMMENT '加购事实表'
PARTITIONED BY (`dt` string)
stored as parquet
row format delimited fields terminated by '\t'
```

```
location '/warehouse/gmall/dwd/dwd_fact_cart_info/'
tblproperties ("parquet.compression"="lzo");
```

2）导入数据

```
hive (gmall)>
insert overwrite table dwd_fact_cart_info partition(dt='2020-03-10')
select
    id,
    user_id,
    sku_id,
    cart_price,
    sku_num,
    sku_name,
    create_time,
    operate_time,
    is_ordered,
    order_time
from ods_cart_info
where dt='2020-03-10';
```

3）查询结果数据

```
hive (gmall)> select * from dwd_fact_cart_info where dt='2020-03-10';
```

6．收藏事实表（周期型快照事实表，每日快照）

收藏事实表采用的同步策略与加购事实表相同。

如表 6-20 所示，收藏事实表与时间、用户、商品三个维度有关。

表 6-20 收藏事实表相关维度

	时间	用户	地区	商品	优惠券	活动	编码	度量值
收藏	√	√		√				个数

1）建表语句

```
hive (gmall)>
drop table if exists dwd_fact_favor_info;
create external table dwd_fact_favor_info(
    `id` string COMMENT '编号',
    `user_id` string COMMENT '用户id',
    `sku_id` string COMMENT '商品id',
    `spu_id` string COMMENT '标准产品单位id',
    `is_cancel` string COMMENT '是否取消',
    `create_time` string COMMENT '收藏时间',
    `cancel_time` string COMMENT '取消时间'
) COMMENT '收藏事实表'
PARTITIONED BY (`dt` string)
stored as parquet
row format delimited fields terminated by '\t'
location '/warehouse/gmall/dwd/dwd_fact_favor_info/'
tblproperties ("parquet.compression"="lzo");
```

2）导入数据

```
hive (gmall)>
insert overwrite table dwd_fact_favor_info partition(dt='2020-03-10')
select
    id,
    user_id,
    sku_id,
    spu_id,
    is_cancel,
    create_time,
    cancel_time
from ods_favor_info
where dt='2020-03-10';
```

3）查询结果数据

```
hive (gmall)> select * from dwd_fact_favor_info where dt='2020-03-10';
```

7. 优惠券领用事实表（累积型快照事实表）

如表 6-21 所示，优惠券领用事实表与时间、用户、优惠券三个维度有关。

表 6-21 优惠券领用事实表相关维度

	时间	用户	地区	商品	优惠券	活动	编码	度量值
优惠券领用	√	√			√			个数

优惠券的使用有一定的生命周期：领取优惠券→使用优惠券下单→优惠券参与支付。所以优惠券领用事实表符合累积型快照事实表的特征，即将优惠券的领用、下单使用、支付使用三个时间节点进行快照记录。

累积型快照事实表可以用来统计优惠券领取次数、使用优惠券下单的次数及优惠券参与支付的次数等数据。

1）建表语句

```
hive (gmall)>
drop table if exists dwd_fact_coupon_use;
create external table dwd_fact_coupon_use(
    `id` string COMMENT '编号',
    `coupon_id` string COMMENT '优惠券id',
    `user_id` string COMMENT 'userid',
    `order_id` string COMMENT '订单id',
    `coupon_status` string COMMENT '优惠券状态',
    `get_time` string COMMENT '领取时间',
    `using_time` string COMMENT '使用(下单)时间',
    `used_time` string COMMENT '使用(支付)时间'
) COMMENT '优惠券领用事实表'
PARTITIONED BY (`dt` string)
stored as parquet
row format delimited fields terminated by '\t'
location '/warehouse/gmall/dwd/dwd_fact_coupon_use/'
```

```
tblproperties ("parquet.compression"="lzo");
```

注意：dt是按照优惠券领取时间get_time进行分区的。

2）导入数据

优惠券领用事实表导入数据的思路如图6-25、图6-26和图6-27所示。

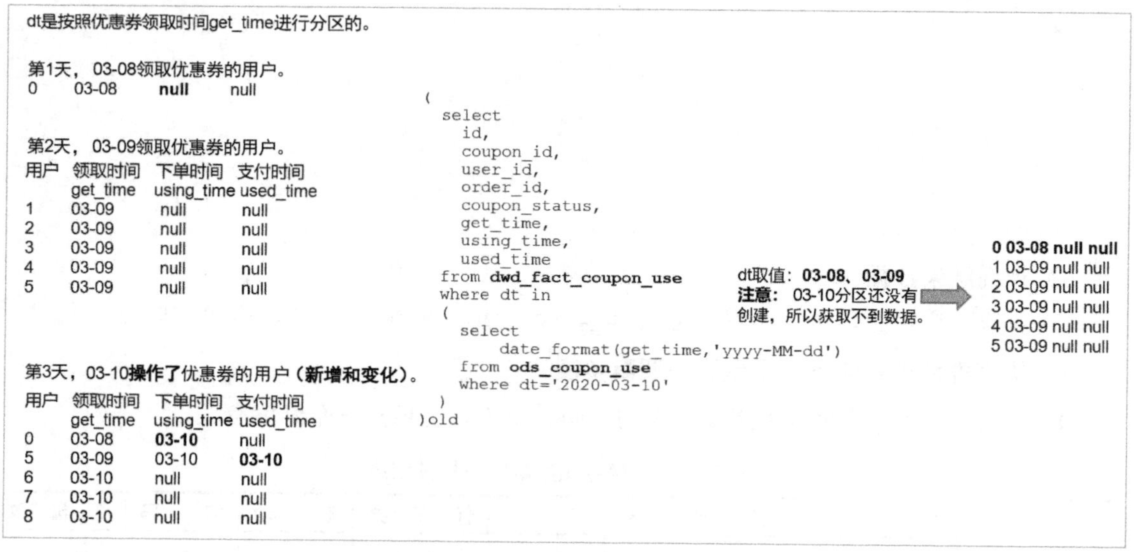

图6-25 优惠券领用事实表导入数据的思路（1）

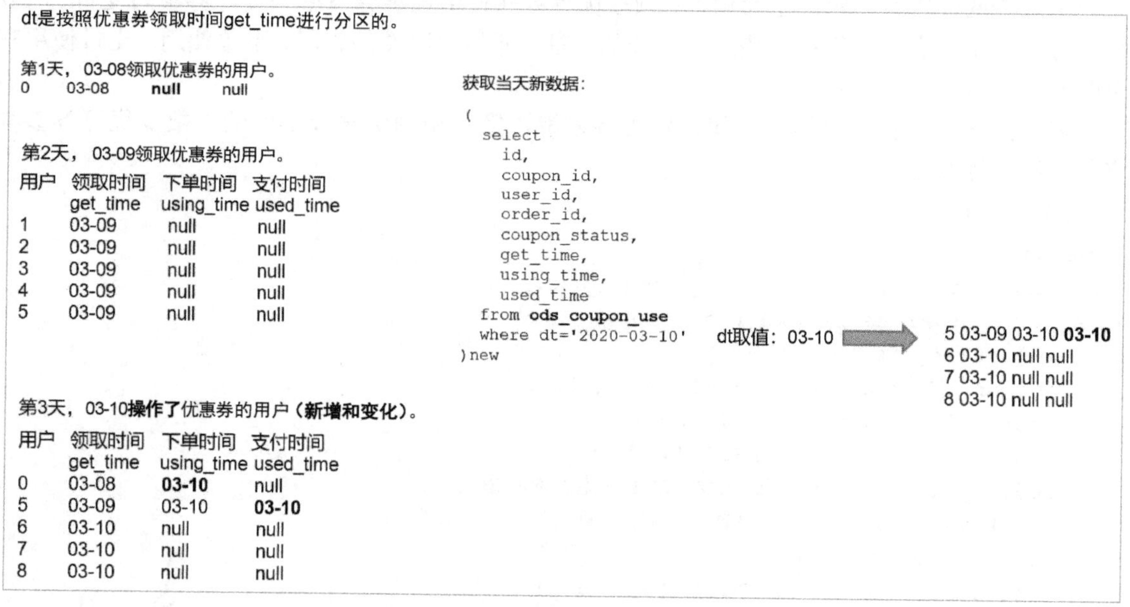

图6-26 优惠券领用事实表导入数据的思路（2）

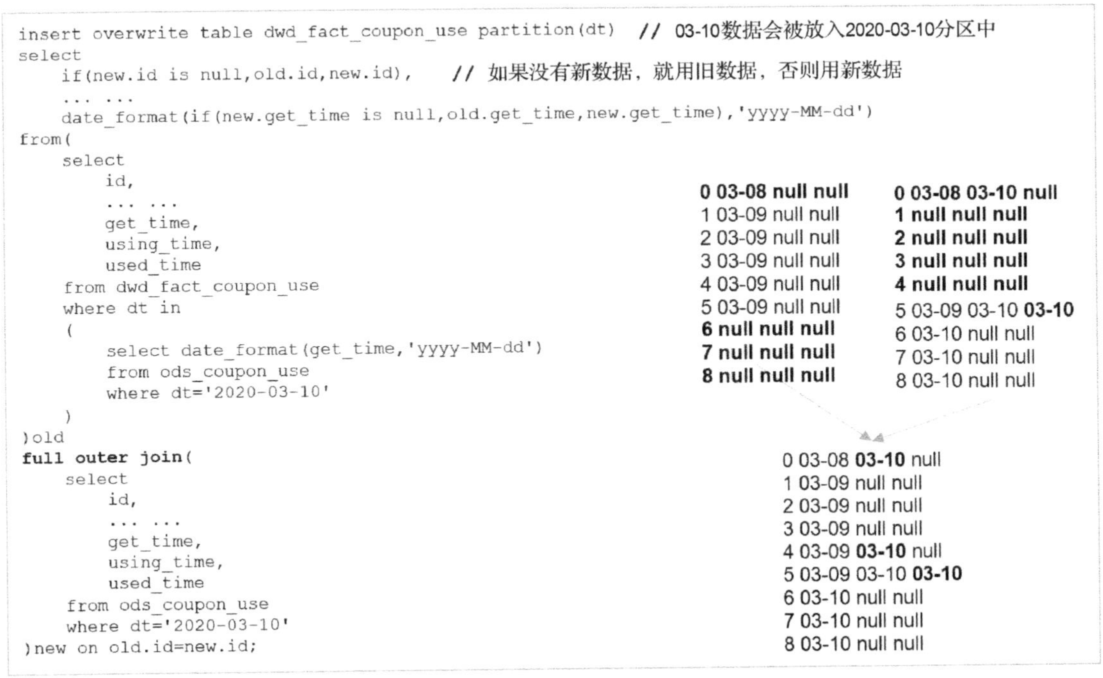

```
insert overwrite table dwd_fact_coupon_use partition(dt)    // 03-10数据会被放入2020-03-10分区中
select
    if(new.id is null,old.id,new.id),    // 如果没有新数据，就用旧数据，否则用新数据
    ... ...
    date_format(if(new.get_time is null,old.get_time,new.get_time),'yyyy-MM-dd')
from(
    select
        id,
        ... ...
        get_time,
        using_time,
        used_time
    from dwd_fact_coupon_use
    where dt in
    (
        select date_format(get_time,'yyyy-MM-dd')
        from ods_coupon_use
        where dt='2020-03-10'
    )
)old
full outer join(
    select
        id,
        ... ...
        get_time,
        using_time,
        used_time
    from ods_coupon_use
    where dt='2020-03-10'
)new on old.id=new.id;
```

图 6-27 优惠券领用事实表导入数据的思路（3）

代码如下。

```
hive (gmall)>
set hive.exec.dynamic.partition.mode=nonstrict;
insert overwrite table dwd_fact_coupon_use partition(dt)
select
    if(new.id is null,old.id,new.id),
    if(new.coupon_id is null,old.coupon_id,new.coupon_id),
    if(new.user_id is null,old.user_id,new.user_id),
    if(new.order_id is null,old.order_id,new.order_id),
    if(new.coupon_status is null,old.coupon_status,new.coupon_status),
    if(new.get_time is null,old.get_time,new.get_time),
    if(new.using_time is null,old.using_time,new.using_time),
    if(new.used_time is null,old.used_time,new.used_time),
    date_format(if(new.get_time is null,old.get_time,new.get_time),'yyyy-MM-dd')
from
(
    select
        id,
        coupon_id,
        user_id,
        order_id,
        coupon_status,
        get_time,
        using_time,
        used_time
```

```
       from dwd_fact_coupon_use
       where dt in
       (
           select
               date_format(get_time,'yyyy-MM-dd')
           from ods_coupon_use
           where dt='2020-03-10'
       )
)old
full outer join
(
    select
        id,
        coupon_id,
        user_id,
        order_id,
        coupon_status,
        get_time,
        using_time,
        used_time
    from ods_coupon_use
    where dt='2020-03-10'
)new
on old.id=new.id;
```

3）查询结果数据

```
hive (gmall)> select * from dwd_fact_coupon_use where dt='2020-03-10';
```

8. 订单事实表（累积型快照事实表）

1）数据导入过程涉及函数

（1）concat()函数。concat()函数用于连接字符串，在连接字符串时，只要其中一个字符串是 NULL，结果就返回 NULL。

```
hive> select concat('a','b');
ab

hive> select concat('a','b',null);
NULL
```

（2）concat_ws()函数。concat_ws()函数同样用于连接字符串，在连接字符串时，只要有一个字符串不是 NULL，结果就不会返回 NULL。concat_ws()函数需要指定分隔符。

```
hive> select concat_ws('-','a','b');
a-b

hive> select concat_ws('-','a','b',null);
a-b

hive> select concat_ws('','a','b',null);
```

ab

（3）str_to_map()函数。
- 语法描述。

str_to_map(VARCHAR text, VARCHAR listDelimiter, VARCHAR keyValueDelimiter)。
- 功能描述。

使用 listDelimiter 将 text 分隔成 key-value 对，然后使用 keyValueDelimiter 分隔每个 key-value 对，并组装成 MAP 返回。默认 listDelimiter 为 ","，keyValueDelimiter 为 "="。
- 案例。

```
str_to_map('1001=2020-03-10,1002=2020-03-10', ',' , '=')
```

输出：

{"1001":"2020-03-10","1002":"2020-03-10"}

2）建表语句

如表 6-22 所示，订单事实表与时间、用户、地区、活动四个维度有关。

表 6-22 订单事实表相关维度

	时间	用户	地区	商品	优惠券	活动	编码	度量值
订单	√	√	√			√		件数/金额

订单从创建到完成同样具有一定的生命周期，这个生命周期为创建→支付→取消→完成→退款→退款完成。

由于 ODS 层的订单表只有创建时间和操作时间两个状态，不能表达所有时间节点，所以需要关联订单状态表。订单事实表中增加了活动 id，所以需要关联活动订单表。

```
hive (gmall)>
drop table if exists dwd_fact_order_info;
create external table dwd_fact_order_info (
    `id` string COMMENT '订单编号',
    `order_status` string COMMENT '订单状态',
    `user_id` string COMMENT '用户id',
    `out_trade_no` string COMMENT '支付流水号',
    `create_time` string COMMENT '创建时间(未支付状态)',
    `payment_time` string COMMENT '支付时间(已支付状态)',
    `cancel_time` string COMMENT '取消时间(已取消状态)',
    `finish_time` string COMMENT '完成时间(已完成状态)',
    `refund_time` string COMMENT '退款时间(退款中状态)',
    `refund_finish_time` string COMMENT '退款完成时间(退款完成状态)',
    `province_id` string COMMENT '省份id',
    `activity_id` string COMMENT '活动id',
    `original_total_amount` string COMMENT '原价金额',
    `benefit_reduce_amount` string COMMENT '优惠金额',
    `feight_fee` string COMMENT '运费',
    `final_total_amount` decimal(10,2) COMMENT '订单金额'
) COMMENT '订单事实表'
PARTITIONED BY (`dt` string)
```

```
stored as parquet
location '/warehouse/gmall/dwd/dwd_fact_order_info/'
tblproperties ("parquet.compression"="lzo");
```

3）导入数据的思路

导入数据的思路如图6-28和图6-29所示。

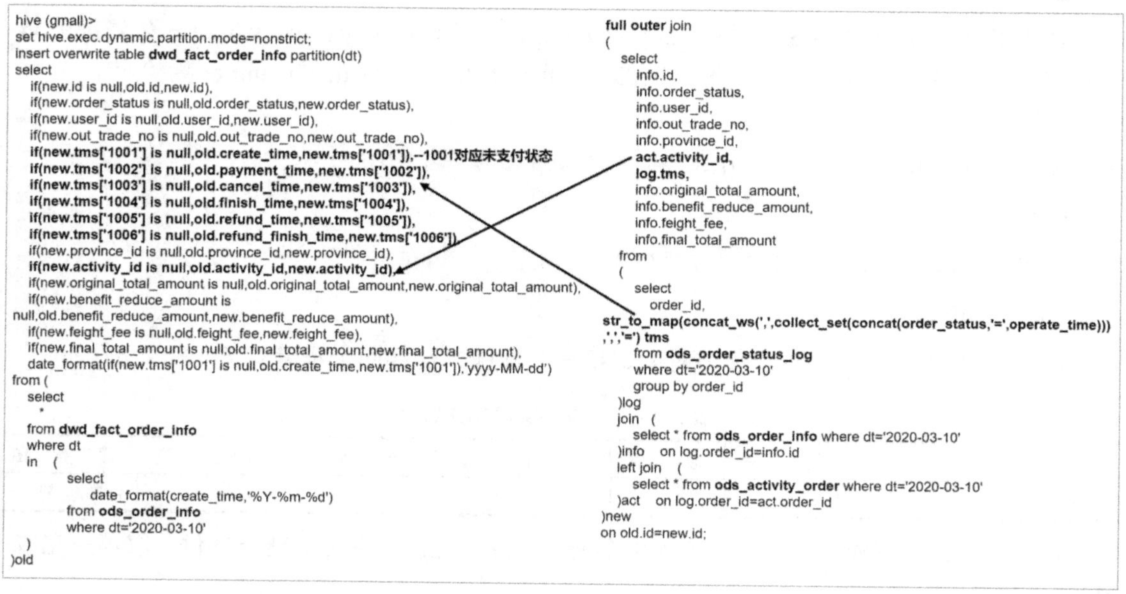

```
hive (gmall)>
set hive.exec.dynamic.partition.mode=nonstrict;
insert overwrite table dwd_fact_order_info partition(dt)
select
    if(new.id is null,old.id,new.id),
    if(new.order_status is null,old.order_status,new.order_status),
    if(new.user_id is null,old.user_id,new.user_id),
    if(new.out_trade_no is null,old.out_trade_no,new.out_trade_no),
    if(new.tms['1001'] is null,old.create_time,new.tms['1001']),--1001对应未支付状态
    if(new.tms['1002'] is null,old.payment_time,new.tms['1002']),
    if(new.tms['1003'] is null,old.cancel_time,new.tms['1003']),
    if(new.tms['1004'] is null,old.finish_time,new.tms['1004']),
    if(new.tms['1005'] is null,old.refund_time,new.tms['1005']),
    if(new.tms['1006'] is null,old.refund_finish_time,new.tms['1006']),
    if(new.province_id is null,old.province_id,new.province_id),
    if(new.activity_id is null,old.activity_id,new.activity_id),
    if(new.original_total_amount is null,old.original_total_amount,new.original_total_amount),
    if(new.benefit_reduce_amount is
null,old.benefit_reduce_amount,new.benefit_reduce_amount),
    if(new.feight_fee is null,old.feight_fee,new.feight_fee),
    if(new.final_total_amount is null,old.final_total_amount,new.final_total_amount),
    date_format(if(new.tms['1001'] is null,old.create_time,new.tms['1001']),'yyyy-MM-dd')
from (
    select
        *
    from dwd_fact_order_info
    where dt
    in  (
        select
            date_format(create_time,'%Y-%m-%d')
        from ods_order_info
        where dt='2020-03-10'
    )
)old

full outer join
(
    select
        info.id,
        info.order_status,
        info.user_id,
        info.out_trade_no,
        info.province_id,
        act.activity_id,
        log.tms,
        info.original_total_amount,
        info.benefit_reduce_amount,
        info.feight_fee,
        info.final_total_amount
    from
    (
        select
            order_id,
            str_to_map(concat_ws(',',collect_set(concat(order_status,'=',operate_time))),',','=') tms
        from ods_order_status_log
        where dt='2020-03-10'
        group by order_id
    )log
    join  (
        select * from ods_order_info where dt='2020-03-10'
    )info  on log.order_id=info.id
    left join  (
        select * from ods_activity_order where dt='2020-03-10'
    )act  on log.order_id=act.order_id
)new
on old.id=new.id;
```

图6-28　订单事实表导入数据的思路（1）

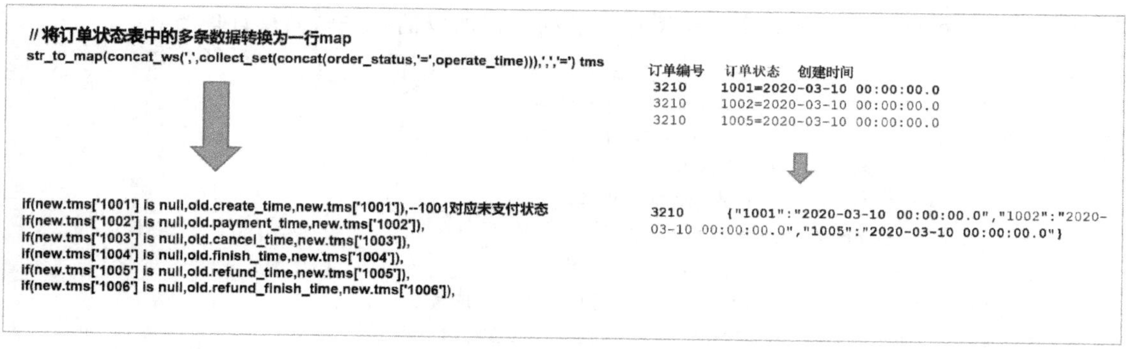

```
//将订单状态表中的多条数据转换为一行map
str_to_map(concat_ws(',',collect_set(concat(order_status,'=',operate_time))),',','=') tms

if(new.tms['1001'] is null,old.create_time,new.tms['1001']),--1001对应未支付状态
if(new.tms['1002'] is null,old.payment_time,new.tms['1002']),
if(new.tms['1003'] is null,old.cancel_time,new.tms['1003']),
if(new.tms['1004'] is null,old.finish_time,new.tms['1004']),
if(new.tms['1005'] is null,old.refund_time,new.tms['1005']),
if(new.tms['1006'] is null,old.refund_finish_time,new.tms['1006']),
```

订单编号　订单状态　创建时间
3210　1001=2020-03-10 00:00:00.0
3210　1002=2020-03-10 00:00:00.0
3210　1005=2020-03-10 00:00:00.0

3210　{"1001":"2020-03-10 00:00:00.0","1002":"2020-03-10 00:00:00.0","1005":"2020-03-10 00:00:00.0"}

图6-29　订单事实表导入数据的思路（2）

4）常用函数在本次数据导入中的使用

```
hive (gmall)> select order_id, concat(order_status,'=', operate_time) from ods_order_status_log where dt='2020-03-10';

3210    1001=2020-03-10 00:00:00.0
3211    1001=2020-03-10 00:00:00.0
3212    1001=2020-03-10 00:00:00.0
3210    1002=2020-03-10 00:00:00.0
3211    1002=2020-03-10 00:00:00.0
3212    1002=2020-03-10 00:00:00.0
```

```
3210    1005=2020-03-10 00:00:00.0
3211    1004=2020-03-10 00:00:00.0
3212    1004=2020-03-10 00:00:00.0
```

```
hive (gmall)> select order_id, collect_set(concat(order_status,'=',operate_
time)) from ods_order_status_log where dt='2020-03-10' group by order_id;

3210    ["1001=2020-03-10 00:00:00.0","1002=2020-03-10 00:00:00.0","1005=2020-
03-10 00:00:00.0"]
3211    ["1001=2020-03-10 00:00:00.0","1002=2020-03-10 00:00:00.0","1004=2020-
03-10 00:00:00.0"]
3212    ["1001=2020-03-10 00:00:00.0","1002=2020-03-10 00:00:00.0","1004=2020-
03-10 00:00:00.0"]

hive (gmall)>
select order_id, concat_ws(',', collect_set(concat(order_status,'=',operate_
time))) from ods_order_status_log where dt='2020-03-10' group by order_id;

3210    1001=2020-03-10 00:00:00.0,1002=2020-03-10 00:00:00.0,1005=2020-03-10
00:00:00.0
3211    1001=2020-03-10 00:00:00.0,1002=2020-03-10 00:00:00.0,1004=2020-03-10
00:00:00.0
3212    1001=2020-03-10 00:00:00.0,1002=2020-03-10 00:00:00.0,1004=2020-03-10
00:00:00.0

hive (gmall)>
select order_id, str_to_map(concat_ws(',',collect_set(concat(order_status,'=',
operate_time))), ',' , '=') from ods_order_status_log where dt='2020-03-10'
group by order_id;

3210    {"1001":"2020-03-10 00:00:00.0","1002":"2020-03-10 00:00:00.0","1005":
"2020-03-10 00:00:00.0"}
3211    {"1001":"2020-03-10 00:00:00.0","1002":"2020-03-10 00:00:00.0","1004":
"2020-03-10 00:00:00.0"}
3212    {"1001":"2020-03-10 00:00:00.0","1002":"2020-03-10 00:00:00.0","1004":
"2020-03-10 00:00:00.0"}
```

5）导入数据

```
hive (gmall)>
set hive.exec.dynamic.partition.mode=nonstrict;
insert overwrite table dwd_fact_order_info partition(dt)
select
    if(new.id is null,old.id,new.id),
    if(new.order_status is null,old.order_status,new.order_status),
    if(new.user_id is null,old.user_id,new.user_id),
    if(new.out_trade_no is null,old.out_trade_no,new.out_trade_no),
    --1001对应未支付状态
```

```sql
        if(new.tms['1001'] is null,old.create_time,new.tms['1001']),
        if(new.tms['1002'] is null,old.payment_time,new.tms['1002']),
        if(new.tms['1003'] is null,old.cancel_time,new.tms['1003']),
        if(new.tms['1004'] is null,old.finish_time,new.tms['1004']),
        if(new.tms['1005'] is null,old.refund_time,new.tms['1005']),
        if(new.tms['1006'] is null,old.refund_finish_time,new.tms['1006']),
        if(new.province_id is null,old.province_id,new.province_id),
        if(new.activity_id is null,old.activity_id,new.activity_id),
        if(new.original_total_amount is null,old.original_total_amount,new.original_total_amount),
        if(new.benefit_reduce_amount is null,old.benefit_reduce_amount,new.benefit_reduce_amount),
        if(new.feight_fee is null,old.feight_fee,new.feight_fee),
        if(new.final_total_amount is null,old.final_total_amount,new.final_total_amount),
        date_format(if(new.tms['1001'] is null,old.create_time,new.tms['1001']),'yyyy-MM-dd')
from
(
    select
        id,
        order_status,
        user_id,
        out_trade_no,
        create_time,
        payment_time,
        cancel_time,
        finish_time,
        refund_time,
        refund_finish_time,
        province_id,
        activity_id,
        original_total_amount,
        benefit_reduce_amount,
        feight_fee,
        final_total_amount
    from dwd_fact_order_info
    where dt
    in
    (
        select
          date_format(create_time,'yyyy-MM-dd')
        from ods_order_info
        where dt='2020-03-10'
    )
)old
full outer join
```

```
(
    select
        info.id,
        info.order_status,
        info.user_id,
        info.out_trade_no,
        info.province_id,
        act.activity_id,
        log.tms,
        info.original_total_amount,
        info.benefit_reduce_amount,
        info.feight_fee,
        info.final_total_amount
    from
    (
        select
            order_id,
            str_to_map(concat_ws(',',collect_set(concat(order_status,'=',operate_time))),',','=') tms
        from ods_order_status_log
        where dt='2020-03-10'
        group by order_id
    )log
    join
    (
        select * from ods_order_info where dt='2020-03-10'
    )info
    on log.order_id=info.id
    left join
    (
        select * from ods_activity_order where dt='2020-03-10'
    )act
    on log.order_id=act.order_id
)new
on old.id=new.id;
```

6）查询结果数据

```
hive (gmall)> select * from dwd_fact_order_info where dt='2020-03-10';
```

6.4.6 拉链表构建之用户维度表

什么是拉链表？

拉链表是维护历史状态及最新状态数据的一种表，用于记录每条信息的生命周期，一旦一条信息的生命周期结束，就重新开始记录一条新的信息，并把当前日期放入生效开始日期，如表 6-23 所示。

如果当前信息至今有效，则在生效结束日期中填入一个极大值（如 9999-99-99）。

表 6-23　用户状态拉链表

用户 id	手 机 号 码	生效开始日期	生效结束日期
1	136****9090	2019-01-01	2019-05-01
1	137****8989	2019-05-02	2019-07-02
1	182****7878	2019-07-03	2019-09-05
1	155****1234	2019-09-06	9999-99-99

如表 6-23 所示的用户状态拉链表，例如，用户想得到 2019-08-01 所有的用户状态信息，则可以通过"生效开始日期≤2019-08-01≤生效结束日期"计算得到。

拉链表适用于如下场景，数据量比较大，且数据部分字段会发生变化，变化的比例不大且频率不高，若采用每日全量同步策略导入数据，则会占用大量内存且会保存很多不变的信息。在此种情况下使用拉链表，既能反映数据的历史状态，又能最大限度地节省存储空间。

比如，用户信息会发生变化，但是变化比例不大。如果用户数量具有一定规模，则按照每日全量同步策略保存，效率会很低。

用户表中的数据每日有可能新增，也有可能修改，但修改频率并不高，属于缓慢变化维度，所以此处采用拉链表存储用户维度数据。

如何制作拉链表？

如图 6-30 所示，将用户当日全部数据和 MySQL 中当日变化数据连接在一起，形成一张新的用户拉链临时表，使用用户拉链临时表中的数据覆盖旧的用户拉链表中的数据，即可解决 Hive 中数据不能更新的问题。

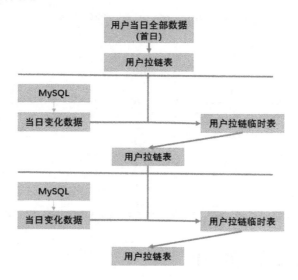

图 6-30　拉链表制作思路

拉链表制作过程如图 6-31 所示。

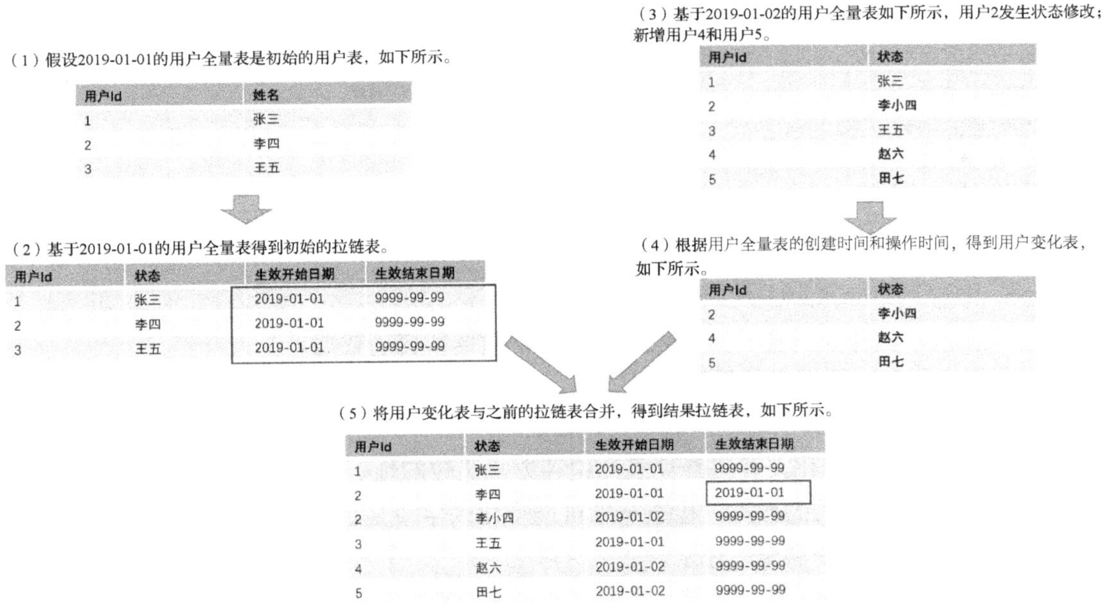

（1）假设2019-01-01的用户全量表是初始的用户表，如下所示。

用户Id	姓名
1	张三
2	李四
3	王五

（2）基于2019-01-01的用户全量表得到初始的拉链表。

用户Id	状态	生效开始日期	生效结束日期
1	张三	2019-01-01	9999-99-99
2	李四	2019-01-01	9999-99-99
3	王五	2019-01-01	9999-99-99

（3）基于2019-01-02的用户全量表如下所示，用户2发生状态修改，新增用户4和用户5。

用户Id	状态
1	张三
2	李小四
3	王五
4	赵六
5	田七

（4）根据用户全量表的创建时间和操作时间，得到用户变化表，如下所示。

用户Id	状态
2	李小四
4	赵六
5	田七

（5）将用户变化表与之前的拉链表合并，得到结果拉链表，如下所示。

用户Id	状态	生效开始日期	生效结束日期
1	张三	2019-01-01	9999-99-99
2	李四	2019-01-01	2019-01-01
2	李小四	2019-01-02	9999-99-99
3	王五	2019-01-01	9999-99-99
4	赵六	2019-01-02	9999-99-99
5	田七	2019-01-02	9999-99-99

图 6-31 拉链表制作过程

（1）步骤1：初始化拉链表（首次独立执行）。

① 创建用户拉链表。

```
hive (gmall)>
drop table if exists dwd_dim_user_info_his;
create external table dwd_dim_user_info_his(
    `id` string COMMENT '用户id',
    `name` string COMMENT '姓名',
    `birthday` string COMMENT '生日',
    `gender` string COMMENT '性别',
    `email` string COMMENT '邮箱',
    `user_level` string COMMENT '用户等级',
    `create_time` string COMMENT '创建时间',
    `operate_time` string COMMENT '操作时间',
    `start_date`  string COMMENT '生效开始日期',
    `end_date`  string COMMENT '生效结束日期'
) COMMENT '用户拉链表'
stored as parquet
location '/warehouse/gmall/dwd/dwd_dim_user_info_his/'
tblproperties ("parquet.compression"="lzo");
```

② 初始化用户拉链表。

```
hive (gmall)>
insert overwrite table dwd_dim_user_info_his
select
    id,
    name,
    birthday,
```

```
    gender,
    email,
    user_level,
    create_time,
    operate_time,
    '2020-03-10',
    '9999-99-99'
from ods_user_info oi
where oi.dt='2020-03-10';
```

（2）步骤 2：制作当日变动数据表，包括新增数据和变动数据。

① 获得当日变动数据表的思路如下。

- 表内最好有创建时间和变动时间。如果没有，则可以利用第三方工具，如 canal，监控 MySQL 的实时变化，并进行记录，这种方式比较麻烦。
- 逐行对比前后两天的数据，检查全部可能变化的字段是否相同。
- 要求业务数据库提供变动流水。

② 因为 ods_user_info 表本身导入进来就是新增变动明细的表，表中有创建时间字段和更改时间字段，所以不用处理，可直接通过查询筛选这两个字段，获取新增和变动数据。

- 通过 Sqoop 把 2020-03-11 的所有数据导入。

```
[atguigu@hadoop102 ~]$ mysqlTohdfs.sh all 2020-03-11
```

- 将 ODS 层数据导入。

```
[atguigu@hadoop102 ~]$ hdfs_to_ods_db.sh all 2020-03-11
```

（3）步骤 3：先合并变动信息，再追加新增信息，插入临时表中。

① 创建用户拉链临时表，与用户拉链表字段完全相同，注意用户拉链临时表以_tmp 结尾。

```
hive (gmall)>
drop table if exists dwd_dim_user_info_his_tmp;
create external table dwd_dim_user_info_his_tmp(
    `id` string COMMENT '用户id',
    `name` string COMMENT '姓名',
    `birthday` string COMMENT '生日',
    `gender` string COMMENT '性别',
    `email` string COMMENT '邮箱',
    `user_level` string COMMENT '用户等级',
    `create_time` string COMMENT '创建时间',
    `operate_time` string COMMENT '操作时间',
    `start_date` string COMMENT '生效开始日期',
    `end_date` string COMMENT '生效结束日期'
) COMMENT '用户拉链临时表'
stored as parquet
location '/warehouse/gmall/dwd/dwd_dim_user_info_his_tmp/'
tblproperties ("parquet.compression"="lzo");
```

② 导入数据。

ods_user_info 表中的 operate_time 字段为当天日期的数据即为当时修改数据，在增加当时修改数据的同时，应该修改原用户拉链表中状态已经发生改变的数据，将生效结束日期修改为前一日，可以通过将原用户拉链表与当日新增及当时修改数据按照 id 进行连接，若新增及当时修改数据中不存在对当前状态的修改，则保留原生效结束日期；若新增及当时修改数据中存在对当前状态的修改，则将生效结束日期修改为前一日。

```
hive (gmall)>
insert overwrite table dwd_dim_user_info_his_tmp
select * from
(
    select
        id,
        name,
        birthday,
        gender,
        email,
        user_level,
        create_time,
        operate_time,
        '2020-03-11' start_date,
        '9999-99-99' end_date
    from ods_user_info where dt='2020-03-11'

    union all
    select
        uh.id,
        uh.name,
        uh.birthday,
        uh.gender,
        uh.email,
        uh.user_level,
        uh.create_time,
        uh.operate_time,
        uh.start_date,
        if(ui.id is not null  and uh.end_date='9999-99-99', date_add(ui.dt,-1),
uh.end_date) end_date
    from dwd_dim_user_info_his uh left join
    (
        select
            *
        from ods_user_info
        where dt='2020-03-11'
    ) ui on uh.id=ui.id
)his
order by his.id, start_date;
```

（4）步骤4：使用用户拉链临时表中的数据覆盖用户拉链表中的数据。

① 导入数据。

```
hive (gmall)>
insert overwrite table dwd_dim_user_info_his
select * from dwd_dim_user_info_his_tmp;
```

② 查询结果数据。

```
hive (gmall)> select id, start_date, end_date from dwd_dim_user_info_his;
```

6.4.7 DWD 层数据导入脚本

将 DWD 层的数据导入过程编写成脚本，方便每日调用执行。

（1）在/home/atguigu/bin 目录下创建脚本 **ods_to_dwd_db.sh**。

```
[atguigu@hadoop102 bin]$ vim ods_to_dwd_db.sh
```

在脚本中编写如下内容。

```bash
#!/bin/bash

APP=gmall
hive=/opt/module/hive/bin/hive

# 如果输入了日期参数，则取输入参数作为日期值；如果没有输入日期参数，则取当前时间的前一天作为日期值
if [ -n "$2" ] ;then
    do_date=$2
else
    do_date=`date -d "-1 day" +%F`
fi

sql1="
set hive.exec.dynamic.partition.mode=nonstrict;

insert overwrite table ${APP}.dwd_dim_sku_info partition(dt='$do_date')
select
    sku.id,
    sku.spu_id,
    sku.price,
    sku.sku_name,
    sku.sku_desc,
    sku.weight,
    sku.tm_id,
    ob.tm_name,
    sku.category3_id,
    c2.id category2_id,
    c1.id category1_id,
    c3.name category3_name,
    c2.name category2_name,
    c1.name category1_name,
    spu.spu_name,
```

```
        sku.create_time
from
(
    select * from ${APP}.ods_sku_info where dt='$do_date'
)sku
join
(
    select * from ${APP}.ods_base_trademark where dt='$do_date'
)ob on sku.tm_id=ob.tm_id
join
(
    select * from ${APP}.ods_spu_info where dt='$do_date'
)spu on spu.id = sku.spu_id
join
(
    select * from ${APP}.ods_base_category3 where dt='$do_date'
)c3 on sku.category3_id=c3.id
join
(
    select * from ${APP}.ods_base_category2 where dt='$do_date'
)c2 on c3.category2_id=c2.id
join
(
    select * from ${APP}.ods_base_category1 where dt='$do_date'
)c1 on c2.category1_id=c1.id;
insert overwrite table ${APP}.dwd_dim_coupon_info partition(dt='$do_date')
select
    id,
    coupon_name,
    coupon_type,
    condition_amount,
    condition_num,
    activity_id,
    benefit_amount,
    benefit_discount,
    create_time,
    range_type,
    spu_id,
    tm_id,
    category3_id,
    limit_num,
    operate_time,
    expire_time
from ${APP}.ods_coupon_info
where dt='$do_date';
```

```sql
insert overwrite table ${APP}.dwd_dim_activity_info partition(dt='$do_date')
select
    info.id,
    info.activity_name,
    info.activity_type,
    rule.condition_amount,
    rule.condition_num,
    rule.benefit_amount,
    rule.benefit_discount,
    rule.benefit_level,
    info.start_time,
    info.end_time,
    info.create_time
from
(
    select * from ${APP}.ods_activity_info where dt='$do_date'
)info
left join
(
    select * from ${APP}.ods_activity_rule where dt='$do_date'
)rule on info.id = rule.activity_id;

insert overwrite table ${APP}.dwd_fact_order_detail partition(dt='$do_date')
select
    od.id,
    od.order_id,
    od.user_id,
    od.sku_id,
    od.sku_name,
    od.order_price,
    od.sku_num,
    od.create_time,
    oi.province_id,
    od.order_price*od.sku_num
from
(
    select * from ${APP}.ods_order_detail where dt='$do_date'
) od
join
(
    select * from ${APP}.ods_order_info where dt='$do_date'
) oi
on od.order_id=oi.id;

insert overwrite table ${APP}.dwd_fact_payment_info partition(dt='$do_date')
select
    pi.id,
```

```sql
        pi.out_trade_no,
        pi.order_id,
        pi.user_id,
        pi.alipay_trade_no,
        pi.total_amount,
        pi.subject,
        pi.payment_type,
        pi.payment_time,
        oi.province_id
from
(
    select * from ${APP}.ods_payment_info where dt='$do_date'
)pi
join
(
    select id, province_id from ${APP}.ods_order_info where dt='$do_date'
)oi
on pi.order_id = oi.id;

insert overwrite table ${APP}.dwd_fact_order_refund_info partition(dt='$do_date')
select
    id,
    user_id,
    order_id,
    sku_id,
    refund_type,
    refund_num,
    refund_amount,
    refund_reason_type,
    create_time
from ${APP}.ods_order_refund_info
where dt='$do_date';

insert overwrite table ${APP}.dwd_fact_comment_info partition(dt='$do_date')
select
    id,
    user_id,
    sku_id,
    spu_id,
    order_id,
    appraise,
    create_time
from ${APP}.ods_comment_info
where dt='$do_date';

insert overwrite table ${APP}.dwd_fact_cart_info partition(dt='$do_date')
```

```sql
select
    id,
    user_id,
    sku_id,
    cart_price,
    sku_num,
    sku_name,
    create_time,
    operate_time,
    is_ordered,
    order_time
from ${APP}.ods_cart_info
where dt='$do_date';

insert overwrite table ${APP}.dwd_fact_favor_info partition(dt='$do_date')
select
    id,
    user_id,
    sku_id,
    spu_id,
    is_cancel,
    create_time,
    cancel_time
from ${APP}.ods_favor_info
where dt='$do_date';

insert overwrite table ${APP}.dwd_fact_coupon_use partition(dt)
select
    if(new.id is null,old.id,new.id),
    if(new.coupon_id is null,old.coupon_id,new.coupon_id),
    if(new.user_id is null,old.user_id,new.user_id),
    if(new.order_id is null,old.order_id,new.order_id),
    if(new.coupon_status is null,old.coupon_status,new.coupon_status),
    if(new.get_time is null,old.get_time,new.get_time),
    if(new.using_time is null,old.using_time,new.using_time),
    if(new.used_time is null,old.used_time,new.used_time),
    date_format(if(new.get_time is null,old.get_time,new.get_time),'yyyy-MM-dd')
from
(
    select
        id,
        coupon_id,
        user_id,
        order_id,
        coupon_status,
        get_time,
        using_time,
```

```
            used_time
        from ${APP}.dwd_fact_coupon_use
        where dt in
        (
            select
                date_format(get_time,'yyyy-MM-dd')
            from ${APP}.ods_coupon_use
            where dt='$do_date'
        )
)old
full outer join
(
    select
        id,
        coupon_id,
        user_id,
        order_id,
        coupon_status,
        get_time,
        using_time,
        used_time
    from ${APP}.ods_coupon_use
    where dt='$do_date'
)new
on old.id=new.id;


insert overwrite table ${APP}.dwd_fact_order_info partition(dt)
select
    if(new.id is null,old.id,new.id),
    if(new.order_status is null,old.order_status,new.order_status),
    if(new.user_id is null,old.user_id,new.user_id),
    if(new.out_trade_no is null,old.out_trade_no,new.out_trade_no),
    --1001 对应未支付状态
    if(new.tms['1001'] is null,old.create_time,new.tms['1001']),
    if(new.tms['1002'] is null,old.payment_time,new.tms['1002']),
    if(new.tms['1003'] is null,old.cancel_time,new.tms['1003']),
    if(new.tms['1004'] is null,old.finish_time,new.tms['1004']),
    if(new.tms['1005'] is null,old.refund_time,new.tms['1005']),
    if(new.tms['1006'] is null,old.refund_finish_time,new.tms['1006']),
    if(new.province_id is null,old.province_id,new.province_id),
    if(new.activity_id is null,old.activity_id,new.activity_id),
    if(new.original_total_amount is null,old.original_total_amount,new.original_total_amount),
    if(new.benefit_reduce_amount is null,old.benefit_reduce_amount,new.benefit_reduce_amount),
    if(new.feight_fee is null,old.feight_fee,new.feight_fee),
```

```sql
        if(new.final_total_amount is null,old.final_total_amount,new.final_total_
amount),
        date_format(if(new.tms['1001'] is null,old.create_time,new.tms['1001']),
'yyyy-MM-dd')
from
(
    select
        id,
        order_status,
        user_id,
        out_trade_no,
        create_time,
        payment_time,
        cancel_time,
        finish_time,
        refund_time,
        refund_finish_time,
        province_id,
        activity_id,
        original_total_amount,
        benefit_reduce_amount,
        feight_fee,
        final_total_amount
    from ${APP}.dwd_fact_order_info
    where dt
    in
    (
        select
          date_format(create_time,'yyyy-MM-dd')
        from ${APP}.ods_order_info
        where dt='$do_date'
    )
)old
full outer join
(
    select
        info.id,
        info.order_status,
        info.user_id,
        info.out_trade_no,
        info.province_id,
        act.activity_id,
        log.tms,
        info.original_total_amount,
        info.benefit_reduce_amount,
        info.feight_fee,
        info.final_total_amount
```

```sql
    from
    (
        select
            order_id,
            str_to_map(concat_ws(',',collect_set(concat(order_status,'=',operate_time))),',','=') tms
        from ${APP}.ods_order_status_log
        where dt='$do_date'
        group by order_id
    )log
    join
    (
        select * from ${APP}.ods_order_info where dt='$do_date'
    )info
    on log.order_id=info.id
    left join
    (
        select * from ${APP}.ods_activity_order where dt='$do_date'
    )act
    on log.order_id=act.order_id
)new
on old.id=new.id;

insert overwrite table ${APP}.dwd_dim_user_info_his_tmp
select * from
(
    select
        id,
        name,
        birthday,
        gender,
        email,
        user_level,
        create_time,
        operate_time,
        '$do_date' start_date,
        '9999-99-99' end_date
    from ${APP}.ods_user_info where dt='$do_date'

    union all
    select
        uh.id,
        uh.name,
        uh.birthday,
        uh.gender,
        uh.email,
```

```
        uh.user_level,
        uh.create_time,
        uh.operate_time,
        uh.start_date,
        if(ui.id is not null  and uh.end_date='9999-99-99', date_add(ui.dt,-1),
uh.end_date) end_date
    from ${APP}.dwd_dim_user_info_his uh left join
    (
        select
            *
        from ${APP}.ods_user_info
        where dt='$do_date'
    ) ui on uh.id=ui.id
)his
order by his.id, start_date;
insert overwrite table ${APP}.dwd_dim_user_info_his select * from ${APP}.dwd_
dim_user_info_his_tmp;
"

sql2="
insert overwrite table ${APP}.dwd_dim_base_province
select
    bp.id,
    bp.name,
    bp.area_code,
    bp.iso_code,
    bp.region_id,
    br.region_name
from ${APP}.ods_base_province bp
join ${APP}.ods_base_region br
on bp.region_id=br.id;
"

case $1 in
"first"){
    $hive -e "$sql1"
    $hive -e "$sql2"
};;
"all"){
    $hive -e "$sql1"
};;
esac
```

（2）增加脚本执行权限。

```
[atguigu@hadoop102 bin]$ chmod 777 ods_to_dwd_db.sh
```

（3）执行脚本，导入数据。

```
[atguigu@hadoop102 bin]$ ods_to_dwd_db.sh all 2020-03-11
```

（4）查询结果数据。

```
hive (gmall)>
select * from dwd_fact_order_info where dt='2020-03-11';
select * from dwd_fact_order_detail where dt='2020-03-11';
select * from dwd_fact_comment_info where dt='2020-03-11';
select * from dwd_fact_order_refund_info where dt='2020-03-11';
```

6.5 数据仓库搭建——DWS 层

DWS 层采用宽表化手段，构建公共指标数据。其站在不同主题的角度，将数据进行汇总和聚合，得到每天每个主题的相关数据。

6.5.1 系统函数

本节在 DWS 层的搭建中，需要用到的重要函数如下所示。

1. collect_set()函数

（1）创建原数据表。

```
hive (gmall)>
drop table if exists stud;
create table stud (name string, area string, course string, score int);
```

（2）向原数据表中插入数据。

```
hive (gmall)>
insert into table stud values('zhang3','bj','math',88);
insert into table stud values('li4','bj','math',99);
insert into table stud values('wang5','sh','chinese',92);
insert into table stud values('zhao6','sh','chinese',54);
insert into table stud values('tian7','bj','chinese',91);
```

（3）查询表中的数据。

```
hive (gmall)> select * from stud;
stud.name        stud.area        stud.course        stud.score
zhang3           bj               math               88
li4              bj               math               99
wang5            sh               chinese            92
zhao6            sh               chinese            54
tian7            bj               chinese            91
```

（4）把同一分组中不同行的数据聚合成一个集合。

```
hive (gmall)> select course, collect_set(area), avg(score) from stud group by course;
```

```
chinese ["sh","bj"]    79.0
math    ["bj"]  93.5
```

（5）使用下标获取聚合结果的某一个值。

```
hive (gmall)> select course, collect_set(area)[0], avg(score) from stud group by course;
chinese sh     79.0
math    bj     93.5
```

2. nvl()函数

基本语法：nvl(表达式 1,表达式 2)。

如果表达式 1 为空值，则 nvl()函数返回表达式 2 的值，否则返回表达式 1 的值。nvl()函数的作用是把一个空值（null）转换成一个实际的值。其表达式的数据类型可以是数字型、字符型和日期型。需要注意的是，表达式 1 和表达式 2 的数据类型必须相同。

3. 日期处理函数

1）date_format()函数（根据格式整理日期）

```
hive (gmall)> select date_format('2020-03-10','yyyy-MM');
2020-03
```

2）date_add()函数（加减日期）

```
hive (gmall)> select date_add('2020-03-10',-1);
2020-03-09
hive (gmall)> select date_add('2020-03-10',1);
2020-03-11
```

3）next_day()函数

（1）获取当前日期的下一个星期一。

```
hive (gmall)> select next_day('2020-03-12','MO');
2020-03-16
```

（2）获取当前周的星期一。

```
hive (gmall)> select date_add(next_day('2020-03-12','MO'),-7);
2020-03-11
```

4）last_day()函数（获取当月最后一天的日期）

```
hive (gmall)> select last_day('2020-03-10');
2020-03-31
```

6.5.2 用户行为数据聚合

出于对后续每日活跃设备、每周活跃设备、每日新增设备等需求的考虑，我们利用用户行为 DWD 层的启动日志表，按照设备 id 进行聚合，得到 DWS 层的设备行为表。

在聚合的过程中，为避免细节数据的丢失，将聚合后的字段使用 concat_ws()函数进行连接，若后期在需求开发过程中需要用到这些细节数据，则可以通过使用爆炸函数 explode()再次获取。

每日设备行为表主要按照设备唯一标识 mid_id 进行分组统计。
1) 建表语句

```
hive (gmall)>
drop table if exists dws_uv_detail_daycount;
create external table dws_uv_detail_daycount
(
    `mid_id` string COMMENT '设备唯一标识',
    `user_id` string COMMENT '用户标识',
    `version_code` string COMMENT '程序版本号',
    `version_name` string COMMENT '程序版本名',
    `lang` string COMMENT '系统语言',
    `source` string COMMENT '渠道号',
    `os` string COMMENT 'Android版本',
    `area` string COMMENT '区域',
    `model` string COMMENT '手机型号',
    `brand` string COMMENT '手机品牌',
    `sdk_version` string COMMENT 'sdkVersion',
    `gmail` string COMMENT 'gmail',
    `height_width` string COMMENT '屏幕宽高',
    `app_time` string COMMENT '客户端日志产生时的时间',
    `network` string COMMENT '网络模式',
    `lng` string COMMENT '经度',
    `lat` string COMMENT '纬度',
    `login_count` bigint COMMENT '活跃次数'
) COMMENT '每日设备行为表'
partitioned by(dt string)
stored as parquet
location '/warehouse/gmall/dws/dws_uv_detail_daycount';
```

2) 导入数据

```
hive (gmall)>
insert overwrite table dws_uv_detail_daycount partition(dt='2020-03-10')
select
    mid_id,
    concat_ws('|', collect_set(user_id)) user_id,
    concat_ws('|', collect_set(version_code)) version_code,
    concat_ws('|', collect_set(version_name)) version_name,
    concat_ws('|', collect_set(lang))lang,
    concat_ws('|', collect_set(source)) source,
    concat_ws('|', collect_set(os)) os,
    concat_ws('|', collect_set(area)) area,
    concat_ws('|', collect_set(model)) model,
    concat_ws('|', collect_set(brand)) brand,
    concat_ws('|', collect_set(sdk_version)) sdk_version,
    concat_ws('|', collect_set(gmail)) gmail,
    concat_ws('|', collect_set(height_width)) height_width,
    concat_ws('|', collect_set(app_time)) app_time,
```

```
    concat_ws('|', collect_set(network)) network,
    concat_ws('|', collect_set(lng)) lng,
    concat_ws('|', collect_set(lat)) lat,
    count(*) login_count
from dwd_start_log
where dt='2020-03-10'
group by mid_id;
```

3）查询结果数据

```
hive (gmall)> select * from dws_uv_detail_daycount where dt='2020-03-10';
```

6.5.3 业务数据聚合

DWS 层的宽表字段是站在不同维度的视角去看事实表的，重点关注事实表的度量值。我们在业务数据 DWD 层主要构建的维度表如图 6-32 所示，其中，编码字典维度表、时间维度表和地区维度表是特殊表（不发生变化）。我们主要关注其余 4 张维度表，分别是用户维度表、商品维度表、优惠券维度表和活动维度表，通过与之关联的事实表，获得不同事实表的度量值。

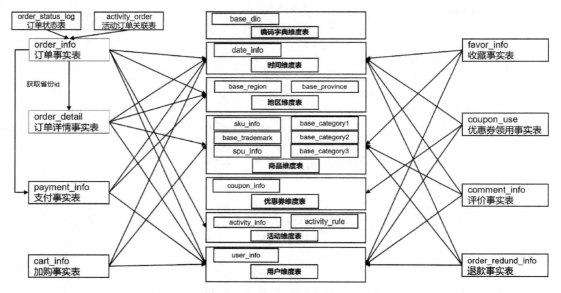

图 6-32　业务数据 DWD 层主要构建的维度表

接下来我们将按照以上思路分别构建 DWS 层的业务数据宽表。

1．每日会员行为

每日会员行为表主要以会员为中心，关注会员的行为，以及该行为对应的度量值。

1）建表语句

```
hive (gmall)>
drop table if exists dws_user_action_daycount;
create external table dws_user_action_daycount
(
```

```sql
    user_id string comment '用户id',
    login_count bigint comment '登录次数',
    cart_count bigint comment '加入购物车次数',
    cart_amount double comment '加入购物车金额',
    order_count bigint comment '下单次数',
    order_amount   decimal(16,2)  comment '下单金额',
    payment_count   bigint     comment '支付次数',
    payment_amount  decimal(16,2) comment '支付金额',
    order_stats array<struct<sku_id:string,sku_num:bigint,order_count:bigint,
order_amount:decimal(20,2)>> comment '下单明细统计'
) COMMENT '每日会员行为表'
PARTITIONED BY (`dt` string)
stored as parquet
location '/warehouse/gmall/dws/dws_user_action_daycount/'
```

2）导入数据

```sql
hive (gmall)>
with
tmp_login as --当日登录统计，统计每个会员的当日登录次数
(
    select
        user_id,
        count(*) login_count --登录次数
    from dwd_start_log
    where dt='2020-03-10'
    and user_id is not null
    group by user_id
),
tmp_cart as --当日加入购物车统计，统计每个会员的当日加入购物车情况
(
    select
        user_id,
        count(*) cart_count, --加入购物车次数
        sum(cart_price*sku_num) cart_amount --加入购物车金额
    from dwd_fact_cart_info
    where dt='2020-03-10'
and user_id is not null
and date_format(create_time,'yyyy-MM-dd')='2020-03-10'
    group by user_id
),
tmp_order as --当日下单统计，统计每个会员的当日下单情况
(
    select
        user_id,
        count(*) order_count, --下单次数
        sum(final_total_amount) order_amount --下单金额
    from dwd_fact_order_info
```

```sql
        where dt='2020-03-10'
        group by user_id
    ),
    tmp_payment as --当日支付统计,统计每个会员的当日支付情况
    (
        select
            user_id,
            count(*) payment_count, --支付次数
            sum(payment_amount) payment_amount --支付金额
        from dwd_fact_payment_info
        where dt='2020-03-10'
        group by user_id
    ),
    tmp_order_detail as --当日订单详情统计,统计每个会员的当日下单详细信息
    (
        select
            user_id,
            --结构为struct数组,每个数组元素对应该会员当日下单的一件商品,包含sku_id、sku_num
            --(下单件数)、order_count(下单次数)、order_amount(下单金额)
            collect_set(named_struct('sku_id',sku_id,'sku_num',sku_num,'order_count',order_count,'order_amount',order_amount)) order_stats
        from
        (
            select
                user_id,
                sku_id,
                sum(sku_num) sku_num,
                count(*) order_count,
                cast(sum(total_amount) as decimal(20,2)) order_amount
            from dwd_fact_order_detail
            where dt='2020-03-10'
            group by user_id,sku_id
        ) tmp
        group by user_id
    )
insert overwrite table dws_user_action_daycount partition(dt='2020-03-10')
select
    coalesce(tmp_login.user_id,tmp_cart.user_id,tmp_order.user_id,tmp_payment.user_id,tmp_order_detail.user_id),
    login_count,
    nvl(cart_count,0),
    nvl(cart_amount,0),
    nvl(order_count,0),
    nvl(order_amount,0),
    nvl(payment_count,0),
    nvl(payment_amount,0),
```

```
    order_stats
from tmp_login
full outer join tmp_cart on tmp_login.user_id=tmp_cart.user_id
full outer join tmp_order on tmp_login.user_id=tmp_order.user_id
full outer join tmp_payment on tmp_login.user_id=tmp_payment.user_id
full outer join tmp_order_detail on tmp_login.user_id=tmp_order_detail.user_id
```

3）查询结果数据

```
hive (gmall)> select * from dws_user_action_daycount where dt='2020-03-10';
```

2. 每日商品行为

每日商品行为表以商品为中心，通过与商品维度有关的事实表获得与商品相关的不同维度的度量值。

1）建表语句

```
hive (gmall)>
drop table if exists dws_sku_action_daycount;
create external table dws_sku_action_daycount
(
    sku_id string comment '商品id',
    order_count bigint comment '被下单次数',
    order_num bigint comment '被下单件数',
    order_amount decimal(16,2) comment '被下单金额',
    payment_count bigint  comment '被支付次数',
    payment_num bigint comment '被支付件数',
    payment_amount decimal(16,2) comment '被支付金额',
    refund_count bigint  comment '被退款次数',
    refund_num bigint comment '被退款件数',
    refund_amount  decimal(16,2) comment '被退款金额',
    cart_count bigint comment '被加入购物车次数',
    cart_num bigint comment '被加入购物车件数',
    favor_count bigint comment '被收藏次数',
    appraise_good_count bigint comment '好评数',
    appraise_mid_count bigint comment '中评数',
    appraise_bad_count bigint comment '差评数',
    appraise_default_count bigint comment '默认评价数'
) COMMENT '每日商品行为表'
PARTITIONED BY (`dt` string)
stored as parquet
location '/warehouse/gmall/dws/dws_sku_action_daycount/'
```

2）导入数据

注意：如果是23:59下单，则支付日期跨天，需要从订单详情中取出支付时间是今天，下单时间是昨天或者今天的订单。

```
hive (gmall)>
with
tmp_order as --下单情况统计，统计每件商品（SKU）当日被下单的情况
```

```sql
(
    select
        sku_id,
        count(*) order_count, --被下单次数
        sum(sku_num) order_num, --被下单件数
        sum(total_amount) order_amount --被下单金额
    from dwd_fact_order_detail
    where dt='2020-03-10'
    group by sku_id
),
tmp_payment as --支付统计,统计每件商品(SKU)当日被支付的情况
(
    select
        sku_id,
        count(*) payment_count, --被支付次数
        sum(sku_num) payment_num, --被支付件数
        sum(total_amount) payment_amount --被支付金额
    from dwd_fact_order_detail
    where dt='2020-03-10'
    and order_id in
    (
        select
           id
        from dwd_fact_order_info
        where (dt='2020-03-10'
        or dt=date_add('2020-03-10',-1))
        and date_format(payment_time,'yyyy-MM-dd')='2020-03-10'
    )
    group by sku_id
),
tmp_refund as --退款统计
(
    select
        sku_id,
        count(*) refund_count, --被退款次数
        sum(refund_num) refund_num, --被退款件数
        sum(refund_amount) refund_amount --被退款金额
    from dwd_fact_order_refund_info
    where dt='2020-03-10'
    group by sku_id
),
tmp_cart as --加入购物车统计
(
    select
        sku_id,
        count(*) cart_count, --被加入购物车次数
        sum(sku_num) cart_num --被加入购物车件数
```

```sql
    from dwd_fact_cart_info
    where dt='2020-03-10'
    and date_format(create_time,'yyyy-MM-dd')='2020-03-10'
    group by sku_id
),
tmp_favor as --收藏统计
(
    select
        sku_id,
        count(*) favor_count --被收藏次数
    from dwd_fact_favor_info
    where dt='2020-03-10'
    and date_format(create_time,'yyyy-MM-dd')='2020-03-10'
    group by sku_id
),
tmp_appraise as
(
select
    sku_id,
    sum(if(appraise='1201',1,0)) appraise_good_count,
    sum(if(appraise='1202',1,0)) appraise_mid_count,
    sum(if(appraise='1203',1,0)) appraise_bad_count,
    sum(if(appraise='1204',1,0)) appraise_default_count
from dwd_fact_comment_info
where dt='2020-03-10'
group by sku_id
)

insert overwrite table dws_sku_action_daycount partition(dt='2020-03-10')
select
    sku_id,
    sum(order_count),
    sum(order_num),
    sum(order_amount),
    sum(payment_count),
    sum(payment_num),
    sum(payment_amount),
    sum(refund_count),
    sum(refund_num),
    sum(refund_amount),
    sum(cart_count),
    sum(cart_num),
    sum(favor_count),
    sum(appraise_good_count),
    sum(appraise_mid_count),
    sum(appraise_bad_count),
    sum(appraise_default_count)
```

```sql
from
(
    select
        sku_id,
        order_count,
        order_num,
        order_amount,
        0 payment_count,
        0 payment_num,
        0 payment_amount,
        0 refund_count,
        0 refund_num,
        0 refund_amount,
        0 cart_count,
        0 cart_num,
        0 favor_count,
        0 appraise_good_count,
        0 appraise_mid_count,
        0 appraise_bad_count,
        0 appraise_default_count
    from tmp_order
    union all
    select
        sku_id,
        0 order_count,
        0 order_num,
        0 order_amount,
        payment_count,
        payment_num,
        payment_amount,
        0 refund_count,
        0 refund_num,
        0 refund_amount,
        0 cart_count,
        0 cart_num,
        0 favor_count,
        0 appraise_good_count,
        0 appraise_mid_count,
        0 appraise_bad_count,
        0 appraise_default_count
    from tmp_payment
    union all
    select
        sku_id,
        0 order_count,
        0 order_num,
        0 order_amount,
```

```
        0 payment_count,
        0 payment_num,
        0 payment_amount,
        refund_count,
        refund_num,
        refund_amount,
        0 cart_count,
        0 cart_num,
        0 favor_count,
        0 appraise_good_count,
        0 appraise_mid_count,
        0 appraise_bad_count,
        0 appraise_default_count
    from tmp_refund
    union all
    select
        sku_id,
        0 order_count,
        0 order_num,
        0 order_amount,
        0 payment_count,
        0 payment_num,
        0 payment_amount,
        0 refund_count,
        0 refund_num,
        0 refund_amount,
        cart_count,
        cart_num,
        0 favor_count,
        0 appraise_good_count,
        0 appraise_mid_count,
        0 appraise_bad_count,
        0 appraise_default_count
    from tmp_cart
    union all
    select
        sku_id,
        0 order_count,
        0 order_num,
        0 order_amount,
        0 payment_count,
        0 payment_num,
        0 payment_amount,
        0 refund_count,
        0 refund_num,
        0 refund_amount,
        0 cart_count,
```

```
            0 cart_num,
            favor_count,
            0 appraise_good_count,
            0 appraise_mid_count,
            0 appraise_bad_count,
            0 appraise_default_count
        from tmp_favor
        union all
        select
            sku_id,
            0 order_count,
            0 order_num,
            0 order_amount,
            0 payment_count,
            0 payment_num,
            0 payment_amount,
            0 refund_count,
            0 refund_num,
            0 refund_amount,
            0 cart_count,
            0 cart_num,
            0 favor_count,
            appraise_good_count,
            appraise_mid_count,
            appraise_bad_count,
            appraise_default_count
        from tmp_appraise
) tmp
group by sku_id;
```

3）查询结果数据

```
hive (gmall)> select * from dws_sku_action_daycount where dt='2020-03-10';
```

3. 每日优惠券统计

每日优惠券统计表以优惠券为中心，统计优惠券的相关行为数，通过优惠券领用事实表与优惠券维度表左连接，获得优惠券的基本信息及领用行为的相关数据。

1）建表语句

```
hive (gmall)>
drop table if exists dws_coupon_use_daycount;
create external table dws_coupon_use_daycount
(
    `coupon_id` string COMMENT '优惠券id',
    `coupon_name` string COMMENT '优惠券名称',
    `coupon_type` string COMMENT '优惠券类型 1 现金券 2 折扣券 3 满减券 4 满件打折券',
    `condition_amount` string COMMENT '满额数',
    `condition_num` string COMMENT '满件数',
    `activity_id` string COMMENT '活动编号',
```

```
    `benefit_amount` string COMMENT '满减金额',
    `benefit_discount` string COMMENT '折扣',
    `create_time` string COMMENT '创建时间',
    `range_type` string COMMENT '范围类型 1 商品 2 品类 3 品牌',
    `spu_id` string COMMENT '标准产品单位id',
    `tm_id` string COMMENT '品牌id',
    `category3_id` string COMMENT '品类id',
    `limit_num` string COMMENT '最多领用次数',
    `get_count` bigint COMMENT '领用次数',
    `using_count` bigint COMMENT '使用(下单)次数',
    `used_count` bigint COMMENT '使用(支付)次数'
) COMMENT '每日优惠券统计表'
PARTITIONED BY (`dt` string)
stored as parquet
location '/warehouse/gmall/dws/dws_coupon_use_daycount/'
```

2) 导入数据

```
hive (gmall)>
insert overwrite table dws_coupon_use_daycount partition(dt='2020-03-10')
select
    cu.coupon_id,
    ci.coupon_name,
    ci.coupon_type,
    ci.condition_amount,
    ci.condition_num,
    ci.activity_id,
    ci.benefit_amount,
    ci.benefit_discount,
    ci.create_time,
    ci.range_type,
    ci.spu_id,
    ci.tm_id,
    ci.category3_id,
    ci.limit_num,
    cu.get_count,
    cu.using_count,
    cu.used_count
from
(
    select
        coupon_id,
        sum(if(date_format(get_time,'yyyy-MM-dd')='2020-03-10',1,0)) get_count,
        sum(if(date_format(using_time,'yyyy-MM-dd')='2020-03-10',1,0)) using_count,
        sum(if(date_format(used_time,'yyyy-MM-dd')='2020-03-10',1,0)) used_count
    from dwd_fact_coupon_use
```

```
        where dt='2020-03-10'
        group by coupon_id
    )cu
    left join
    (
        select
            *
        from dwd_dim_coupon_info
        where dt='2020-03-10'
    )ci on cu.coupon_id=ci.id;
```

3）查询结果数据

```
hive (gmall)> select * from dws_coupon_use_daycount where dt='2020-03-10';
```

4. 每日活动统计

每日活动统计表主要以活动为中心，通过订单事实表 dwd_fact_order_info 获得与活动有关的度量值。

1）建表语句

```
hive (gmall)>
drop table if exists dws_activity_info_daycount;
create external table dws_activity_info_daycount(
    `id` string COMMENT '编号',
    `activity_name` string COMMENT '活动名称',
    `activity_type` string COMMENT '活动类型',
    `start_time` string COMMENT '开始时间',
    `end_time` string COMMENT '结束时间',
    `create_time` string COMMENT '创建时间',
    `order_count` bigint COMMENT '下单次数',
    `payment_count` bigint COMMENT '支付次数'
) COMMENT '每日活动统计表'
PARTITIONED BY (`dt` string)
row format delimited fields terminated by '\t'
location '/warehouse/gmall/dws/dws_activity_info_daycount/'
```

2）导入数据

```
hive (gmall)>
insert overwrite table dws_activity_info_daycount partition(dt='2020-03-10')
select
    oi.activity_id,
    ai.activity_name,
    ai.activity_type,
    ai.start_time,
    ai.end_time,
    ai.create_time,
    oi.order_count,
    oi.payment_count
from
```

```sql
(
    select
        activity_id,
        sum(if(date_format(create_time,'yyyy-MM-dd')='2020-03-10',1,0)) order_count,
        sum(if(date_format(payment_time,'yyyy-MM-dd')='2020-03-10',1,0)) payment_count
    from dwd_fact_order_info
    where (dt='2020-03-10' or dt=date_add('2020-03-10',-1))
    and activity_id is not null
    group by activity_id
)oi
join
(
    select
        *
    from dwd_dim_activity_info
    where dt='2020-03-10'
)ai
on oi.activity_id=ai.id;
```

3）查询结果数据

```
hive (gmall)> select * from dws_activity_info_daycount where dt='2020-03-10';
```

6.5.4 DWS 层数据导入脚本

将 DWS 层的数据导入过程编写成数据导入脚本。

（1）在/home/atguigu/bin 目录下创建脚本 dwd_to_dws.sh。

```
[atguigu@hadoop102 bin]$ vim dwd_to_dws.sh
```

在脚本中编写如下内容。

```bash
#!/bin/bash

APP=gmall
hive=/opt/module/hive/bin/hive

# 如果输入了日期参数，则取输入参数作为日期值；如果没有输入日期参数，则取当前时间的前一天作为日期值
if [ -n "$1" ] ;then
    do_date=$1
else
    do_date=`date -d "-1 day" +%F`
fi

sql="
insert overwrite table ${APP}.dws_uv_detail_daycount partition(dt='$do_date')
select
```

```sql
    mid_id,
    concat_ws('|', collect_set(user_id)) user_id,
    concat_ws('|', collect_set(version_code)) version_code,
    concat_ws('|', collect_set(version_name)) version_name,
    concat_ws('|', collect_set(lang))lang,
    concat_ws('|', collect_set(source)) source,
    concat_ws('|', collect_set(os)) os,
    concat_ws('|', collect_set(area)) area,
    concat_ws('|', collect_set(model)) model,
    concat_ws('|', collect_set(brand)) brand,
    concat_ws('|', collect_set(sdk_version)) sdk_version,
    concat_ws('|', collect_set(gmail)) gmail,
    concat_ws('|', collect_set(height_width)) height_width,
    concat_ws('|', collect_set(app_time)) app_time,
    concat_ws('|', collect_set(network)) network,
    concat_ws('|', collect_set(lng)) lng,
    concat_ws('|', collect_set(lat)) lat,
    count(*) login_count
from ${APP}.dwd_start_log
where dt='$do_date'
group by mid_id;


with
tmp_login as
(
    select
        user_id,
        count(*) login_count
    from ${APP}.dwd_start_log
    where dt='$do_date'
    and user_id is not null
    group by user_id
),
tmp_cart as
(
    select
        user_id,
        count(*) cart_count,
        sum(cart_price*sku_num) cart_amount
    from ${APP}.dwd_fact_cart_info
    where dt='$do_date'
    and user_id is not null
    and date_format(create_time,'yyyy-MM-dd')='$do_date'
    group by user_id
),
tmp_order as
```

```sql
(
    select
        user_id,
        count(*) order_count,
        sum(final_total_amount) order_amount
    from ${APP}.dwd_fact_order_info
    where dt='$do_date'
    group by user_id
),
tmp_payment as
(
    select
        user_id,
        count(*) payment_count,
        sum(payment_amount) payment_amount
    from ${APP}.dwd_fact_payment_info
    where dt='$do_date'
    group by user_id
),
tmp_order_detail as
(
    select
        user_id,
        collect_set(named_struct('sku_id',sku_id,'sku_num',sku_num,'order_count',order_count,'order_amount',order_amount)) order_stats
    from
    (
        select
            user_id,
            sku_id,
            sum(sku_num) sku_num,
            count(*) order_count,
            cast(sum(total_amount) as decimal(20,2)) order_amount
        from ${APP}.dwd_fact_order_detail
        where dt='$do_date'
        group by user_id,sku_id
    ) tmp
    group by user_id
)

insert overwrite table ${APP}.dws_user_action_daycount partition(dt='$do_date')
select
    coalesce(tmp_login.user_id,tmp_cart.user_id,tmp_order.user_id,tmp_payment.user_id,tmp_order_detail.user_id),
    login_count,
    nvl(cart_count,0),
    nvl(cart_amount,0),
```

```sql
        nvl(order_count,0),
        nvl(order_amount,0),
        nvl(payment_count,0),
        nvl(payment_amount,0),
        order_stats
    from tmp_login
    full outer join tmp_cart on tmp_login.user_id=tmp_cart.user_id
    full outer join tmp_order on tmp_login.user_id=tmp_order.user_id
    full outer join tmp_payment on tmp_login.user_id=tmp_payment.user_id
    full outer join tmp_order_detail on tmp_login.user_id=tmp_order_detail.user_id;

with
tmp_order as
(
    select
        sku_id,
        count(*) order_count,
        sum(sku_num) order_num,
        sum(total_amount) order_amount
    from ${APP}.dwd_fact_order_detail
    where dt='$do_date'
    group by sku_id
),
tmp_payment as
(
    select
        sku_id,
        count(*) payment_count,
        sum(sku_num) payment_num,
        sum(total_amount) payment_amount
    from ${APP}.dwd_fact_order_detail
    where dt='$do_date'
    and order_id in
    (
        select
            id
        from ${APP}.dwd_fact_order_info
        where (dt='$do_date' or dt=date_add('$do_date',-1))
        and date_format(payment_time,'yyyy-MM-dd')='$do_date'
    )
    group by sku_id
),
tmp_refund as
(
    select
        sku_id,
        count(*) refund_count,
```

```sql
        sum(refund_num) refund_num,
        sum(refund_amount) refund_amount
    from ${APP}.dwd_fact_order_refund_info
    where dt='$do_date'
    group by sku_id
),
tmp_cart as
(
    select
        sku_id,
        count(*) cart_count,
        sum(sku_num) cart_num
    from ${APP}.dwd_fact_cart_info
    where dt='$do_date'
    and date_format(create_time,'yyyy-MM-dd')='$do_date'
    group by sku_id
),
tmp_favor as
(
    select
        sku_id,
        count(*) favor_count
    from ${APP}.dwd_fact_favor_info
    where dt='$do_date'
    and date_format(create_time,'yyyy-MM-dd')='$do_date'
    group by sku_id
),
tmp_appraise as
(
    select
        sku_id,
        sum(if(appraise='1201',1,0)) appraise_good_count,
        sum(if(appraise='1202',1,0)) appraise_mid_count,
        sum(if(appraise='1203',1,0)) appraise_bad_count,
        sum(if(appraise='1204',1,0)) appraise_default_count
    from ${APP}.dwd_fact_comment_info
    where dt='$do_date'
    group by sku_id
)

insert overwrite table ${APP}.dws_sku_action_daycount partition(dt='$do_date')
select
    sku_id,
    sum(order_count),
    sum(order_num),
    sum(order_amount),
    sum(payment_count),
```

```sql
        sum(payment_num),
        sum(payment_amount),
        sum(refund_count),
        sum(refund_num),
        sum(refund_amount),
        sum(cart_count),
        sum(cart_num),
        sum(favor_count),
        sum(appraise_good_count),
        sum(appraise_mid_count),
        sum(appraise_bad_count),
        sum(appraise_default_count)
from
(
    select
        sku_id,
        order_count,
        order_num,
        order_amount,
        0 payment_count,
        0 payment_num,
        0 payment_amount,
        0 refund_count,
        0 refund_num,
        0 refund_amount,
        0 cart_count,
        0 cart_num,
        0 favor_count,
        0 appraise_good_count,
        0 appraise_mid_count,
        0 appraise_bad_count,
        0 appraise_default_count
    from tmp_order
    union all
    select
        sku_id,
        0 order_count,
        0 order_num,
        0 order_amount,
        payment_count,
        payment_num,
        payment_amount,
        0 refund_count,
        0 refund_num,
        0 refund_amount,
        0 cart_count,
        0 cart_num,
```

```sql
        0 favor_count,
        0 appraise_good_count,
        0 appraise_mid_count,
        0 appraise_bad_count,
        0 appraise_default_count
from tmp_payment
union all
select
        sku_id,
        0 order_count,
        0 order_num,
        0 order_amount,
        0 payment_count,
        0 payment_num,
        0 payment_amount,
        refund_count,
        refund_num,
        refund_amount,
        0 cart_count,
        0 cart_num,
        0 favor_count,
        0 appraise_good_count,
        0 appraise_mid_count,
        0 appraise_bad_count,
        0 appraise_default_count
from tmp_refund
union all
select
        sku_id,
        0 order_count,
        0 order_num,
        0 order_amount,
        0 payment_count,
        0 payment_num,
        0 payment_amount,
        0 refund_count,
        0 refund_num,
        0 refund_amount,
        cart_count,
        cart_num,
        0 favor_count,
        0 appraise_good_count,
        0 appraise_mid_count,
        0 appraise_bad_count,
        0 appraise_default_count
from tmp_cart
union all
```

```sql
    select
        sku_id,
        0 order_count,
        0 order_num,
        0 order_amount,
        0 payment_count,
        0 payment_num,
        0 payment_amount,
        0 refund_count,
        0 refund_num,
        0 refund_amount,
        0 cart_count,
        0 cart_num,
        favor_count,
        0 appraise_good_count,
        0 appraise_mid_count,
        0 appraise_bad_count,
        0 appraise_default_count
    from tmp_favor
    union all
    select
        sku_id,
        0 order_count,
        0 order_num,
        0 order_amount,
        0 payment_count,
        0 payment_num,
        0 payment_amount,
        0 refund_count,
        0 refund_num,
        0 refund_amount,
        0 cart_count,
        0 cart_num,
        0 favor_count,
        appraise_good_count,
        appraise_mid_count,
        appraise_bad_count,
        appraise_default_count
    from tmp_appraise
)tmp
group by sku_id;


insert overwrite table ${APP}.dws_coupon_use_daycount partition(dt='$do_date')
select
    cu.coupon_id,
    ci.coupon_name,
```

```sql
    ci.coupon_type,
    ci.condition_amount,
    ci.condition_num,
    ci.activity_id,
    ci.benefit_amount,
    ci.benefit_discount,
    ci.create_time,
    ci.range_type,
    ci.spu_id,
    ci.tm_id,
    ci.category3_id,
    ci.limit_num,
    cu.get_count,
    cu.using_count,
    cu.used_count
from
(
    select
        coupon_id,
        sum(if(date_format(get_time,'yyyy-MM-dd')='$do_date',1,0)) get_count,
        sum(if(date_format(using_time,'yyyy-MM-dd')='$do_date',1,0)) using_count,
        sum(if(date_format(used_time,'yyyy-MM-dd')='$do_date',1,0)) used_count
    from ${APP}.dwd_fact_coupon_use
    where dt='$do_date'
    group by coupon_id
)cu
left join
(
    select
        *
    from ${APP}.dwd_dim_coupon_info
    where dt='$do_date'
)ci on cu.coupon_id=ci.id;

insert overwrite table ${APP}.dws_activity_info_daycount partition(dt='$do_date')
select
    oi.activity_id,
    ai.activity_name,
    ai.activity_type,
    ai.start_time,
    ai.end_time,
    ai.create_time,
    oi.order_count,
    oi.payment_count
from
(
```

```
    select
        activity_id,
        sum(if(date_format(create_time,'yyyy-MM-dd')='$do_date',1,0)) order_count,
        sum(if(date_format(payment_time,'yyyy-MM-dd')='$do_date',1,0)) payment_count
    from ${APP}.dwd_fact_order_info
    where (dt='$do_date' or dt=date_add('$do_date',-1))
    and activity_id is not null
    group by activity_id
)oi
join
(
    select
        *
    from ${APP}.dwd_dim_activity_info
    where dt='$do_date'
)ai
on oi.activity_id=ai.id;
"

$hive -e "$sql"
```

（2）增加脚本执行权限。

```
[atguigu@hadoop102 bin]$ chmod 777 dwd_to_dws.sh
```

（3）执行脚本，导入数据。

```
[atguigu@hadoop102 bin]$ dwd_to_dws.sh 2020-03-11
```

（4）查询结果数据。

```
hive (gmall)>
select * from dws_uv_detail_daycount where dt='2020-03-11';
select * from dws_user_action_daycount where dt='2020-03-11';
select * from dws_sku_action_daycount where dt='2020-03-11';
select * from dws_sale_detail_daycount where dt='2020-03-11';
select * from dws_coupon_use_daycount where dt='2020-03-11';
select * from dws_activity_info_daycount where dt='2020-03-11';
```

6.6 数据仓库搭建——DWT 层

在 DWS 层的搭建中，我们把不同的主题按照天进行了聚合，获得了每天每个主题的相关事实度量数据。在 DWT 层中，我们将会把这些不同的主题进行进一步汇总，获得每个主题的全量数据表。

DWT 层主题宽表记录的字段包括每个维度关联的不同事实表度量值、累计某个时间段的度量值，以及首次时间、末次时间、累计至今的度量值。

6.6.1 设备主题宽表

DWT 层的设备主题宽表将在每日设备行为表的基础上进行进一步汇总，获得每台设备对应的详细信息，每天将新增加的设备信息增加到设备主题宽表中，并添加首次活跃时间、末次活跃时间、当日活跃次数和累计活跃天数信息，方便后续实现与设备相关的需求，如图 6-33 所示。

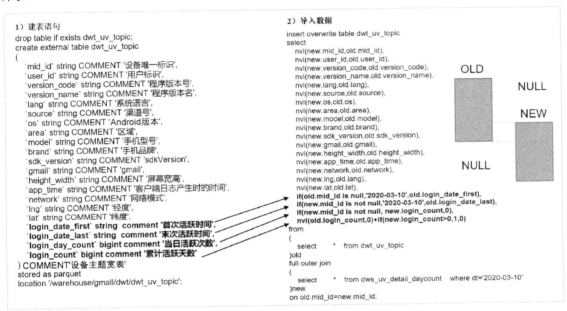

图 6-33 设备主题宽表数据导入思路

1）建表语句

```
hive (gmall)>
drop table if exists dwt_uv_topic;
create external table dwt_uv_topic
(
    `mid_id` string COMMENT '设备唯一标识',
    `user_id` string COMMENT '用户标识',
    `version_code` string COMMENT '程序版本号',
    `version_name` string COMMENT '程序版本名',
    `lang` string COMMENT '系统语言',
    `source` string COMMENT '渠道号',
    `os` string COMMENT 'Android版本',
    `area` string COMMENT '区域',
    `model` string COMMENT '手机型号',
    `brand` string COMMENT '手机品牌',
    `sdk_version` string COMMENT 'sdkVersion',
    `gmail` string COMMENT 'gmail',
    `height_width` string COMMENT '屏幕宽高',
    `app_time` string COMMENT '客户端日志产生时的时间',
```

```
    `network` string COMMENT '网络模式',
    `lng` string COMMENT '经度',
    `lat` string COMMENT '纬度',
    `login_date_first` string comment '首次活跃时间',
    `login_date_last` string comment '末次活跃时间',
    `login_day_count` bigint comment '当日活跃次数',
    `login_count` bigint comment '累计活跃天数'
) COMMENT '设备主题宽表'
stored as parquet
location '/warehouse/gmall/dwt/dwt_uv_topic';
```

2）导入数据

```
hive (gmall)>
insert overwrite table dwt_uv_topic
select
   nvl(new.mid_id,old.mid_id),
   nvl(new.user_id,old.user_id),
   nvl(new.version_code,old.version_code),
   nvl(new.version_name,old.version_name),
   nvl(new.lang,old.lang),
   nvl(new.source,old.source),
   nvl(new.os,old.os),
   nvl(new.area,old.area),
   nvl(new.model,old.model),
   nvl(new.brand,old.brand),
   nvl(new.sdk_version,old.sdk_version),
   nvl(new.gmail,old.gmail),
   nvl(new.height_width,old.height_width),
   nvl(new.app_time,old.app_time),
   nvl(new.network,old.network),
   nvl(new.lng,old.lng),
   nvl(new.lat,old.lat),
   if(old.mid_id is null,'2020-03-10',old.login_date_first),
   if(new.mid_id is not null,'2020-03-10',old.login_date_last),
   if(new.mid_id is not null, new.login_count,0),
   nvl(old.login_count,0)+if(new.login_count>0,1,0)
from
(
   select
       *
   from dwt_uv_topic
)old
full outer join
(
   select
       *
   from dws_uv_detail_daycount
```

```
       where dt='2020-03-10'
)new
on old.mid_id=new.mid_id;
```

3）查询结果数据

```
hive (gmall)> select * from dwt_uv_topic limit 5;
```

6.6.2 会员主题宽表

DWT 层会员主题宽表与多张事实表有关联，需要获取多个事实行为的首次时间、末次时间和累计度量值。会员主题宽表数据导入思路如图 6-34 所示。

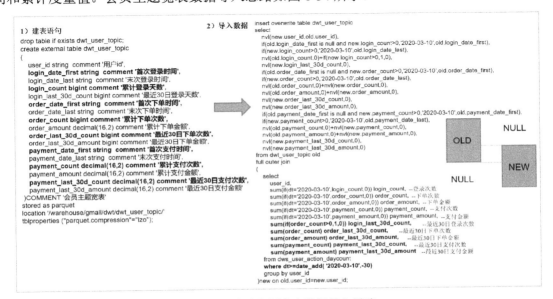

图 6-34 会员主题宽表数据导入思路

1）建表语句

```
hive (gmall)>
drop table if exists dwt_user_topic;
create external table dwt_user_topic
(
    user_id string  comment '用户id',
    login_date_first string  comment '首次登录时间',
    login_date_last string  comment '末次登录时间',
    login_count bigint comment '累计登录天数',
    login_last_30d_count bigint comment '最近30日登录天数',
    order_date_first string  comment '首次下单时间',
    order_date_last string  comment '末次下单时间',
    order_count bigint comment '累计下单次数',
    order_amount decimal(16,2) comment '累计下单金额',
    order_last_30d_count bigint comment '最近30日下单次数',
    order_last_30d_amount bigint comment '最近30日下单金额',
```

```
    payment_date_first string  comment '首次支付时间',
    payment_date_last string  comment '末次支付时间',
    payment_count decimal(16,2) comment '累计支付次数',
    payment_amount decimal(16,2) comment '累计支付金额',
    payment_last_30d_count decimal(16,2) comment '最近30日支付次数',
    payment_last_30d_amount decimal(16,2) comment '最近30日支付金额'
)COMMENT '会员主题宽表'
stored as parquet
location '/warehouse/gmall/dwt/dwt_user_topic/'
```

2）导入数据

```
hive (gmall)>
insert overwrite table dwt_user_topic
select
    nvl(new.user_id,old.user_id),
    if(old.login_date_first is null and new.login_count>0,'2020-03-10',old.login_date_first),
    if(new.login_count>0,'2020-03-10',old.login_date_last),
    nvl(old.login_count,0)+if(new.login_count>0,1,0),
    nvl(new.login_last_30d_count,0),
    if(old.order_date_first is null and new.order_count>0,'2020-03-10',old.order_date_first),
    if(new.order_count>0,'2020-03-10',old.order_date_last),
    nvl(old.order_count,0)+nvl(new.order_count,0),
    nvl(old.order_amount,0)+nvl(new.order_amount,0),
    nvl(new.order_last_30d_count,0),
    nvl(new.order_last_30d_amount,0),
    if(old.payment_date_first is null and new.payment_count>0,'2020-03-10',old.payment_date_first),
    if(new.payment_count>0,'2020-03-10',old.payment_date_last),
    nvl(old.payment_count,0)+nvl(new.payment_count,0),
    nvl(old.payment_amount,0)+nvl(new.payment_amount,0),
    nvl(new.payment_last_30d_count,0),
    nvl(new.payment_last_30d_amount,0)
from
dwt_user_topic old
full outer join
(
    select
        user_id,
        sum(if(dt='2020-03-10',login_count,0)) login_count,
        sum(if(dt='2020-03-10',order_count,0)) order_count,
        sum(if(dt='2020-03-10',order_amount,0)) order_amount,
        sum(if(dt='2020-03-10',payment_count,0)) payment_count,
        sum(if(dt='2020-03-10',payment_amount,0)) payment_amount,
        sum(if(login_count>0,1,0)) login_last_30d_count,
        sum(order_count) order_last_30d_count,
```

```
        sum(order_amount) order_last_30d_amount,
        sum(payment_count) payment_last_30d_count,
        sum(payment_amount) payment_last_30d_amount
    from dws_user_action_daycount
    where dt>=date_add( '2020-03-10',-30)
    group by user_id
)new
on old.user_id=new.user_id;
```

3）查询结果数据

```
hive (gmall)> select * from dwt_user_topic limit 5;
```

6.6.3 商品主题宽表

商品主题宽表与会员主题宽表稍有不同，商品的首次被购买时间和末次被购买时间数据没有太大的意义，重点需要获取多个事实行为的累计度量值和累计行为次数。

1）建表语句

```
hive (gmall)>
drop table if exists dwt_sku_topic;
create external table dwt_sku_topic
(
    sku_id string comment '商品id',
    spu_id string comment '标准产品单位id',
    order_last_30d_count bigint comment '最近30日被下单次数',
    order_last_30d_num bigint comment '最近30日被下单件数',
    order_last_30d_amount decimal(16,2)  comment '最近30日被下单金额',
    order_count bigint comment '累计被下单次数',
    order_num bigint comment '累计被下单件数',
    order_amount decimal(16,2) comment '累计被下单金额',
    payment_last_30d_count  bigint  comment '最近30日被支付次数',
    payment_last_30d_num bigint comment '最近30日被支付件数',
    payment_last_30d_amount decimal(16,2) comment '最近30日被支付金额',
    payment_count  bigint  comment '累计被支付次数',
    payment_num bigint comment '累计被支付件数',
    payment_amount  decimal(16,2) comment '累计被支付金额',
    refund_last_30d_count bigint comment '最近30日退款次数',
    refund_last_30d_num bigint comment '最近30日退款件数',
    refund_last_30d_amount decimal(10,2) comment '最近30日退款金额',
    refund_count bigint comment '累计退款次数',
    refund_num bigint comment '累计退款件数',
    refund_amount decimal(10,2) comment '累计退款金额',
    cart_last_30d_count bigint comment '最近30日被加入购物车次数',
    cart_last_30d_num bigint comment '最近30日被加入购物车件数',
    cart_count bigint comment '累计被加入购物车次数',
    cart_num bigint comment '累计被加入购物车件数',
    favor_last_30d_count bigint comment '最近30日被收藏次数',
```

```
    favor_count bigint comment '累计被收藏次数',
    appraise_last_30d_good_count bigint comment '最近30日好评数',
    appraise_last_30d_mid_count bigint comment '最近30日中评数',
    appraise_last_30d_bad_count bigint comment '最近30日差评数',
    appraise_last_30d_default_count bigint comment '最近30日默认评价数',
    appraise_good_count bigint comment '累计好评数',
    appraise_mid_count bigint comment '累计中评数',
    appraise_bad_count bigint comment '累计差评数',
    appraise_default_count bigint comment '累计默认评价数'
) COMMENT '商品主题宽表'
stored as parquet
location '/warehouse/gmall/dwt/dwt_sku_topic/'
```

2）导入数据

```
hive (gmall)>
insert overwrite table dwt_sku_topic
select
    nvl(new.sku_id,old.sku_id),
    sku_info.spu_id,
    nvl(new.order_count30,0),
    nvl(new.order_num30,0),
    nvl(new.order_amount30,0),
    nvl(old.order_count,0) + nvl(new.order_count,0),
    nvl(old.order_num,0) + nvl(new.order_num,0),
    nvl(old.order_amount,0) + nvl(new.order_amount,0),
    nvl(new.payment_count30,0),
    nvl(new.payment_num30,0),
    nvl(new.payment_amount30,0),
    nvl(old.payment_count,0) + nvl(new.payment_count,0),
    nvl(old.payment_num,0) + nvl(new.payment_count,0),
    nvl(old.payment_amount,0) + nvl(new.payment_count,0),
    nvl(new.refund_count30,0),
    nvl(new.refund_num30,0),
    nvl(new.refund_amount30,0),
    nvl(old.refund_count,0) + nvl(new.refund_count,0),
    nvl(old.refund_num,0) + nvl(new.refund_num,0),
    nvl(old.refund_amount,0) + nvl(new.refund_amount,0),
    nvl(new.cart_count30,0),
    nvl(new.cart_num30,0),
    nvl(old.cart_count,0) + nvl(new.cart_count,0),
    nvl(old.cart_num,0) + nvl(new.cart_num,0),
    nvl(new.favor_count30,0),
    nvl(old.favor_count,0) + nvl(new.favor_count,0),
    nvl(new.appraise_good_count30,0),
    nvl(new.appraise_mid_count30,0),
    nvl(new.appraise_bad_count30,0),
    nvl(new.appraise_default_count30,0)  ,
    nvl(old.appraise_good_count,0) + nvl(new.appraise_good_count,0),
```

```sql
        nvl(old.appraise_mid_count,0) + nvl(new.appraise_mid_count,0),
        nvl(old.appraise_bad_count,0) + nvl(new.appraise_bad_count,0),
        nvl(old.appraise_default_count,0) + nvl(new.appraise_default_count,0)
from
(
    select
        sku_id,
        spu_id,
        order_last_30d_count,
        order_last_30d_num,
        order_last_30d_amount,
        order_count,
        order_num,
        order_amount  ,
        payment_last_30d_count,
        payment_last_30d_num,
        payment_last_30d_amount,
        payment_count,
        payment_num,
        payment_amount,
        refund_last_30d_count,
        refund_last_30d_num,
        refund_last_30d_amount,
        refund_count,
        refund_num,
        refund_amount,
        cart_last_30d_count,
        cart_last_30d_num,
        cart_count,
        cart_num,
        favor_last_30d_count,
        favor_count,
        appraise_last_30d_good_count,
        appraise_last_30d_mid_count,
        appraise_last_30d_bad_count,
        appraise_last_30d_default_count,
        appraise_good_count,
        appraise_mid_count,
        appraise_bad_count,
        appraise_default_count
    from dwt_sku_topic
)old
full outer join
(
    select
        sku_id,
        sum(if(dt='2020-03-10', order_count,0 )) order_count,
```

```sql
        sum(if(dt='2020-03-10',order_num ,0 ))  order_num,
        sum(if(dt='2020-03-10',order_amount,0 ))  order_amount ,
        sum(if(dt='2020-03-10',payment_count,0 ))  payment_count,
        sum(if(dt='2020-03-10',payment_num,0 ))  payment_num,
        sum(if(dt='2020-03-10',payment_amount,0 ))  payment_amount,
        sum(if(dt='2020-03-10',refund_count,0 ))  refund_count,
        sum(if(dt='2020-03-10',refund_num,0 ))  refund_num,
        sum(if(dt='2020-03-10',refund_amount,0 ))  refund_amount,
        sum(if(dt='2020-03-10',cart_count,0 ))  cart_count,
        sum(if(dt='2020-03-10',cart_num,0 ))  cart_num,
        sum(if(dt='2020-03-10',favor_count,0 ))  favor_count,
        sum(if(dt='2020-03-10',appraise_good_count,0 ))  appraise_good_count,
        sum(if(dt='2020-03-10',appraise_mid_count,0 ) )  appraise_mid_count ,
        sum(if(dt='2020-03-10',appraise_bad_count,0 ))  appraise_bad_count,
        sum(if(dt='2020-03-10',appraise_default_count,0 ))  appraise_default_count,
        sum(order_count) order_count30 ,
        sum(order_num) order_num30,
        sum(order_amount) order_amount30,
        sum(payment_count) payment_count30,
        sum(payment_num) payment_num30,
        sum(payment_amount) payment_amount30,
        sum(refund_count) refund_count30,
        sum(refund_num) refund_num30,
        sum(refund_amount) refund_amount30,
        sum(cart_count) cart_count30,
        sum(cart_num) cart_num30,
        sum(favor_count) favor_count30,
        sum(appraise_good_count) appraise_good_count30,
        sum(appraise_mid_count) appraise_mid_count30,
        sum(appraise_bad_count) appraise_bad_count30,
        sum(appraise_default_count) appraise_default_count30
    from dws_sku_action_daycount
    where dt >= date_add ('2020-03-10', -30)
    group by sku_id
)new
on new.sku_id = old.sku_id
left join
(select * from dwd_dim_sku_info where dt='2020-03-10') sku_info
on nvl(new.sku_id,old.sku_id)= sku_info.id;
```

3）查询结果数据

```
hive (gmall)> select * from dwt_sku_topic limit 5;
```

6.6.4 优惠券主题宽表

优惠券主题宽表主要获取优惠券的领用、下单使用、支付使用行为的累计发生次数和当日累计发生次数。优惠券主题宽表数据导入思路如图 6-35 所示。

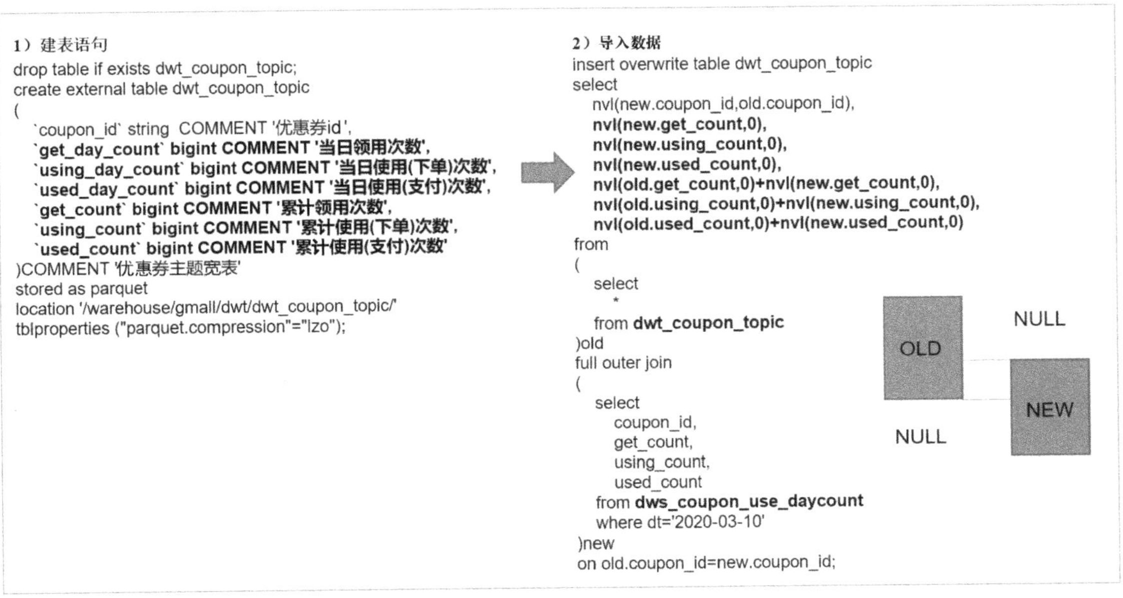

```
1）建表语句
drop table if exists dwt_coupon_topic;
create external table dwt_coupon_topic
(
    `coupon_id` string COMMENT '优惠券id',
    `get_day_count` bigint COMMENT '当日领用次数',
    `using_day_count` bigint COMMENT '当日使用(下单)次数',
    `used_day_count` bigint COMMENT '当日使用(支付)次数',
    `get_count` bigint COMMENT '累计领用次数',
    `using_count` bigint COMMENT '累计使用(下单)次数',
    `used_count` bigint COMMENT '累计使用(支付)次数'
)COMMENT '优惠券主题宽表'
stored as parquet
location '/warehouse/gmall/dwt/dwt_coupon_topic/'
tblproperties ("parquet.compression"="lzo");
```

```
2）导入数据
insert overwrite table dwt_coupon_topic
select
    nvl(new.coupon_id,old.coupon_id),
    nvl(new.get_count,0),
    nvl(new.using_count,0),
    nvl(new.used_count,0),
    nvl(old.get_count,0)+nvl(new.get_count,0),
    nvl(old.using_count,0)+nvl(new.using_count,0),
    nvl(old.used_count,0)+nvl(new.used_count,0)
from
(
    select
        *
    from dwt_coupon_topic
)old
full outer join
(
    select
        coupon_id,
        get_count,
        using_count,
        used_count
    from dws_coupon_use_daycount
    where dt='2020-03-10'
)new
on old.coupon_id=new.coupon_id;
```

图 6-35　优惠券主题宽表数据导入思路

1）建表语句

```
hive (gmall)>
drop table if exists dwt_coupon_topic;
create external table dwt_coupon_topic
(
    `coupon_id` string COMMENT '优惠券id',
    `get_day_count` bigint COMMENT '当日领用次数',
    `using_day_count` bigint COMMENT '当日使用(下单)次数',
    `used_day_count` bigint COMMENT '当日使用(支付)次数',
    `get_count` bigint COMMENT '累计领用次数',
    `using_count` bigint COMMENT '累计使用(下单)次数',
    `used_count` bigint COMMENT '累计使用(支付)次数'
)COMMENT '优惠券主题宽表'
stored as parquet
location '/warehouse/gmall/dwt/dwt_coupon_topic/'
```

2）导入数据

```
hive (gmall)>
insert overwrite table dwt_coupon_topic
select
    nvl(new.coupon_id,old.coupon_id),
    nvl(new.get_count,0),
    nvl(new.using_count,0),
    nvl(new.used_count,0),
    nvl(old.get_count,0)+nvl(new.get_count,0),
    nvl(old.using_count,0)+nvl(new.using_count,0),
```

```
        nvl(old.used_count,0)+nvl(new.used_count,0)
from
(
    select
        *
    from dwt_coupon_topic
)old
full outer join
(
    select
        coupon_id,
        get_count,
        using_count,
        used_count
    from dws_coupon_use_daycount
    where dt='2020-03-10'
)new
on old.coupon_id=new.coupon_id;
```

3）查询结果数据

```
hive (gmall)> select * from dwt_coupon_topic limit 5;
```

6.6.5 活动主题宽表

活动主题宽表与优惠券主题宽表类似，主要获取下单、支付行为的当日行为次数和累计行为次数。活动主题宽表数据导入思路如图 6-36 所示。

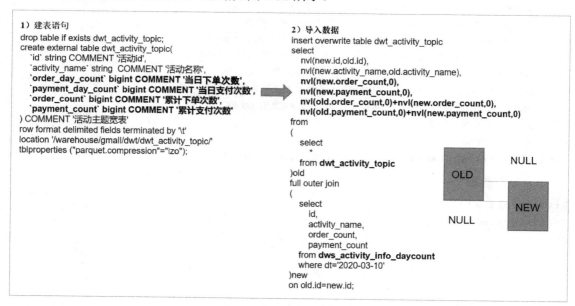

图 6-36 活动主题宽表数据导入思路

1）建表语句

```sql
hive (gmall)>
drop table if exists dwt_activity_topic;
create external table dwt_activity_topic(
    `id` string COMMENT '活动id',
    `activity_name` string  COMMENT '活动名称',
    `order_day_count` bigint COMMENT '当日下单次数',
    `payment_day_count` bigint COMMENT '当日支付次数',
    `order_count` bigint COMMENT '累计下单次数',
    `payment_count` bigint COMMENT '累计支付次数'
) COMMENT '活动主题宽表'
stored as parquet
location '/warehouse/gmall/dwt/dwt_activity_topic/'
```

2）导入数据

```sql
hive (gmall)>
insert overwrite table dwt_activity_topic
select
    nvl(new.id,old.id),
    nvl(new.activity_name,old.activity_name),
    nvl(new.order_count,0),
    nvl(new.payment_count,0),
    nvl(old.order_count,0)+nvl(new.order_count,0),
    nvl(old.payment_count,0)+nvl(new.payment_count,0)
from
(
    select
        *
    from dwt_activity_topic
)old
full outer join
(
    select
        id,
        activity_name,
        order_count,
        payment_count
    from dws_activity_info_daycount
    where dt='2020-03-10'
)new
on old.id=new.id;
```

3）查询结果数据

```sql
hive (gmall)> select * from dwt_activity_topic limit 5;
```

6.6.6 DWT层数据导入脚本

将DWT层加载数据的过程编写成脚本，方便每日调用执行。

（1）在/home/atguigu/bin目录下创建脚本dws_to_dwt.sh。

```
[atguigu@hadoop102 bin]$ vim dws_to_dwt.sh
```

在脚本中编写如下内容。

```
#!/bin/bash

APP=gmall
hive=/opt/module/hive/bin/hive

# 如果输入了日期参数,则取输入参数作为日期值;如果没有输入日期参数,则取当前时间的前一天作为日期值
if [ -n "$1" ] ;then
    do_date=$1
else
    do_date=`date -d "-1 day" +%F`
fi

sql="
insert overwrite table ${APP}.dwt_uv_topic
select
   nvl(new.mid_id,old.mid_id),
   nvl(new.user_id,old.user_id),
   nvl(new.version_code,old.version_code),
   nvl(new.version_name,old.version_name),
   nvl(new.lang,old.lang),
   nvl(new.source,old.source),
   nvl(new.os,old.os),
   nvl(new.area,old.area),
   nvl(new.model,old.model),
   nvl(new.brand,old.brand),
   nvl(new.sdk_version,old.sdk_version),
   nvl(new.gmail,old.gmail),
   nvl(new.height_width,old.height_width),
   nvl(new.app_time,old.app_time),
   nvl(new.network,old.network),
   nvl(new.lng,old.lng),
   nvl(new.lat,old.lat),
   nvl(old.login_date_first,'$do_date'),
   if(new.login_count>0,'$do_date',old.login_date_last),
   nvl(new.login_count,0),
   nvl(new.login_count,0)+nvl(old.login_count,0)
from
(
   select
```

```sql
        *
    from ${APP}.dwt_uv_topic
)old
full outer join
(
    select
        *
    from ${APP}.dws_uv_detail_daycount
    where dt='$do_date'
)new
on old.mid_id=new.mid_id;

insert overwrite table ${APP}.dwt_user_topic
select
    nvl(new.user_id,old.user_id),
    if(old.login_date_first is null and new.login_count>0,'$do_date',old.login_date_first),
    if(new.login_count>0,'$do_date',old.login_date_last),
    nvl(old.login_count,0)+if(new.login_count>0,1,0),
    nvl(new.login_last_30d_count,0),
    if(old.order_date_first is null and new.order_count>0,'$do_date',old.order_date_first),
    if(new.order_count>0,'$do_date',old.order_date_last),
    nvl(old.order_count,0)+nvl(new.order_count,0),
    nvl(old.order_amount,0)+nvl(new.order_amount,0),
    nvl(new.order_last_30d_count,0),
    nvl(new.order_last_30d_amount,0),
    if(old.payment_date_first is null and new.payment_count>0,'$do_date',old.payment_date_first),
    if(new.payment_count>0,'$do_date',old.payment_date_last),
    nvl(old.payment_count,0)+nvl(new.payment_count,0),
    nvl(old.payment_amount,0)+nvl(new.payment_amount,0),
    nvl(new.payment_last_30d_count,0),
    nvl(new.payment_last_30d_amount,0)
from
(
    select
        *
    from ${APP}.dwt_user_topic
)old
full outer join
(
    select
        user_id,
        sum(if(dt='$do_date',login_count,0)) login_count,
        sum(if(dt='$do_date',order_count,0)) order_count,
        sum(if(dt='$do_date',order_amount,0)) order_amount,
```

```sql
        sum(if(dt='$do_date',payment_count,0)) payment_count,
        sum(if(dt='$do_date',payment_amount,0)) payment_amount,
        sum(if(order_count>0,1,0)) login_last_30d_count,
        sum(order_count) order_last_30d_count,
        sum(order_amount) order_last_30d_amount,
        sum(payment_count) payment_last_30d_count,
        sum(payment_amount) payment_last_30d_amount
    from ${APP}.dws_user_action_daycount
    where dt>=date_add( '$do_date',-30)
    group by user_id
)new
on old.user_id=new.user_id;

with
sku_act as
(
select
    sku_id,
    sum(if(dt='$do_date', order_count,0 )) order_count,
    sum(if(dt='$do_date',order_num ,0 )) order_num,
    sum(if(dt='$do_date',order_amount,0 )) order_amount ,
    sum(if(dt='$do_date',payment_count,0 )) payment_count,
    sum(if(dt='$do_date',payment_num,0 )) payment_num,
    sum(if(dt='$do_date',payment_amount,0 )) payment_amount,
    sum(if(dt='$do_date',refund_count,0 )) refund_count,
    sum(if(dt='$do_date',refund_num,0 )) refund_num,
    sum(if(dt='$do_date',refund_amount,0 )) refund_amount,
    sum(if(dt='$do_date',cart_count,0 )) cart_count,
    sum(if(dt='$do_date',cart_num,0 )) cart_num,
    sum(if(dt='$do_date',favor_count,0 )) favor_count,
    sum(if(dt='$do_date',appraise_good_count,0 )) appraise_good_count,
    sum(if(dt='$do_date',appraise_mid_count,0 ) ) appraise_mid_count ,
    sum(if(dt='$do_date',appraise_bad_count,0 )) appraise_bad_count,
    sum(if(dt='$do_date',appraise_default_count,0 )) appraise_default_count,
    sum( order_count ) order_count30 ,
    sum( order_num ) order_num30,
    sum(order_amount ) order_amount30,
    sum(payment_count ) payment_count30,
    sum(payment_num ) payment_num30,
    sum(payment_amount ) payment_amount30,
    sum(refund_count ) refund_count30,
    sum(refund_num ) refund_num30,
    sum(refund_amount ) refund_amount30,
    sum(cart_count ) cart_count30,
    sum(cart_num ) cart_num30,
    sum(favor_count ) favor_count30,
    sum(appraise_good_count ) appraise_good_count30,
```

```sql
    sum(appraise_mid_count ) appraise_mid_count30,
    sum(appraise_bad_count ) appraise_bad_count30,
    sum(appraise_default_count ) appraise_default_count30
from ${APP}.dws_sku_action_daycount
where dt>=date_add ( '$do_date',-30)
group by sku_id
),
sku_topic
as
(
select
    sku_id,
    spu_id,
    order_last_30d_count,
    order_last_30d_num,
    order_last_30d_amount,
    order_count,
    order_num,
    order_amount ,
    payment_last_30d_count,
    payment_last_30d_num,
    payment_last_30d_amount,
    payment_count,
    payment_num,
    payment_amount,
    refund_last_30d_count,
    refund_last_30d_num,
    refund_last_30d_amount ,
    refund_count ,
    refund_num ,
    refund_amount ,
    cart_last_30d_count ,
    cart_last_30d_num ,
    cart_count ,
    cart_num ,
    favor_last_30d_count ,
    favor_count ,
    appraise_last_30d_good_count ,
    appraise_last_30d_mid_count ,
    appraise_last_30d_bad_count ,
    appraise_last_30d_default_count ,
    appraise_good_count ,
    appraise_mid_count ,
    appraise_bad_count ,
    appraise_default_count
from ${APP}.dwt_sku_topic
)
```

```sql
insert overwrite table ${APP}.dwt_sku_topic
select
    nvl(sku_act.sku_id,sku_topic.sku_id) ,
    sku_info.spu_id,
    nvl (sku_act.order_count30,0) ,
    nvl (sku_act.order_num30,0) ,
    nvl (sku_act.order_amount30,0) ,
    nvl(sku_topic.order_count,0)+ nvl (sku_act.order_count,0) ,
    nvl(sku_topic.order_num,0)+ nvl (sku_act.order_num,0) ,
    nvl(sku_topic.order_amount,0)+ nvl (sku_act.order_amount,0),
    nvl (sku_act.payment_count30,0),
    nvl (sku_act.payment_num30,0),
    nvl (sku_act.payment_amount30,0),
    nvl(sku_topic.payment_count,0)+ nvl (sku_act.payment_count,0) ,
    nvl(sku_topic.payment_num,0)+ nvl (sku_act.payment_count,0) ,
    nvl(sku_topic.payment_amount,0)+ nvl (sku_act.payment_count,0) ,
    nvl (refund_count30,0),
    nvl (sku_act.refund_num30,0),
    nvl (sku_act.refund_amount30,0),
    nvl(sku_topic.refund_count,0)+ nvl (sku_act.refund_count,0),
    nvl(sku_topic.refund_num,0)+ nvl (sku_act.refund_num,0),
    nvl(sku_topic.refund_amount,0)+ nvl (sku_act.refund_amount,0),
    nvl(sku_act.cart_count30,0) ,
    nvl(sku_act.cart_num30,0) ,
    nvl(sku_topic.cart_count ,0)+ nvl (sku_act.cart_count,0),
    nvl( sku_topic.cart_num ,0)+ nvl (sku_act.cart_num,0),
    nvl(sku_act.favor_count30 ,0) ,
    nvl (sku_topic.favor_count ,0)+ nvl (sku_act.favor_count,0),
    nvl (sku_act.appraise_good_count30 ,0) ,
    nvl (sku_act.appraise_mid_count30 ,0) ,
    nvl (sku_act.appraise_bad_count30 ,0) ,
    nvl (sku_act.appraise_default_count30 ,0) ,
    nvl (sku_topic.appraise_good_count ,0)+ nvl (sku_act.appraise_good_count,0) ,
    nvl (sku_topic.appraise_mid_count ,0)+ nvl (sku_act.appraise_mid_count,0) ,
    nvl (sku_topic.appraise_bad_count ,0)+ nvl (sku_act.appraise_bad_count,0) ,
    nvl (sku_topic.appraise_default_count ,0)+ nvl (sku_act.appraise_default_count,0)
from sku_act
full outer join sku_topic
on sku_act.sku_id =sku_topic.sku_id
left join
(select * from ${APP}.dwd_dim_sku_info where dt='$do_date') sku_info
on nvl(sku_topic.sku_id,sku_act.sku_id)= sku_info.id;

insert overwrite table ${APP}.dwt_coupon_topic
select
    nvl(new.coupon_id,old.coupon_id),
```

```sql
        nvl(new.get_count,0),
        nvl(new.using_count,0),
        nvl(new.used_count,0),
        nvl(old.get_count,0)+nvl(new.get_count,0),
        nvl(old.using_count,0)+nvl(new.using_count,0),
        nvl(old.used_count,0)+nvl(new.used_count,0)
from
(
    select
         *
    from ${APP}.dwt_coupon_topic
)old
full outer join
(
    select
        coupon_id,
        get_count,
        using_count,
        used_count
    from ${APP}.dws_coupon_use_daycount
    where dt='$do_date'
)new
on old.coupon_id=new.coupon_id;

insert overwrite table ${APP}.dwt_activity_topic
select
    nvl(new.id,old.id),
    nvl(new.activity_name,old.activity_name),
    nvl(new.order_count,0),
    nvl(new.payment_count,0),
    nvl(old.order_count,0)+nvl(new.order_count,0),
    nvl(old.payment_count,0)+nvl(new.payment_count,0)
from
(
    select
         *
    from ${APP}.dwt_activity_topic
)old
full outer join
(
    select
        id,
        activity_name,
        order_count,
        payment_count
    from ${APP}.dws_activity_info_daycount
    where dt='$do_date'
```

```
)new
on old.id=new.id;
"

$hive -e "$sql"
```

(2) 增加脚本执行权限。

```
[atguigu@hadoop102 bin]$ chmod 777 dws_to_dwt.sh
```

(3) 执行脚本,导入数据。

```
[atguigu@hadoop102 bin]$ dws_to_dwt.sh 2020-03-11
```

(4) 查询结果数据。

```
hive (gmall)>
select * from dwt_uv_topic limit 5;
select * from dwt_user_topic limit 5;
select * from dwt_sku_topic limit 5;
select * from dwt_coupon_topic limit 5;
select * from dwt_activity_topic limit 5;
```

6.7 数据仓库搭建——ADS 层

前面已完成 ODS、DWD、DWS、DWT 层数据仓库的搭建,本节主要实现具体需求。

6.7.1 设备主题

本节主要实现与设备主题相关的需求,可通过对设备主题宽表进行适当维度的聚合得到结果。

1. 活跃设备数(日、周、月)

需求定义如下。
- 日活:当日活跃的设备数。
- 周活:当周活跃的设备数。
- 月活:当月活跃的设备数。

1)建表语句

```
hive (gmall)>
drop table if exists ads_uv_count;
create external table ads_uv_count(
    `dt` string COMMENT '统计日期',
    `day_count` bigint COMMENT '当日活跃设备数量',
    `wk_count` bigint COMMENT '当周活跃设备数量',
    `mn_count` bigint COMMENT '当月活跃设备数量',
    `is_weekend` string COMMENT 'Y 或 N 表示是否周末,用于得到本周最终结果',
    `is_monthend` string COMMENT 'Y 或 N 表示是否月末,用于得到本月最终结果'
```

```
) COMMENT '活跃设备数表'
row format delimited fields terminated by '\t'
location '/warehouse/gmall/ads/ads_uv_count/';
```

2）导入数据

```
hive (gmall)>
insert into table ads_uv_count
select
    '2020-03-10' dt,
    daycount.ct,
    wkcount.ct,
    mncount.ct,
    if(date_add(next_day('2020-03-10','MO'),-1)='2020-03-10','Y','N') ,
    if(last_day('2020-03-10')='2020-03-10','Y','N')
from
(
    select
        '2020-03-10' dt,
        count(*) ct
    from dwt_uv_topic
    where login_date_last='2020-03-10'
)daycount join
(
    select
        '2020-03-10' dt,
        count (*) ct
    from dwt_uv_topic
    where login_date_last>=date_add(next_day('2020-03-10','MO'),-7)
    and login_date_last<= date_add(next_day('2020-03-10','MO'),-1)
) wkcount on daycount.dt=wkcount.dt
join
(
    select
        '2020-03-10' dt,
        count (*) ct
    from dwt_uv_topic
    where date_format(login_date_last,'yyyy-MM')=date_format('2020-03-10',
'yyyy-MM')
)mncount on daycount.dt=mncount.dt;
```

3）查询结果数据

```
hive (gmall)> select * from ads_uv_count;
```

2．每日新增设备数

1）建表语句

```
hive (gmall)>
drop table if exists ads_new_mid_count;
```

```
create external table ads_new_mid_count
(
    `create_date`       string comment '创建时间',
    `new_mid_count`     BIGINT comment '新增设备数量'
) COMMENT '每日新增设备数表'
row format delimited fields terminated by '\t'
location '/warehouse/gmall/ads/ads_new_mid_count/';
```

2）导入数据

```
hive (gmall)>
insert into table ads_new_mid_count
select
    login_date_first,
    count(*)
from dwt_uv_topic
where login_date_first='2020-03-10'
group by login_date_first;
```

3）查询结果数据

```
hive (gmall)> select * from ads_new_mid_count;
```

3. 沉默设备数

需求定义如下。

沉默设备：只在安装当天启动过，且启动时间在 7 天前。

1）建表语句

```
hive (gmall)>
drop table if exists ads_silent_count;
create external table ads_silent_count(
    `dt` string COMMENT '统计日期',
    `silent_count` bigint COMMENT '沉默设备数'
)COMMENT '沉默设备数表'
row format delimited fields terminated by '\t'
location '/warehouse/gmall/ads/ads_silent_count';
```

2）导入 2020-03-15 的数据

```
hive (gmall)>
insert into table ads_silent_count
select
    '2020-03-15',
    count(*)
from dwt_uv_topic
where login_date_first=login_date_last
and login_date_last<=date_add('2020-03-15',-7);
```

3）查询结果数据

```
hive (gmall)> select * from ads_silent_count;
```

4．本周回流设备数

需求定义如下。

本周回流设备：上周未活跃，而本周活跃的设备，且不是本周新增的设备。

1）建表语句

```
hive (gmall)>
drop table if exists ads_back_count;
create external table ads_back_count(
    `dt` string COMMENT '统计日期',
    `wk_dt` string COMMENT '统计日期所在周',
    `wastage_count` bigint COMMENT '回流设备数'
) COMMENT '本周回流设备数表'
row format delimited fields terminated by '\t'
location '/warehouse/gmall/ads/ads_back_count';
```

2）导入数据

```
hive (gmall)>
insert into table ads_back_count
select
    '2020-03-15',
    count(*)
from
(
    select
        mid_id
    from dwt_uv_topic
    where login_date_last>=date_add(next_day('2020-03-15','MO'),-7)
    and login_date_last<= date_add(next_day('2020-03-15','MO'),-1)
    and login_date_first<date_add(next_day('2020-03-15','MO'),-7)
)current_wk
left join
(
    select
        mid_id
    from dws_uv_detail_daycount
    where dt>=date_add(next_day('2020-03-15','MO'),-7*2)
    and dt<= date_add(next_day('2020-03-15','MO'),-7-1)
    group by mid_id
)last_wk
on current_wk.mid_id=last_wk.mid_id
where last_wk.mid_id is null;
```

3）查询结果数据

```
hive (gmall)> select * from ads_back_count;
```

5．流失设备数

需求定义如下。

流失设备：最近 7 天未活跃的设备。

1）建表语句

```
hive (gmall)>
drop table if exists ads_wastage_count;
create external table ads_wastage_count(
    `dt` string COMMENT '统计日期',
    `wastage_count` bigint COMMENT '流失设备数'
) COMMENT '流失设备数表'
row format delimited fields terminated by '\t'
location '/warehouse/gmall/ads/ads_wastage_count';
```

2）导入 2020-03-20 的数据

```
hive (gmall)>
insert into table ads_wastage_count
select
    '2020-03-20',
    count(*)
from
(
    select
        mid_id
    from dwt_uv_topic
    where login_date_last<=date_add('2020-03-20',-7)
    group by mid_id
)t1;
```

3）查询结果数据

```
hive (gmall)> select * from ads_wastage_count;
```

6. 留存率

1）概念

留存设备是指某段时间内的新增设备（活跃设备），经过一段时间后，又被继续使用的设备；留存设备占当时新增设备（活跃设备）的比例是留存率。

例如，2019-02-10 新增设备 100 台，在这 100 台设备上，2019-02-11 启动过应用的有 30 台，2019-02-12 启动过应用的有 25 台，2019-02-13 启动过应用的有 32 台，则 2019-02-10 新增设备的次日留存率是 30/100 = 30%，两日留存率是 25/100=25%，三日留存率是 32/100=32%，如图 6-37 所示。

时间	新增设备	1天后	2天后	3天后
2019-02-10	100	30% (2019-02-11)	25% (2019-02-12)	32% (2019-02-13)
2019-02-11	200	20% (2019-02-12)	15% (2019-02-13)	
2019-02-12	100	25% (2019-02-13)		
2019-02-13				

图 6-37 留存率计算示例

2）需求描述

需求：每天计算前 1、2、3、…n 天的留存率。

分析：假设今天是 2 月 11 日，统计前 1 天也就是 2 月 10 日新增设备的留存率，公式如下：

2 月 10 日新增设备的留存率=2 月 10 日的新增设备且 2 月 11 日活跃的设备数/2 月 10 日的新增设备数

3）思路分析

在 ADS 层中，对设备创建时间和留存天数进行汇总，即可获取每天新增设备的后期留存情况，再将 1、2、3 天留存率数据进行汇总。

4）建表语句

```
hive (gmall)>
drop table if exists ads_user_retention_day_rate;
create external table ads_user_retention_day_rate
(
    `stat_date`         string comment '统计日期',
    `create_date`       string comment '设备新增日期',
    `retention_day`     int comment '截至当前日期留存天数',
    `retention_count`   bigint comment '留存数量',
    `new_mid_count`     bigint comment '设备新增数量',
    `retention_ratio`   decimal(10,2) comment '留存率'
) COMMENT '每日设备留存率表'
row format delimited fields terminated by '\t'
location '/warehouse/gmall/ads/ads_user_retention_day_rate/';
```

5）导入数据

```
hive (gmall)>
insert into table ads_user_retention_day_rate
select
    '2020-03-10',--统计日期
    date_add('2020-03-10',-1),--新增日期
    1,--留存天数
    sum(if(login_date_first=date_add('2020-03-10',-1) and login_date_last='2020-03-10',1,0)),--2020-03-09 的 1 日留存数
    sum(if(login_date_first=date_add('2020-03-10',-1),1,0)),--2020-03-09 新增
    sum(if(login_date_first=date_add('2020-03-10',-1) and login_date_last='2020-03-10',1,0))/sum(if(login_date_first=date_add('2020-03-10',-1),1,0))*100
from dwt_uv_topic

union all

select
    '2020-03-10',--统计日期
    date_add('2020-03-10',-2),--新增日期
    2,--留存天数
    sum(if(login_date_first=date_add('2020-03-10',-2) and login_date_last='2020-03-10',1,0)),--2020-03-08 的 2 日留存数
```

```sql
    sum(if(login_date_first=date_add('2020-03-10',-2),1,0)),--2020-03-08 新增
    sum(if(login_date_first=date_add('2020-03-10',-2) and login_date_last=
'2020-03-10',1,0))/sum(if(login_date_first=date_add('2020-03-10',-2),1,0))*100
from dwt_uv_topic

union all

select
    '2020-03-10',--统计日期
    date_add('2020-03-10',-3),--新增日期
    3,--留存天数
    sum(if(login_date_first=date_add('2020-03-10',-3) and login_date_last=
'2020-03-10',1,0)),--2020-03-07 的 3 日留存数
    sum(if(login_date_first=date_add('2020-03-10',-3),1,0)),--2020-03-07 新增
    sum(if(login_date_first=date_add('2020-03-10',-3) and login_date_last=
'2020-03-10',1,0))/sum(if(login_date_first=date_add('2020-03-10',-3),1,0))*100
from dwt_uv_topic;
```

6）查询结果数据

```
hive (gmall)>select * from ads_user_retention_day_rate;
```

7. 最近连续三周活跃设备数

1）建表语句

```
hive (gmall)>
drop table if exists ads_continuity_wk_count;
create external table ads_continuity_wk_count(
    `dt` string COMMENT '统计日期,一般用结束周周日日期,如果每天计算一次,则可用当天日期',
    `wk_dt` string COMMENT '持续时间',
    `continuity_count` bigint COMMENT '活跃次数'
)COMMENT '最近连续三周活跃设备数表'
row format delimited fields terminated by '\t'
location '/warehouse/gmall/ads/ads_continuity_wk_count';
```

2）导入 2020-03-10 所在周的数据

```
hive (gmall)>
insert into table ads_continuity_wk_count
select
    '2020-03-10',
    concat(date_add(next_day('2020-03-10','MO'),-
7*3),'_',date_add(next_day('2020-03-10','MO'),-1)),
    count(*)
from
(
    select
        mid_id
    from
    (
```

```
    Select --查找本周活跃设备
        mid_id
    from dws_uv_detail_daycount
    where dt>=date_add(next_day('2020-03-10','monday'),-7)
    and dt<=date_add(next_day('2020-03-10','monday'),-1)
    group by mid_id

    union all

    select --查找上周活跃设备
        mid_id
    from dws_uv_detail_daycount
    where dt>=date_add(next_day('2020-03-10','monday'),-7*2)
    and dt<=date_add(next_day('2020-03-10','monday'),-7-1)
    group by mid_id

    union all

    select --查找上上周活跃设备
        mid_id
    from dws_uv_detail_daycount
    where dt>=date_add(next_day('2020-03-10','monday'),-7*3)
    and dt<=date_add(next_day('2020-03-10','monday'),-7*2-1)
    group by mid_id
)t1
group by mid_id   --对三周内的所有活跃设备进行分组
having count(*)=3 --分组后,mid_id个数为3的设备为最近连续三周活跃设备
)t2
```

3) 查询结果数据

```
hive (gmall)> select * from ads_continuity_wk_count;
```

8. 最近七天内连续三天活跃设备数

1) 建表语句

```
hive (gmall)>
drop table if exists ads_continuity_uv_count;
create external table ads_continuity_uv_count(
    `dt` string COMMENT '统计日期',
    `wk_dt` string COMMENT '最近七天日期',
    `continuity_count` bigint
) COMMENT '最近七天内连续三天活跃设备数表'
row format delimited fields terminated by '\t'
location '/warehouse/gmall/ads/ads_continuity_uv_count';
```

2) 导入数据

```
hive (gmall)>
insert into table ads_continuity_uv_count
```

```sql
select
    '2020-03-10',
    concat(date_add('2020-03-10',-6),'_','2020-03-10'),
    count(*)
from
(
    select mid_id
    from
    (
        select mid_id
        from
        (
            select
                mid_id,
                date_sub(dt,rank) date_dif --取排序值与日期值之差作为连续标志
            from
            (
                select
                    mid_id,
                    dt,
                    --对七天内登录过的设备按照登录日期进行排序
                    rank() over(partition by mid_id order by dt) rank
                from dws_uv_detail_daycount
                where dt>=date_add('2020-03-10',-6) and dt<='2020-03-10'
            )t1
        )t2
        group by mid_id,date_dif --按照连续标志和设备的mid_id进行分组
        --分组后,mid_id个数大于或等于3的设备为最近七天内连续三天活跃的设备
        having count(*)>=3
    )t3
    group by mid_id
)t4;
```

3）查询结果数据

```
hive (gmall)> select * from ads_continuity_uv_count;
```

6.7.2 会员主题

本节主要实现与会员主题相关的需求，大部分需求可通过 DWT 层的会员主题宽表实现。

1. 会员主题信息

在创建了 DWT 层的会员主题宽表之后，大部分需求的实现都比较简单，我们将这些比较简单的需求汇总到一起实现。

1）建表语句

```
hive (gmall)>
drop table if exists ads_user_topic;
```

```sql
create external table ads_user_topic(
    `dt` string COMMENT '统计日期',
    `day_users` string COMMENT '活跃会员数',
    `day_new_users` string COMMENT '新增会员数',
    `day_new_payment_users` string COMMENT '新增付费会员数',
    `payment_users` string COMMENT '总付费会员数',
    `users` string COMMENT '总会员数',
    `day_users2users` decimal(10,2) COMMENT '会员活跃率',
    `payment_users2users` decimal(10,2) COMMENT '会员付费率',
    `day_new_users2users` decimal(10,2) COMMENT '会员新鲜度'
) COMMENT '会员主题信息表'
row format delimited fields terminated by '\t'
location '/warehouse/gmall/ads/ads_user_topic';
```

2）导入数据

```
hive (gmall)>
insert into table ads_user_topic
select
    '2020-03-10',
    sum(if(login_date_last='2020-03-10',1,0)),
    sum(if(login_date_first='2020-03-10',1,0)),
    sum(if(payment_date_first='2020-03-10',1,0)),
    sum(if(payment_count>0,1,0)),
    count(*),
    sum(if(login_date_last='2020-03-10',1,0))/count(*),
    sum(if(payment_count>0,1,0))/count(*),
    sum(if(login_date_first='2020-03-10',1,0))/sum(if(login_date_last='2020-03-10',1,0))
from dwt_user_topic
```

3）查询结果数据

```
hive (gmall)> select * from ads_user_topic;
```

2．用户行为漏斗分析

用户行为漏斗分析也称为转化率，具体求何种转化率视具体需求而定，比如，消费用户转化率指的是单日日活中最终有多少用户下单消费，即消费用户转化率=单日消费用户数/日活数。再如：

新访问用户转化率=单日新访问用户数/日活数
新注册用户转化率=单日新注册用户数/日活数
新付费用户转化率=单日新付费用户数/日活数

在实际业务中，我们通常关注一些特定页面和行为的转化率，图 6-38 显示了首页访问到商品详情页浏览的转化率，商品详情页浏览到加入购物车的转化率，加入购物车到提交订单的转化率，提交订单到最终支付成功的转化率。通过这些数字，我们可以得知网页设计是否存在缺陷，以及在哪一步损失了用户，从而对网页设计进行改进。

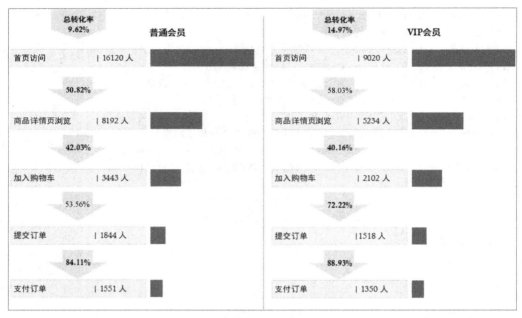

图 6-38　转化率示意

在本需求中，主要计算三个转化率：访问到加入购物车的转化率、加入购物车到下单的转化率和下单到支付的转化率。访问人数可以从 ADS 层的活跃设备数表 ads_uv_count 中获取，加入购物车人数、下单人数和支付人数从每日会员行为表中获取，再进一步计算转化率，即用户行为漏斗分析。

1）建表语句

```
hive (gmall)>
drop table if exists ads_user_action_convert_day;
create external table ads_user_action_convert_day(
    `dt` string COMMENT '统计日期',
    `total_visitor_m_count`  bigint COMMENT '总访问人数',
    `cart_u_count` bigint COMMENT '加入购物车的人数',
    `visitor2cart_convert_ratio` decimal(10,2) COMMENT '访问到加入购物车的转化率',
    `order_u_count` bigint     COMMENT '下单人数',
    `cart2order_convert_ratio`  decimal(10,2) COMMENT '加入购物车到下单的转化率',
    `payment_u_count` bigint    COMMENT '支付人数',
    `order2payment_convert_ratio` decimal(10,2) COMMENT '下单到支付的转化率'
) COMMENT '用户行为漏斗分析表'
row format delimited  fields terminated by '\t'
location '/warehouse/gmall/ads/ads_user_action_convert_day/';
```

2）导入数据

```
hive (gmall)>
insert into table ads_user_action_convert_day
select
    '2020-03-10',
    uv.day_count,
```

```
    ua.cart_count,
    cast(ua.cart_count/uv.day_count as decimal(10,2)) visitor2cart_convert_ratio,
    ua.order_count,
    cast(ua.order_count/ua.cart_count as decimal(10,2)) visitor2order_convert_
ratio,
    ua.payment_count,
    cast(ua.payment_count/ua.order_count as decimal(10,2)) order2payment_
convert_ratio
from
(
    select
        dt,
        sum(if(cart_count>0,1,0)) cart_count,
        sum(if(order_count>0,1,0)) order_count,
        sum(if(payment_count>0,1,0)) payment_count
    from dws_user_action_daycount
where dt='2020-03-10'
group by dt
)ua join ads_uv_count uv on uv.dt=ua.dt;
```

3）查询结果数据

```
hive (gmall)> select * from ads_user_action_convert_day;
```

6.7.3 商品主题

本节主要实现与商品主题相关的需求，大部分需求可通过 DWT 层的商品主题宽表实现。

1．商品个数信息

1）建表语句

```
hive (gmall)>
drop table if exists ads_product_info;
create external table ads_product_info(
    `dt` string COMMENT '统计日期',
    `sku_num` string COMMENT '商品个数',
    `spu_num` string COMMENT '标准产品单位个数'
) COMMENT '商品个数信息表'
row format delimited fields terminated by '\t'
location '/warehouse/gmall/ads/ads_product_info';
```

2）导入数据

```
hive (gmall)>
insert into table ads_product_info
select
    '2020-03-10' dt,
    sku_num,
    spu_num
```

```
from
(
    select
        '2020-03-10' dt,
        count(*) sku_num
    from
        dwt_sku_topic
) tmp_sku_num
join
(
    select
        '2020-03-10' dt,
        count(*) spu_num
    from
    (
        select
            spu_id
        from
            dwt_sku_topic
        group by
            spu_id
    ) tmp_spu_id
) tmp_spu_num
on
    tmp_sku_num.dt=tmp_spu_num.dt;
```

3）查询结果数据

```
hive (gmall)> select * from ads_product_info;
```

2. 商品销量排名

1）建表语句

```
hive (gmall)>
drop table if exists ads_product_sale_topN;
create external table ads_product_sale_topN(
    `dt` string COMMENT '统计日期',
    `sku_id` string COMMENT '商品id',
    `payment_amount` bigint COMMENT '销量'
) COMMENT '商品销量排名表'
row format delimited fields terminated by '\t'
location '/warehouse/gmall/ads/ads_product_sale_topN';
```

2）导入数据

```
hive (gmall)>
insert into table ads_product_sale_topN
select
    '2020-03-10' dt,
    sku_id,
```

```
    payment_amount
from
    dws_sku_action_daycount
where
    dt='2020-03-10'
order by payment_amount desc
limit 10;
```

3) 查询结果数据

```
hive (gmall)> select * from ads_product_sale_topN;
```

3. 商品收藏排名

1) 建表语句

```
hive (gmall)>
drop table if exists ads_product_favor_topN;
create external table ads_product_favor_topN(
    `dt` string COMMENT '统计日期',
    `sku_id` string COMMENT '商品id',
    `favor_count` bigint COMMENT '收藏量'
) COMMENT '商品收藏排名表'
row format delimited fields terminated by '\t'
location '/warehouse/gmall/ads/ads_product_favor_topN';
```

2) 导入数据

```
hive (gmall)>
insert into table ads_product_favor_topN
select
    '2020-03-10' dt,
    sku_id,
    favor_count
from
    dws_sku_action_daycount
where
    dt='2020-03-10'
order by favor_count desc
limit 10;
```

3) 查询结果数据

```
hive (gmall)> select * from ads_product_favor_topN;
```

4. 商品加入购物车排名

1) 建表语句

```
hive (gmall)>
drop table if exists ads_product_cart_topN;
create external table ads_product_cart_topN(
    `dt` string COMMENT '统计日期',
    `sku_id` string COMMENT '商品id',
```

```
    `cart_num` bigint COMMENT '加入购物车数量'
) COMMENT '商品加入购物车排名表'
row format delimited fields terminated by '\t'
location '/warehouse/gmall/ads/ads_product_cart_topN';
```

2）导入数据

```
hive (gmall)>
insert into table ads_product_cart_topN
select
    '2020-03-10' dt,
    sku_id,
    cart_num
from
    dws_sku_action_daycount
where
    dt='2020-03-10'
order by cart_num desc
limit 10;
```

3）查询结果数据

```
hive (gmall)> select * from ads_product_cart_topN;
```

5. 商品退款率排名（最近 30 天）

1）建表语句

```
hive (gmall)>
drop table if exists ads_product_refund_topN;
create external table ads_product_refund_topN(
    `dt` string COMMENT '统计日期',
    `sku_id` string COMMENT '商品id',
    `refund_ratio` decimal(10,2) COMMENT '退款率'
) COMMENT '商品退款率排名表'
row format delimited fields terminated by '\t'
location '/warehouse/gmall/ads/ads_product_refund_topN';
```

2）导入数据

```
hive (gmall)>
insert into table ads_product_refund_topN
select
    '2020-03-10',
    sku_id,
    refund_last_30d_count/payment_last_30d_count*100 refund_ratio
from dwt_sku_topic
order by refund_ratio desc
limit 10;
```

3）查询结果数据

```
hive (gmall)> select * from ads_product_refund_topN;
```

6. 商品差评率

1）建表语句

```
hive (gmall)>
drop table if exists ads_appraise_bad_topN;
create external table ads_appraise_bad_topN(
    `dt` string COMMENT '统计日期',
    `sku_id` string COMMENT '商品id',
    `appraise_bad_ratio` decimal(10,2) COMMENT '差评率'
) COMMENT '商品差评率排名表'
row format delimited fields terminated by '\t'
location '/warehouse/gmall/ads/ads_appraise_bad_topN';
```

2）导入数据

```
hive (gmall)>
insert into table ads_appraise_bad_topN
select
    '2020-03-10' dt,
    sku_id,
appraise_bad_count/(appraise_good_count+appraise_mid_count+appraise_bad_count
+appraise_default_count) appraise_bad_ratio
from
    dws_sku_action_daycount
where
    dt='2020-03-10'
order by appraise_bad_ratio desc
limit 10;
```

3）查询结果数据

```
hive (gmall)> select * from ads_appraise_bad_topN;
```

6.7.4 营销主题

本节主要实现同时涉及用户和商品 2 个主题的需求，在实际工作中，这种需求很常见，有时我们需要整合多张主题宽表来实现。

1．下单数目统计

需求分析：统计每日下单笔数、下单金额及下单用户数。

1）建表语句

```
hive (gmall)>
drop table if exists ads_order_daycount;
create external table ads_order_daycount(
    dt string comment '统计日期',
    order_count bigint comment '每日下单笔数',
    order_amount bigint comment '每日下单金额',
    order_users bigint comment '每日下单用户数'
```

```
) comment '每日订单总计表'
row format delimited fields terminated by '\t'
location '/warehouse/gmall/ads/ads_order_daycount';
```

2)导入数据

```
hive (gmall)>
insert into table ads_order_daycount
select
    '2020-03-10',
    sum(order_count),
    sum(order_amount),
    sum(if(order_count>0,1,0))
from dws_user_action_daycount
where dt='2020-03-10';
```

3)查询结果数据

```
hive (gmall)> select * from ads_order_daycount;
```

2. 支付信息统计

需求分析:统计每日支付金额、支付人数、支付商品数、支付笔数,以及下单到支付的平均时长(取 DWD 层的数据)。

1)建表语句

```
hive (gmall)>
drop table if exists ads_payment_daycount;
create external table ads_payment_daycount(
    dt string comment '统计日期',
    payment_count bigint comment '每日支付笔数',
    payment_amount bigint comment '每日支付金额',
    payment_user_count bigint comment '每日支付人数',
    payment_sku_count bigint comment '每日支付商品数',
    payment_avg_time double comment '下单到支付的平均时长,取分钟数'
) comment '每日支付总计表'
row format delimited fields terminated by '\t'
location '/warehouse/gmall/ads/ads_payment_daycount';
```

2)导入数据

```
hive (gmall)>
insert into table ads_payment_daycount
select
    tmp_payment.dt,
    tmp_payment.payment_count,
    tmp_payment.payment_amount,
    tmp_payment.payment_user_count,
    tmp_skucount.payment_sku_count,
    tmp_time.payment_avg_time
from
(
```

```sql
    select
        '2020-03-10' dt,
        sum(payment_count) payment_count,
        sum(payment_amount) payment_amount,
        sum(if(payment_count>0,1,0)) payment_user_count
    from dws_user_action_daycount
    where dt='2020-03-10'
)tmp_payment
join
(
    select
        '2020-03-10' dt,
        sum(if(payment_count>0,1,0)) payment_sku_count
    from dws_sku_action_daycount
    where dt='2020-03-10'
)tmp_skucount on tmp_payment.dt=tmp_skucount.dt
join
(
    select
        '2020-03-10' dt,
        sum(unix_timestamp(payment_time)-
unix_timestamp(create_time))/count(*)/60 payment_avg_time
    from dwd_fact_order_info
    where dt='2020-03-10'
    and payment_time is not null
)tmp_time on tmp_payment.dt=tmp_time.dt
```

3）查询结果数据

```
hive (gmall)> select * from ads_payment_daycount;
```

3. 复购率

1）建表语句

```
hive (gmall)>
drop table ads_sale_tm_category1_stat_mn;
create external table ads_sale_tm_category1_stat_mn
(
    tm_id string comment '品牌id',
    category1_id string comment '一级品类id ',
    category1_name string comment '一级品类名称 ',
    buycount    bigint comment '购买人数',
    buy_twice_last bigint  comment '两次以上购买人数',
    buy_twice_last_ratio decimal(10,2)  comment '单次复购率',
    buy_3times_last   bigint comment  '三次以上购买人数',
    buy_3times_last_ratio decimal(10,2)  comment '多次复购率',
    stat_mn string comment '统计月份',
    stat_date string comment '统计日期'
) COMMENT '复购率统计表'
```

```
row format delimited fields terminated by '\t'
location '/warehouse/gmall/ads/ads_sale_tm_category1_stat_mn/';
```

2）导入数据

```
hive (gmall)>
with
tmp_order as
(
    select
        user_id,
        order_stats_struct.sku_id sku_id,
        order_stats_struct.order_count order_count
    from dws_user_action_daycount lateral view explode(order_stats) tmp as order_stats_struct
    where date_format(dt,'yyyy-MM')=date_format('2020-03-10','yyyy-MM')
),
tmp_sku as
(
    select
        id,
        tm_id,
        category1_id,
        category1_name
    from dwd_dim_sku_info
    where dt='2020-03-10'
)
insert into table ads_sale_tm_category1_stat_mn
select
    tm_id,
    category1_id,
    category1_name,
    sum(if(order_count>=1,1,0)) buycount,
    sum(if(order_count>=2,1,0)) buyTwiceLast,
    sum(if(order_count>=2,1,0))/sum( if(order_count>=1,1,0)) buyTwiceLastRatio,
    sum(if(order_count>=3,1,0))  buy3timeLast  ,
    sum(if(order_count>=3,1,0))/sum( if(order_count>=1,1,0)) buy3timeLastRatio,
    date_format('2020-03-10' ,'yyyy-MM') stat_mn,
    '2020-03-10' stat_date
from
(
    select
        tmp_order.user_id,
        tmp_sku.category1_id,
        tmp_sku.category1_name,
        tmp_sku.tm_id,
        sum(order_count) order_count
    from tmp_order
```

```
    join tmp_sku
    on tmp_order.sku_id=tmp_sku.id
    group by tmp_order.user_id,tmp_sku.category1_id,tmp_sku.category1_name,tmp_
sku.tm_id
)tmp
group by tm_id, category1_id, category1_name
```

6.7.5 ADS 层数据导入脚本

(1)在/home/atguigu/bin 目录下创建脚本 dwt_to_ads.sh。

```
[atguigu@hadoop102 bin]$ vim dwt_to_ads.sh
```

在脚本中编写如下内容。

```
#!/bin/bash

hive=/opt/module/hive/bin/hive

# 如果输入了日期参数,则取输入参数作为日期值;如果没有输入日期参数,则取当前时间的前一天作为日期值
if [ -n "$1" ] ;then
    do_date=$1
else
    do_date=`date -d "-1 day" +%F`
fi

sql="use gmall;
insert into table ads_uv_count
select
    '$do_date',
    sum(if(login_date_last='$do_date',1,0)),
    sum(if(login_date_last>=date_add(next_day('$do_date','monday'),-7) and
login_date_last<=date_add(next_day('$do_date','monday'),-1) ,1,0)),
    sum(if(date_format(login_date_last,'yyyy-MM')=date_format('$do_date','
yyyy-MM'),1,0)),
    if('$do_date'=date_add(next_day('$do_date','monday'),-1),'Y','N'),
    if('$do_date'=last_day('$do_date'),'Y','N')
from dwt_uv_topic;

insert into table ads_new_mid_count
select
    '$do_date',
    count(*)
from dwt_uv_topic
where login_date_first='$do_date';

insert into table ads_silent_count
select
```

```sql
    '$do_date',
    count(*)
from dwt_uv_topic
where login_date_first=login_date_last
and login_date_last<=date_add('$do_date',-7);

insert into table ads_back_count
select
    '$do_date',
    concat(date_add(next_day('2020-03-10','MO'),-7),'_',date_add(next_day('2020-03-10','MO'),-1)),
    count(*)
from
(
    select
        mid_id
    from dwt_uv_topic
    where login_date_last>=date_add(next_day('$do_date','MO'),-7)
    and login_date_last<= date_add(next_day('$do_date','MO'),-1)
    and login_date_first<date_add(next_day('$do_date','MO'),-7)
)current_wk
left join
(
    select
        mid_id
    from dws_uv_detail_daycount
    where dt>=date_add(next_day('$do_date','MO'),-7*2)
    and dt<= date_add(next_day('$do_date','MO'),-7-1)
    group by mid_id
)last_wk
on current_wk.mid_id=last_wk.mid_id
where last_wk.mid_id is null;


insert into table ads_wastage_count
select
    '$do_date',
    count(*)
from dwt_uv_topic
where login_date_last<=date_add('$do_date',-7);

insert into table ads_user_retention_day_rate
select
    '$do_date',
    date_add('$do_date',-3),
    3,
    sum(if(login_date_first=date_add('$do_date',-3) and login_date_last='$do_
```

```sql
date',1,0)),
    sum(if(login_date_first=date_add('$do_date',-3),1,0)),
    sum(if(login_date_first=date_add('$do_date',-3) and login_date_last='$do_
date',1,0))/sum(if(login_date_first=date_add('$do_date',-3),1,0))*100
from dwt_uv_topic
union all
select
    '$do_date',
    date_add('$do_date',-2),
    2,
    sum(if(login_date_first=date_add('$do_date',-2) and login_date_last='$do_
date',1,0)),
    sum(if(login_date_first=date_add('$do_date',-2),1,0)),
    sum(if(login_date_first=date_add('$do_date',-2) and login_date_last='$do_
date',1,0))/sum(if(login_date_first=date_add('$do_date',-2),1,0))*100
from dwt_uv_topic
union all
select
    '$do_date',
    date_add('$do_date',-1),
    1,
    sum(if(login_date_first=date_add('$do_date',-1) and login_date_last='$do_
date',1,0)),
    sum(if(login_date_first=date_add('$do_date',-1),1,0)),
    sum(if(login_date_first=date_add('$do_date',-1) and login_date_last='$do_
date',1,0))/sum(if(login_date_first=date_add('$do_date',-1),1,0))*100
from dwt_uv_topic;


insert into table ads_continuity_wk_count
select
    '$do_date',
    concat(date_add(next_day('$do_date','MO'),-7*3),'_',date_add(next_day('$do_
date','MO'),-1)),
    count(*)
from
(
    select
        mid_id
    from
    (
        select
            mid_id
        from dws_uv_detail_daycount
        where dt>=date_add(next_day('$do_date','monday'),-7)
        and dt<=date_add(next_day('$do_date','monday'),-1)
        group by mid_id
```

```sql
        union all

        select
            mid_id
        from dws_uv_detail_daycount
        where dt>=date_add(next_day('$do_date','monday'),-7*2)
        and dt<=date_add(next_day('$do_date','monday'),-7-1)
        group by mid_id

        union all

        select
            mid_id
        from dws_uv_detail_daycount
        where dt>=date_add(next_day('$do_date','monday'),-7*3)
        and dt<=date_add(next_day('$do_date','monday'),-7*2-1)
        group by mid_id
    )t1
    group by mid_id
    having count(*)=3
)t2;
insert into table ads_continuity_uv_count
select
    '$do_date',
    concat(date_add('$do_date',-6),'_','$do_date'),
    count(*)
from
(
    select mid_id
    from
    (
        select mid_id
        from
        (
            select
                mid_id,
                date_sub(dt,rank) date_dif
            from
            (
                select
                    mid_id,
                    dt,
                    rank() over(partition by mid_id order by dt) rank
                from dws_uv_detail_daycount
                where dt>=date_add('$do_date',-6) and dt<='$do_date'
```

```sql
            )t1
        )t2
        group by mid_id,date_dif
        having count(*)>=3
    )t3
    group by mid_id
)t4;

insert into table ads_user_topic
select
    '$do_date',
    sum(if(login_date_last='$do_date',1,0)),
    sum(if(login_date_first='$do_date',1,0)),
    sum(if(payment_date_first='$do_date',1,0)),
    sum(if(payment_count>0,1,0)),
    count(*),
    sum(if(login_date_last='$do_date',1,0))/count(*),
    sum(if(payment_count>0,1,0))/count(*),
    sum(if(login_date_first='$do_date',1,0))/sum(if(login_date_last='$do_date',
1,0))
from dwt_user_topic;

insert into table ads_user_action_convert_day
select
    '$do_date',
    uv.day_count,
    ua.cart_count,
    ua.cart_count/uv.day_count*100 visitor2cart_convert_ratio,
    ua.order_count,
    ua.order_count/ua.cart_count*100  visitor2order_convert_ratio,
    ua.payment_count,
    ua.payment_count/ua.order_count*100 order2payment_convert_ratio
from
(
    select
        '$do_date' dt,
        sum(if(cart_count>0,1,0)) cart_count,
        sum(if(order_count>0,1,0)) order_count,
        sum(if(payment_count>0,1,0)) payment_count
    from dws_user_action_daycount
    where dt='$do_date'
)ua join ads_uv_count uv on uv.dt=ua.dt;


insert into table ads_product_info
select
    '$do_date' dt,
```

```
        sku_num,
        spu_num
    from
    (
        select
            '$do_date' dt,
            count(*) sku_num
        from
            dwt_sku_topic
    ) tmp_sku_num
    join
    (
        select
            '$do_date' dt,
            count(*) spu_num
        from
        (
            select
                spu_id
            from
                dwt_sku_topic
            group by
                spu_id
        ) tmp_spu_id
    ) tmp_spu_num
    on tmp_sku_num.dt=tmp_spu_num.dt;


insert into table ads_product_sale_topN
select
    '$do_date',
    sku_id,
    payment_amount
from dws_sku_action_daycount
where dt='$do_date'
order by payment_amount desc
limit 10;


insert into table ads_product_favor_topN
select
    '$do_date',
    sku_id,
    favor_count
from dws_sku_action_daycount
where dt='$do_date'
order by favor_count
```

```sql
limit 10;


insert into table ads_product_cart_topN
select
    '$do_date' dt,
    sku_id,
    cart_num
from dws_sku_action_daycount
where dt='$do_date'
order by cart_num
limit 10;

insert into table ads_product_refund_topN
select
    '$do_date',
    sku_id,
    refund_last_30d_count/payment_last_30d_count*100 refund_ratio
from dwt_sku_topic
order by refund_ratio desc
limit 10;

insert into table ads_appraise_bad_topN
select
    '$do_date' dt,
    sku_id,
    appraise_bad_count/(appraise_bad_count+appraise_good_count+appraise_mid_count+appraise_default_count)*100 appraise_bad_ratio
from dws_sku_action_daycount
where dt='$do_date'
order by appraise_bad_ratio desc
limit 10;


insert into table ads_order_daycount
select
    '$do_date',
    sum(order_count),
    sum(order_amount),
    sum(if(order_count>0,1,0))
from dws_user_action_daycount
where dt='$do_date';

insert into table ads_payment_daycount
select
    tmp_payment.dt,
    tmp_payment.payment_count,
```

```sql
    tmp_payment.payment_amount,
    tmp_payment.payment_user_count,
    tmp_skucount.payment_sku_count,
    tmp_time.payment_avg_time
from
(
    select
        '$do_date' dt,
        sum(payment_count) payment_count,
        sum(payment_amount) payment_amount,
        sum(if(payment_count>0,1,0)) payment_user_count
    from dws_user_action_daycount
    where dt='$do_date'
)tmp_payment
join
(
    select
        '$do_date' dt,
        sum(if(payment_count>0,1,0)) payment_sku_count
    from dws_sku_action_daycount
    where dt='$do_date'
)tmp_skucount on tmp_payment.dt=tmp_skucount.dt
join
(
    select
        '$do_date' dt,
        sum(unix_timestamp(payment_time)-
unix_timestamp(create_time))/count(*)/60 payment_avg_time
    from dwd_fact_order_info
    where dt='$do_date'
    and payment_time is not null
)tmp_time on tmp_payment.dt=tmp_time.dt;


with
tmp_order as
(
    select
        user_id,
        order_stats_struct.sku_id sku_id,
        order_stats_struct.order_count order_count
    from dws_user_action_daycount lateral view explode(order_stats) tmp as order_stats_struct
    where date_format(dt,'yyyy-MM')=date_format('$do_date','yyyy-MM')
),
tmp_sku as
(
```

```
    select
        id,
        tm_id,
        category1_id,
        category1_name
    from dwd_dim_sku_info
    where dt='$do_date'
)
insert into table ads_sale_tm_category1_stat_mn
select
    tm_id,
    category1_id,
    category1_name,
    sum(if(order_count>=1,1,0)) buycount,
    sum(if(order_count>=2,1,0)) buyTwiceLast,
    sum(if(order_count>=2,1,0))/sum( if(order_count>=1,1,0)) buyTwiceLastRatio,
    sum(if(order_count>=3,1,0))  buy3timeLast  ,
    sum(if(order_count>=3,1,0))/sum( if(order_count>=1,1,0)) buy3timeLastRatio ,
    date_format('$do_date' ,'yyyy-MM') stat_mn,
    '$do_date' stat_date
from
(
    select
        tmp_order.user_id,
        tmp_sku.category1_id,
        tmp_sku.category1_name,
        tmp_sku.tm_id,
        sum(order_count) order_count
    from tmp_order
    join tmp_sku
    on tmp_order.sku_id=tmp_sku.id
    group by tmp_order.user_id,tmp_sku.category1_id,tmp_sku.category1_name,tmp_sku.tm_id
)tmp
group by tm_id, category1_id, category1_name;
"

$hive -e "$sql"
```

6.8 结果数据导出脚本

想要对数据仓库中的结果数据进行可视化，需要使用 Sqoop 将结果数据导出到关系型数据库（MySQL）中。使用 Sqoop 可以方便地在关系型数据库和大数据存储系统之间实现数据传输。通过编写数据导出脚本即可实现。脚本编写步骤如下所示。

1. 编写结果数据的 Sqoop 导出脚本

在 /home/atguigu/bin 目录下创建脚本 hdfs_to_mysql.sh。

```
[atguigu@hadoop102 bin]$ vim hdfs_to_mysql.sh
```

在脚本中编写如下内容。

```bash
#!/bin/bash

hive_db_name=gmall
mysql_db_name=gmall_report

export_data() {
/opt/module/sqoop/bin/sqoop export \
--connect "jdbc:mysql://hadoop102:3306/${mysql_db_name}?useUnicode=true&characterEncoding=utf-8" \
--username root \
--password 000000 \
--table $1 \
--num-mappers 1 \
--export-dir /warehouse/$hive_db_name/ads/$1 \
--input-fields-terminated-by "\t" \
--update-mode allowinsert \
--update-key $2 \
--input-null-string '\\N'    \
--input-null-non-string '\\N'
}

case $1 in
  "ads_uv_count")
     export_data "ads_uv_count" "dt"
;;
  "ads_user_action_convert_day")
     export_data "ads_user_action_convert_day" "dt"
;;
  "ads_user_topic")
     export_data "ads_user_topic" "dt"
;;
  "all")
     export_data "ads_uv_count" "dt"
     export_data "ads_user_action_convert_day" "dt"
     export_data "ads_user_topic" "dt"
;;
esac
```

导出脚本中的指令参数说明如下。

- --connect、--username、--password 参数与导入指令中的参数含义相同，参见 5.2.5 节相关内容。

- --table：导出到 MySQL 中的表名。
- --num-mappers：导出数据的作业个数。
- --export-dir：导出数据所在目录。
- --input-fields-terminated-by：指定分隔符，需要与创建 Hive 表时使用的分隔符一致。
- --update-mode：若传入 updateonly，则导出数据后只根据 update-key 进行更新，不允许插入新数据；若传入 allowinsert，则可以根据 update-key 进行更新，并允许插入新数据。
- --update-key：在允许更新的情况下，指定一个或多个字段，在进行数据导出时，若这些字段全部相同，则说明是同一条数据，该条数据将进行更新操作，而不是插入一条新数据。多个字段用逗号分隔。例如，一个表中存在 10 个字段，我们可以指定其中 2 个字段为 update-key，在进行数据导出时，新旧数据间这 2 个字段完全相同，则视新数据与旧数据为同一条数据，就会执行更新操作，否则执行插入操作。
- --input-null-string 和--input-null-non-string：将字符串列与非字符串列的空串和 "NULL" 转换成'\\N'。

Hive 中的 NULL 在底层是以 "\N" 存储的，而 MySQL 中的 NULL 在底层就是 NULL，为了保证数据两端的一致性，在导出数据时使用--input-null-string 和--input-null-non-string 两个参数，导入数据时则使用--null-string 和--null-non-string 两个参数，这样可避免导出数据时出错。

2．Sqoop 导出脚本用法

```
[atguigu@hadoop102 bin]$ chmod 777 hdfs_to_mysql.sh
[atguigu@hadoop102 bin]$ hdfs_to_mysql.sh all
```

6.9 会员主题指标获取的全调度流程

6.9.1 Azkaban 安装

Azkaban 是由 Linkedin 公司推出的一个批量工作流任务调度器，主要用于在一个工作流内以一个特定的顺序运行一组工作和流程。它通过简单的 key-value 对的方式进行配置，通过配置中的 dependencies 来设置依赖关系。Azkaban 使用 job 描述文件建立任务之间的依赖关系，并提供一个易于使用的 Web 界面维护和跟踪用户的工作流。

1．安装前准备

（1）将 Azkaban Web 服务器（azkaban-web-server-2.5.0.tar.gz）、Azkaban 执行服务器（azkaban-executor-server-2.5.0.tar.gz）、Azkaban 的 SQL 执行脚本（azkaban-sql-script-2.5.0.tar.gz）及 MySQL 安装包（mysql-libs.zip）复制到 hadoop102 虚拟机的/opt/software 目录下。

（2）Azkaban 建立了一些 MySQL 连接增强功能，所以选择 MySQL 作为 Azkaban 数据库，以方便 Azkaban 的设置，并可增强服务可靠性。

2．安装 Azkaban

（1）在/opt/module 目录下创建 azkaban 目录。

```
[atguigu@hadoop102 module]$ mkdir azkaban
```

（2）解压 azkaban-web-server-2.5.0.tar.gz、azkaban-executor-server-2.5.0.tar.gz、azkaban-sql-script-2.5.0.tar.gz 到/opt/module/azkaban 目录下。

```
[atguigu@hadoop102 software]$ tar -zxvf azkaban-web-server-2.5.0.tar.gz -C /opt/module/azkaban
[atguigu@hadoop102 software]$ tar -zxvf azkaban-executor-server-2.5.0.tar.gz -C /opt/module/azkaban
[atguigu@hadoop102 software]$ tar -zxvf azkaban-sql-script-2.5.0.tar.gz -C /opt/module/azkaban
```

（3）对解压后的 azkaban-web-server-2.5.0 和 azkaban-executor-server-2.5.0 文件重新命名。

```
[atguigu@hadoop102 azkaban]$ mv azkaban-web-server-2.5.0 server
[atguigu@hadoop102 azkaban]$ mv azkaban-executor-server-2.5.0 executor
```

（4）Azkaban 脚本导入。

进入 MySQL，创建 azkaban 数据库，并将解压的脚本导入 azkaban 数据库。

```
[atguigu@hadoop102 azkaban]$ mysql -uroot -p000000
mysql> create database azkaban;
mysql> use azkaban;
mysql> source /opt/module/azkaban/azkaban-2.5.0/create-all-sql-2.5.0.sql
```

注意：source 后跟.sql 文件，用于批量处理.sql 文件中的 SQL 语句。

3．生成密钥库

- keytool：Java 数据证书的管理工具，使用户能够管理自己的公钥/私钥对及相关证书。
- -keystore：指定密钥库的名称及位置。
- -genkey：在用户主目录中创建一个默认文件 ".keystore"。
- -alias：对生成的 ".keystore" 文件指定别名；如果没有，则默认是 mykey。
- -keyalg：指定密钥的算法 RSA/DSA，默认是 DSA。

（1）生成密钥库的口令及相应信息。

```
[atguigu@hadoop102 azkaban]$ keytool -keystore keystore -alias jetty -genkey -keyalg RSA
输入密钥库口令：
再次输入新口令：
您的名字与姓氏是什么？
  [Unknown]:
您的组织单位名称是什么？
  [Unknown]:
您的组织名称是什么？
  [Unknown]:
您所在的城市或区域名称是什么？
  [Unknown]:
您所在的省/市/自治区名称是什么？
  [Unknown]:
该单位的双字母国家/地区代码是什么？
```

```
[Unknown]:
CN=Unknown, OU=Unknown, O=Unknown, L=Unknown, ST=Unknown, C=Unknown 是否正确?
  [否]: y
```

输入 <jetty> 的密钥口令
            (如果和密钥库口令相同，则按 Enter 键):
再次输入新口令:

注意:

密钥库的口令至少是 6 个字符，可以是纯数字、字母或者数字和字母的组合等。

密钥库的口令最好与<jetty> 的密钥口令相同，方便记忆。

（2）将密钥库所在文件夹 keystore 复制到 Azkaban Web 服务器根目录下。

```
[atguigu@hadoop102 azkaban]$ mv keystore /opt/module/azkaban/server/
```

4．时间同步配置

配置节点服务器上的时区。

（1）如果在/usr/share/zoneinfo 目录下不存在时区配置文件 Asia/Shanghai，则通过执行 tzselect 命令来生成。

```
[atguigu@hadoop102 azkaban]$ tzselect
Please identify a location so that time zone rules can be set correctly.
Please select a continent or ocean.
 1) Africa
 2) Americas
 3) Antarctica
 4) Arctic Ocean
 5) Asia
 6) Atlantic Ocean
 7) Australia
 8) Europe
 9) Indian Ocean
10) Pacific Ocean
11) none - I want to specify the time zone using the Posix TZ format.
#? 5
Please select a country.
 1) Afghanistan           18) Israel              35) Palestine
 2) Armenia               19) Japan               36) Philippines
 3) Azerbaijan            20) Jordan              37) Qatar
 4) Bahrain               21) Kazakhstan          38) Russia
 5) Bangladesh            22) Korea (North)       39) Saudi Arabia
 6) Bhutan                23) Korea (South)       40) Singapore
 7) Brunei                24) Kuwait              41) Sri Lanka
 8) Cambodia              25) Kyrgyzstan          42) Syria
 9) China                 26) Laos                43) Taiwan
10) Cyprus                27) Lebanon             44) Tajikistan
11) East Timor            28) Macau               45) Thailand
```

```
12) Georgia             29) Malaysia            46) Turkmenistan
13) Hong Kong           30) Mongolia            47) United Arab Emirates
14) India               31) Myanmar (Burma)     48) Uzbekistan
15) Indonesia           32) Nepal               49) Vietnam
16) Iran                33) Oman                50) Yemen
17) Iraq                34) Pakistan
#? 9
Please select one of the following time zone regions.
1) Beijing Time
2) Xinjiang Time
#? 1

The following information has been given:

        China
        Beijing Time

Therefore TZ='Asia/Shanghai' will be used.
Local time is now:      Thu Oct 18 16:24:23 CST 2018.
Universal Time is now:  Thu Oct 18 08:24:23 UTC 2018.
Is the above information OK?
1) Yes
2) No
#? 1

You can make this change permanent for yourself by appending the line
       TZ='Asia/Shanghai'; export TZ
to the file '.profile' in your home directory; then log out and log in again.

Here is that TZ value again, this time on standard output so that you
can use the /usr/bin/tzselect command in shell scripts:
Asia/Shanghai
```

（2）复制该时区文件，覆盖系统本地时区的配置。

```
[atguigu@hadoop102 azkaban]$ cp /usr/share/zoneinfo/Asia/Shanghai /etc/localtime
```

（3）集群时间同步配置。

集群时间同步配置可参考 3.3.1 节中的内容。

5．Web 服务器配置

（1）进入 Azkaban Web 服务器安装目录 conf，打开 azkaban.properties 文件。

```
[atguigu@hadoop102 conf]$ pwd
/opt/module/azkaban/server/conf
[atguigu@hadoop102 conf]$ vim azkaban.properties
```

（2）按照如下配置，修改 azkaban.properties 文件。

```
#Azkaban 个性化设置
#服务器UI名称，在服务器上方显示
```

```
azkaban.name=Test
#描述
azkaban.label=My Local Azkaban
#UI 颜色
azkaban.color=#FF3601
azkaban.default.servlet.path=/index
#默认 Web 服务器存放 Web 文件的目录
web.resource.dir=/opt/module/azkaban/server/web/
#默认时区为美国，已改为亚洲/上海
default.timezone.id=Asia/Shanghai

#Azkaban 用户管理定制
user.manager.class=azkaban.user.XmlUserManager
#用户权限管理默认类（绝对路径）
user.manager.xml.file=/opt/module/azkaban/server/conf/azkaban-users.xml

#项目加载配置
#global 配置文件所在位置（绝对路径）
executor.global.properties=/opt/module/azkaban/executor/conf/global.properties
azkaban.project.dir=projects

#数据库类型
database.type=mysql
#端口
mysql.port=3306
#数据库连接 IP 地址
mysql.host=hadoop102
#数据库实例名
mysql.database=azkaban
#数据库用户名
mysql.user=root
#数据库密码
mysql.password=000000
#最大连接数
mysql.numconnections=100

velocity.dev.mode=false

#Jetty 服务器属性
#最大线程数
jetty.maxThreads=25
#Jetty SSL 端口
jetty.ssl.port=8443
#Jetty 端口
jetty.port=8081

#密钥库配置
```

```
#SSL 文件名（绝对路径）
jetty.keystore=/opt/module/azkaban/server/keystore
#SSL 文件密码
jetty.password=000000
#Jetty 主密码与.keystore 文件相同
jetty.keypassword=000000

#可信密钥库配置
#SSL 文件名（绝对路径）
jetty.truststore=/opt/module/azkaban/server/keystore
#SSL 文件密码
jetty.trustpassword=000000

#Azkaban Executor 配置
executor.port=12321

#邮件发送配置
mail.sender=
mail.host=
job.failure.email=
job.success.email=

lockdown.create.projects=false

cache.directory=cache
```

（3）Web 服务器用户配置。在 Azkaban Web 服务器安装目录 conf 下，按照如下配置修改 azkaban-users.xml 文件，增加管理员用户。

```
[atguigu@hadoop102 conf]$ vim azkaban-users.xml
<azkaban-users>
 <user username="azkaban" password="azkaban" roles="admin" groups="azkaban" />
 <user username="metrics" password="metrics" roles="metrics"/>
 <user username="admin" password="admin" roles="admin" />
 <role name="admin" permissions="ADMIN" />
 <role name="metrics" permissions="METRICS"/>
</azkaban-users>
```

6. 执行服务器配置

（1）进入服务器安装目录 conf，打开 azkaban.properties 文件。

```
[atguigu@hadoop102 conf]$ pwd
/opt/module/azkaban/executor/conf
[atguigu@hadoop102 conf]$ vim azkaban.properties
```

（2）按照如下配置，修改 azkaban.properties 文件。

```
#Azkaban
#时区
default.timezone.id=Asia/Shanghai
```

```
#Azkaban 作业类型插件配置
#JobTypes 插件所在位置
azkaban.jobtype.plugin.dir=plugins/jobtypes

#Loader for projects
executor.global.properties=/opt/module/azkaban/executor/conf/global.properties
azkaban.project.dir=projects

database.type=mysql
mysql.port=3306
mysql.host=hadoop102
mysql.database=azkaban
mysql.user=root
mysql.password=000000
mysql.numconnections=100

# Azkaban Executor 配置
#最大线程数
executor.maxThreads=50
#端口（如需修改，要与 Web 服务器中的端口保持一致）
executor.port=12321
#线程数
executor.flow.threads=30
```

7. 启动 Executor 服务器

在 Executor 服务器目录下执行启动命令。

```
[atguigu@hadoop102 executor]$ pwd
/opt/module/azkaban/executor
[atguigu@hadoop102 executor]$ bin/azkaban-executor-start.sh
```

8. 启动 Web 服务器

在 Azkaban Web 服务器目录下执行启动命令。

```
[atguigu@hadoop102 server]$ pwd
/opt/module/azkaban/server
[atguigu@hadoop102 server]$ bin/azkaban-web-start.sh
```

注意：先启动 Executor 服务器，再启动 Web 服务器，避免 Web 服务器因找不到执行器而启动失败。

执行 jps 命令，查看进程。

```
[atguigu@hadoop102 server]$ jps
3601 AzkabanExecutorServer
5880 Jps
3661 AzkabanWebServer
```

启动完成后，在浏览器（建议使用谷歌浏览器）地址栏中输入 https://服务器 IP 地址:8443，即可访问 Azkaban 服务。

在登录页面中，输入刚才在 azkaban-users.xml 文件中新添加的用户名及密码，单击"Login"按钮，如图 6-39 所示。

图 6-39　Azkaban 登录

Azkaban 登录成功的页面如图 6-40 所示。

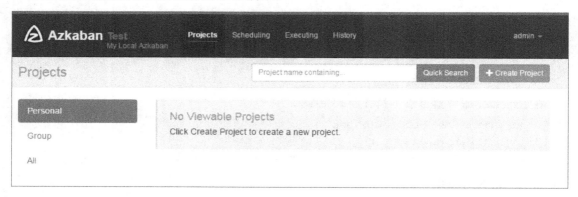

图 6-40　Azkaban 登录成功的页面

6.9.2　创建可视化的 MySQL 数据库和表

（1）创建 gmall_report 数据库，如图 6-41 所示。

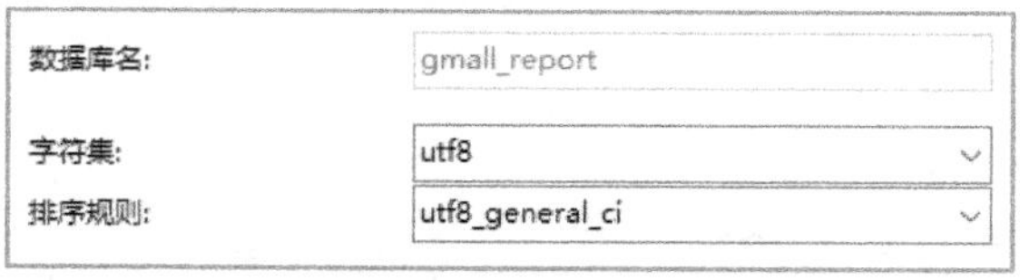

图 6-41　创建 gmall_report 数据库

创建 gmall_report 数据库的 SQL 语句如下。

```
CREATE DATABASE `gmall_report` CHARACTER SET 'utf8' COLLATE 'utf8_general_ci';
```

（2）创建表，语句如下。
```sql
DROP TABLE IF EXISTS `ads_user_topic`;
CREATE TABLE `ads_user_topic` (
  `dt` date NOT NULL,
  `day_users` bigint(255) NULL DEFAULT NULL,
  `day_new_users` bigint(255) NULL DEFAULT NULL,
  `day_new_payment_users` bigint(255) NULL DEFAULT NULL,
  `payment_users` bigint(255) NULL DEFAULT NULL,
  `users` bigint(255) NULL DEFAULT NULL,
  `day_users2users` double(255, 2) NULL DEFAULT NULL,
  `payment_users2users` double(255, 2) NULL DEFAULT NULL,
  `day_new_users2users` double(255, 2) NULL DEFAULT NULL,
  PRIMARY KEY (`dt`) USING BTREE
) ENGINE = InnoDB CHARACTER SET = utf8 COLLATE = utf8_general_ci ROW_FORMAT = Compact;
```

6.9.3 编写指标获取调度流程

Azkaban 安装完成之后，即可编写 job 描述文件，然后将 job 描述文件打包成压缩包提交到 Azkaban 并执行。

Azkaban 内置的任务类型支持 command、Java，任务类型可以通过 type 指定，若该任务执行时需要传入参数，则可以通过 &{PARAMETER} 的形式传入，若任务之间存在依赖关系，则可以通过 dependencies 指定。

（1）生成 2020-03-13 的数据。

（2）编写 Azkaban 程序，运行 job 描述文件。

① mysql_to_hdfs.job 文件。
```
type=command
command=/home/atguigu/bin/mysql_to_hdfs.sh all ${dt}
```

② hdfs_to_ods_log.job 文件。
```
type=command
command=/home/atguigu/bin/hdfs_to_ods_log.sh ${dt}
```

③ hdfs_to_ods_db.job 文件。
```
type=command
command=/home/atguigu/bin/hdfs_to_ods_db.sh all ${dt}
dependencies=mysql_to_hdfs
```

④ ods_to_dwd_start_log.job 文件。
```
type=command
command=/home/atguigu/bin/ods_to_dwd_start_log.sh ${dt}
dependencies=hdfs_to_ods_log
```

⑤ ods_to_dwd_db.job 文件。
```
type=command
```

```
command=/home/atguigu/bin/ods_to_dwd_db.sh ${dt}
dependencies=hdfs_to_ods_db
```

⑥ dwd_to_dws.job 文件。

```
type=command
command=/home/atguigu/bin/dwd_to_dws.sh ${dt}
dependencies=ods_to_dwd_db,ods_to_dwd_start_log
```

⑦ dws_to_dwt.job 文件。

```
type=command
command=/home/atguigu/bin/dws_to_dwt.sh ${dt}
dependencies=dwd_to_dws
```

⑧ dwt_to_ads.job 文件。

```
type=command
command=/home/atguigu/bin/dwt_to_ads.sh ${dt}
dependencies=dws_to_dwt
```

⑨ hdfs_to_mysql.job 文件。

```
type=command
command=/home/atguigu/bin/hdfs_to_mysql.sh ads_user_topic
dependencies=dwt_to_ads
```

⑩ 将以上 9 个 job 描述文件压缩成 gmall.zip 文件，如图 6-42 所示。

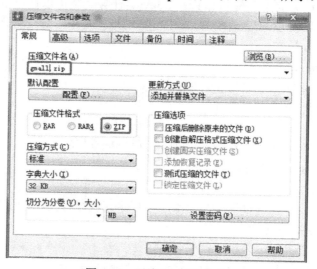

图 6-42 压缩 job 描述文件

（3）任务调度执行。

① 在浏览器地址栏中输入 https://hadoop102:8443，登录 Azkaban，登录成功后，在页面上单击"Create Project"按钮，创建工程，如图 6-43 所示。

图 6-43　创建工程入口

② 对该调度工程进行命名并添加描述信息，然后单击"Create Project"按钮，如图 6-44 所示。

图 6-44　命名并添加描述信息

③ 在打开的页面上单击"Upload"按钮，上传 job 描述 zip 包，如图 6-45 所示。

图 6-45　上传 job 描述 zip 包

④ 上传 job 描述 zip 包之后，工程即创建成功，用户可以单击"Execute Flow"按钮对调度流程进行配置并执行，如图 6-46 所示。

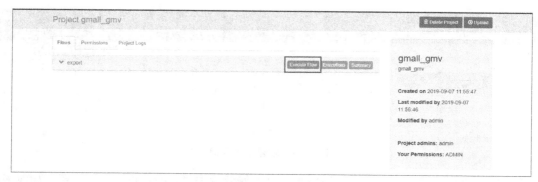

图 6-46 工程创建成功

⑤ 单击"Flow Parameters"按钮,传入参数,如图 6-47 和图 6-48 所示。

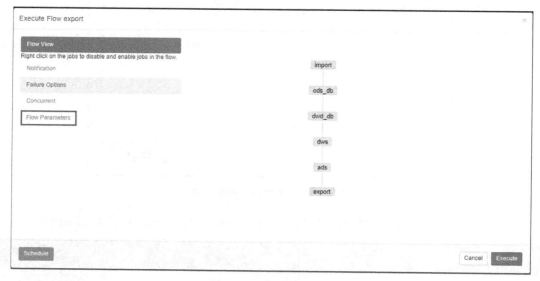

图 6-47 传入参数入口

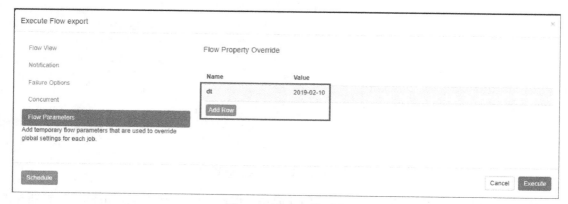

图 6-48 传入参数成功

⑥ 参数设置完毕后,可以单击"Schedule"按钮,进行定时任务调度(企业中通常采用此方案),如图 6-49 所示,也可以单击"Execute"按钮,使任务立即执行,如图 6-50 所示。

图 6-49　提交定时任务入口

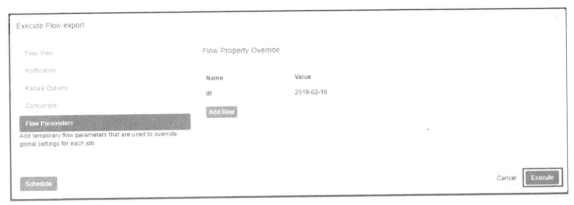

图 6-50　任务立即执行入口

⑦ 任务执行过程中的页面显示如图 6-51 所示，任务执行成功的页面显示如图 6-52 所示。

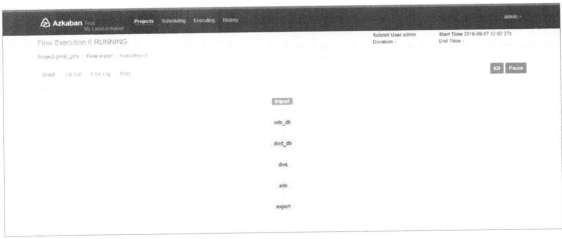

图 6-51　任务执行过程中的页面显示

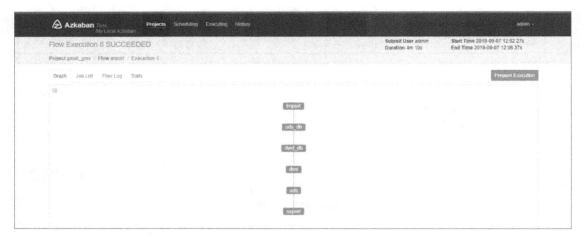

图 6-52　任务执行成功的页面显示

⑧ 任务执行成功后，在 MySQL 中查看结果。

```
select * from ads_gmv_sum_day;
```

6.10　本章总结

　　本章主要对数据仓库搭建模块进行了讲解，从数据仓库搭建所需的理论基础开始，讲解了数据仓库的建模分层理念，以及本数据仓库项目对理论的具体应用，然后讲解了数据仓库搭建所需的基本环境，接着重点讲解了数据仓库如何按照分层的理念一层层地搭建并实现需求，最后为读者简单介绍了数据仓库的任务调度流程。本章重点讲解了数据仓库的分层理论及分层搭建过程，这是无论搭建多大规模的数据仓库都需要首先考虑的问题。

第7章 数据可视化模块

将需求实现,获取到最终的结果数据之后,仅仅让结果数据存放于数据仓库中是远远不够的,还需要将数据进行可视化。通常可视化的思路是从大数据的存储系统中将数据导出到关系型数据库中,再使用 Java 程序进行展示。

7.1 模拟可视化数据

在 MySQL 中,根据可视化的需求进行建表,要求与 ADS 层的表结构完全一致。本章只对会员主题中的 ads_user_topic 表和地区主题中的 ads_area_topic 表进行可视化,为了能尽快看到展示效果,我们在创建表的同时向表中插入若干条数据,建表及数据插入语句如下。

7.1.1 会员主题

1. 创建会员主题结果表 ads_user_topic

```sql
DROP TABLE IF EXISTS `ads_user_topic`;
CREATE TABLE `ads_user_topic` (
  `dt` date NOT NULL,
  `day_users` bigint(255) NULL DEFAULT NULL,
  `day_new_users` bigint(255) NULL DEFAULT NULL,
  `day_new_payment_users` bigint(255) NULL DEFAULT NULL,
  `payment_users` bigint(255) NULL DEFAULT NULL,
  `users` bigint(255) NULL DEFAULT NULL,
  `day_users2users` double(255, 2) NULL DEFAULT NULL,
  `payment_users2users` double(255, 2) NULL DEFAULT NULL,
  `day_new_users2users` double(255, 2) NULL DEFAULT NULL,
  PRIMARY KEY (`dt`) USING BTREE
) ENGINE = InnoDB CHARACTER SET = utf8 COLLATE = utf8_general_ci ROW_FORMAT = Compact;
```

2. 导入数据

```sql
INSERT INTO `ads_user_topic` VALUES ('2020-03-12', 761, 52, 0, 8, 989, 0.77,
```

```
0.01, 0.07);
INSERT INTO `ads_user_topic` VALUES ('2020-03-13', 840, 69, 10, 5, 1000, 0.59, 0.01, 0.05);
INSERT INTO `ads_user_topic` VALUES ('2020-03-14', 900, 69, 10, 5, 1000, 0.59, 0.01, 0.05);
INSERT INTO `ads_user_topic` VALUES ('2020-03-15', 890, 69, 10, 5, 1000, 0.59, 0.01, 0.05);
INSERT INTO `ads_user_topic` VALUES ('2020-03-16', 607, 69, 10, 5, 1000, 0.59, 0.01, 0.05);
INSERT INTO `ads_user_topic` VALUES ('2020-03-17', 812, 69, 10, 5, 1000, 0.59, 0.01, 0.05);
INSERT INTO `ads_user_topic` VALUES ('2020-03-18', 640, 69, 10, 5, 1000, 0.59, 0.01, 0.05);
INSERT INTO `ads_user_topic` VALUES ('2020-03-19', 740, 69, 10, 5, 1000, 0.59, 0.01, 0.05);
INSERT INTO `ads_user_topic` VALUES ('2020-03-20', 540, 69, 10, 5, 1000, 0.59, 0.01, 0.05);
INSERT INTO `ads_user_topic` VALUES ('2020-03-21', 940, 69, 10, 5, 1000, 0.59, 0.01, 0.05);
INSERT INTO `ads_user_topic` VALUES ('2020-03-22', 840, 69, 10, 5, 1000, 0.59, 0.01, 0.05);
INSERT INTO `ads_user_topic` VALUES ('2020-03-23', 1000, 32, 32, 32, 23432, 22.00, 0.11, 0.55);
```

7.1.2 地区主题

1. 创建地区主题结果表 ads_area_topic

```
DROP TABLE IF EXISTS `ads_area_topic`;
CREATE TABLE `ads_area_topic` (
  `dt` date NOT NULL,
  `iso_code` varchar(255) CHARACTER SET utf8 COLLATE utf8_general_ci NOT NULL,
  `province_name` varchar(255) CHARACTER SET utf8 COLLATE utf8_general_ci NULL DEFAULT NULL,
  `area_name` varchar(255) CHARACTER SET utf8 COLLATE utf8_general_ci NULL DEFAULT NULL,
  `order_count` bigint(255) NULL DEFAULT NULL,
  `order_amount` double(255, 2) NULL DEFAULT NULL,
  `payment_count` bigint(255) NULL DEFAULT NULL,
  `payment_amount` double(255, 2) NULL DEFAULT NULL,
  PRIMARY KEY (`dt`, `iso_code`) USING BTREE
) ENGINE = InnoDB CHARACTER SET = utf8 COLLATE = utf8_general_ci ROW_FORMAT = Compact;
```

2. 导入数据

```
INSERT INTO `ads_area_topic` VALUES ('2020-03-10', 'CN-11', '北京市', '华北', 652, 652.00, 652, 652.00);
```

```sql
INSERT INTO `ads_area_topic` VALUES ('2020-03-10', 'CN-12', '天津市', '华北', 42, 42.00, 42, 42.00);
INSERT INTO `ads_area_topic` VALUES ('2020-03-10', 'CN-13', '河北', '华北', 3435, 3435.00, 3435, 3435.00);
INSERT INTO `ads_area_topic` VALUES ('2020-03-10', 'CN-14', '山西', '华北', 4, 4.00, 4, 4.00);
INSERT INTO `ads_area_topic` VALUES ('2020-03-10', 'CN-15', '内蒙古', '华北', 44, 44.00, 44, 44.00);
INSERT INTO `ads_area_topic` VALUES ('2020-03-10', 'CN-21', '辽宁', '东北', 335, 335.00, 335, 335.00);
INSERT INTO `ads_area_topic` VALUES ('2020-03-10', 'CN-22', '吉林', '东北', 44, 44.00, 44, 44.00);
INSERT INTO `ads_area_topic` VALUES ('2020-03-10', 'CN-23', '黑龙江', '东北', 337, 337.00, 337, 337.00);
INSERT INTO `ads_area_topic` VALUES ('2020-03-10', 'CN-31', '上海', '华东', 4, 4.00, 4, 4.00);
INSERT INTO `ads_area_topic` VALUES ('2020-03-10', 'CN-32', '江苏', '华东', 4545, 4545.00, 4545, 4545.00);
INSERT INTO `ads_area_topic` VALUES ('2020-03-10', 'CN-33', '浙江', '华东', 43, 43.00, 43, 43.00);
INSERT INTO `ads_area_topic` VALUES ('2020-03-10', 'CN-34', '安徽', '华东', 12345, 2134.00, 324, 252.00);
INSERT INTO `ads_area_topic` VALUES ('2020-03-10', 'CN-35', '福建', '华东', 435, 435.00, 435, 435.00);
INSERT INTO `ads_area_topic` VALUES ('2020-03-10', 'CN-36', '江西', '华东', 4453, 4453.00, 4453, 4453.00);
INSERT INTO `ads_area_topic` VALUES ('2020-03-10', 'CN-37', '山东', '华东', 34, 34.00, 34, 34.00);
INSERT INTO `ads_area_topic` VALUES ('2020-03-10', 'CN-41', '河南', '华中', 34, 34.00, 34, 34.00);
INSERT INTO `ads_area_topic` VALUES ('2020-03-10', 'CN-42', '湖北', '华中', 4, 4.00, 4, 4.00);
INSERT INTO `ads_area_topic` VALUES ('2020-03-10', 'CN-43', '湖南', '华中', 54, 54.00, 54, 54.00);
INSERT INTO `ads_area_topic` VALUES ('2020-03-10', 'CN-44', '广东', '华南', 24, 24.00, 24, 24.00);
INSERT INTO `ads_area_topic` VALUES ('2020-03-10', 'CN-45', '广西', '华南', 4, 4.00, 4, 4.00);
INSERT INTO `ads_area_topic` VALUES ('2020-03-10', 'CN-46', '海南', '华南', 42, 42.00, 42, 42.00);
INSERT INTO `ads_area_topic` VALUES ('2020-03-10', 'CN-50', '重庆', '西南', 4532, 4532.00, 4532, 4532.00);
INSERT INTO `ads_area_topic` VALUES ('2020-03-10', 'CN-51', '四川', '西南', 3435, 3435.00, 3435, 3435.00);
INSERT INTO `ads_area_topic` VALUES ('2020-03-10', 'CN-52', '贵州', '西南', 5725, 5725.00, 5725, 5725.00);
INSERT INTO `ads_area_topic` VALUES ('2020-03-10', 'CN-53', '云南', '西南', 4357,
```

```
4357.00, 4357, 4357.00);
INSERT INTO `ads_area_topic` VALUES ('2020-03-10', 'CN-54', '西藏', '西南', 54,
54.00, 54, 54.00);
INSERT INTO `ads_area_topic` VALUES ('2020-03-10', 'CN-61', '陕西', '西北', 44,
44.00, 44, 44.00);
INSERT INTO `ads_area_topic` VALUES ('2020-03-10', 'CN-62', '甘肃', '西北', 78,
78.00, 78, 78.00);
INSERT INTO `ads_area_topic` VALUES ('2020-03-10', 'CN-63', '青海', '西北', 3444,
3444.00, 3444, 3444.00);
INSERT INTO `ads_area_topic` VALUES ('2020-03-10', 'CN-64', '宁夏', '西北', 445,
445.00, 445, 445.00);
INSERT INTO `ads_area_topic` VALUES ('2020-03-10', 'CN-65', '新疆', '西北', 4442,
4442.00, 4442, 4442.00);
INSERT INTO `ads_area_topic` VALUES ('2020-03-10', 'CN-71', '台湾', '华东', 343,
343.00, 343, 343.00);
INSERT INTO `ads_area_topic` VALUES ('2020-03-10', 'CN-91', '香港', '华南', 44,
44.00, 44, 44.00);
INSERT INTO `ads_area_topic` VALUES ('2020-03-10', 'CN-92', '澳门', '华南', 34,
34.00, 34, 34.00);
```

7.2 Superset 部署

Apache Superset 是一个开源的、现代的、轻量级的 BI 分析工具，能够对接多种数据源，拥有丰富的图标展示形式，支持自定义仪表盘，且拥有友好的用户界面，十分易用。

7.2.1 环境准备

Superset 是由 Python 编写的 Web 应用，要求使用 Python3.6 的环境，但是 CentOS 系统自带的 Python 环境是 2.x 版本的，所以我们需要先安装 Python3 环境。

1. 安装 Miniconda

Conda 是一个开源的包和环境管理器，可以用于在同一台机器上安装不同版本的 Python 软件包及依赖，并能在不同的 Python 环境之间切换。Anaconda 和 Miniconda 都集成了 Conda，而 Anaconda 包括更多的工具包，如 numpy、pandas，Miniconda 则只包括 Conda 和 Python。

此处，我们不需要太多的工具包，故选择 Miniconda。

1）下载 Miniconda（Python3 版本）

2）安装 Miniconda

（1）执行以下命令，安装 Miniconda，并按照提示进行操作，直到安装完成。

```
[atguigu@hadoop102 lib]$ bash Miniconda3-latest-Linux-x86_64.sh
```

（2）在安装过程中，出现如图 7-1 所示的提示时，需指定安装路径。

```
Miniconda3 will now be installed into this location:
/home/atguigu/miniconda3

  - Press ENTER to confirm the location
  - Press CTRL-C to abort the installation
  - Or specify a different location below

[/home/atguigu/miniconda3] >>> /opt/module/miniconda3
```

图 7-1 Miniconda 安装提示

（3）出现如图 7-2 所示的提示时，表示安装完成。

```
Thank you for installing Miniconda3!
```

图 7-2 Miniconda 安装完成提示

3）加载环境变量文件

```
[atguigu@hadoop102 ~]$ source ~/.bashrc
```

4）禁止激活默认的 base 环境

Miniconda 安装完成后，每次打开终端都会激活其默认的 base 环境，我们可通过以下命令，禁止激活默认的 base 环境。

```
[atguigu@hadoop102 ~]$ conda config --set auto_activate_base false
```

2．使用 Python 3.6 环境

（1）配置 Conda 国内镜像。

```
[atguigu@hadoop102 ~]$ conda config --add channels https://mirrors.tuna.tsinghua.edu.cn/anaconda/pkgs/free
[atguigu@hadoop102 ~]$ conda config --add channels https://mirrors.tuna.tsinghua.edu.cn/anaconda/pkgs/main
[atguigu@hadoop102 ~]$ conda config --set show_channel_urls yes
```

（2）创建 Python 3.6 环境。

```
[atguigu@hadoop102 ~]$ conda create --name superset python=3.6
```

Conda 环境管理器常用命令如下。

- 创建环境：conda create -n env_name。
- 查看所有环境：conda info --envs。
- 删除一个环境：conda remove -n env_name --all。

（3）激活 Superset 环境。

```
[atguigu@hadoop102 ~]$ conda activate superset
```

激活后的效果如图 7-3 所示。

```
(superset) [atguigu@hadoop102 ~]$
```

图 7-3 Superset 环境激活后的效果

（4）退出当前环境。

```
[atguigu@hadoop102 ~]$ conda deactivate
```

（5）执行 python 命令，查看 Python 版本，如图 7-4 所示。

```
(superset) [atguigu@hadoop102 ~]$ python
Python 3.6.10 |Anaconda, Inc.| (default, Jan  7 2020, 21:14:29)
[GCC 7.3.0] on linux
Type "help", "copyright", "credits" or "license" for more information.
>>> quit();
```

图 7-4　查看 Python 版本

7.2.2　Superset 安装

1．安装依赖

在安装 Superset 之前，需要先安装以下所需的依赖。

```
(superset) [atguigu@hadoop102 ~]$ sudo yum install -y python-setuptools
(superset) [atguigu@hadoop102 ~]$ sudo yum install -y gcc gcc-c++ libffi-devel python-devel python-pip python-wheel openssl-devel cyrus-sasl-devel openldap-devel
```

2．安装 Superset

1）安装（更新）setuptools 和 pip

```
(superset) [atguigu@hadoop102 ~]$ pip install --upgrade setuptools pip -i https://pypi.douban.com/simple/
```

说明：pip 是 Python 的包管理工具，与 CentOS 中的 yum 类似。

2）安装 Superset

```
(superset) [atguigu@hadoop102 ~]$ pip install apache-superset -i https://pypi.douban.com/simple/
```

说明：-i 的作用是指定镜像，这里选择国内镜像。

3）初始化 superset 数据库

```
(superset) [atguigu@hadoop102 ~]$ superset db upgrade
```

4）创建管理员用户

```
(superset) [atguigu@hadoop102 ~]$ export FLASK_APP=superset
(superset) [atguigu@hadoop102 ~]$ flask fab create-admin
```

说明：flask 是一个 Python Web 框架，Superset 使用的就是 flask。

5）初始化 Superset

```
(superset) [atguigu@hadoop102 ~]$ superset init
```

3．操作 Superset

1）安装 gunicorn

```
(superset) [atguigu@hadoop102 ~]$ pip install gunicorn -i https://pypi.douban.com/simple/
```

说明：gunicorn 是一个 Python Web Server，与 Java 中的 TomCat 类似。

2）启动 Superset

（1）确保当前 Conda 的环境为 Superset。

（2）启动。

```
(superset) [atguigu@hadoop102 ~]$ gunicorn --workers 5 --timeout 120 --bind hadoop102:8787 --daemon "superset.app:create_app()"
```

参数说明如下。

- --workers：指定进程个数。
- --timeout：Worker 进程超时时间，超时会自动重启。
- --bind：绑定本机地址，即 Superset 的访问地址。
- --daemon：后台运行。

3）停止运行 Superset

（1）停掉 gunicorn 进程。

```
(superset) [atguigu@hadoop102 ~]$ ps -ef | awk '/gunicorn/ && !/awk/{print $2}' | xargs kill -9
```

（2）退出 Superset 环境。

```
(superset) [atguigu@hadoop102 ~]$ conda deactivate
```

4）启动后登录 Superset

访问 http://hadoop102:8787，进入 Superset 登录页面，如图 7-5 所示，并使用前面创建的管理员账号进行登录。

图 7-5 Superset 登录页面

7.3 Superset 使用

7.3.1 对接 MySQL 数据源

1．安装依赖

```
(superset) [atguigu@hadoop102 ~]$ conda install mysqlclient
```

说明：对接不同的数据源，需要安装不同的依赖。

2．重启 Superset

1）停掉 gunicorn 进程

```
(superset) [atguigu@hadoop102 ~]$ ps -ef | awk '/gunicorn/ && !/awk/{print $2}' | xargs kill -9
```

2）启动 Superset

```
(superset) [atguigu@hadoop102 ~]$ gunicorn --workers 5 --timeout 120 --bind hadoop102:8787  --daemon "superset.app:create_app()"
```

3．数据源配置

1）Database 配置

Step1：选择"Sources"→"Databases"选项，如图 7-6 所示。

图 7-6　Database 配置入口

Step2：单击 ❸ 图标添加数据库，如图 7-7 所示。

图 7-7　添加数据库操作

Step3：填写"Database"及"SQLAlchemy URI"，如图 7-8 所示。

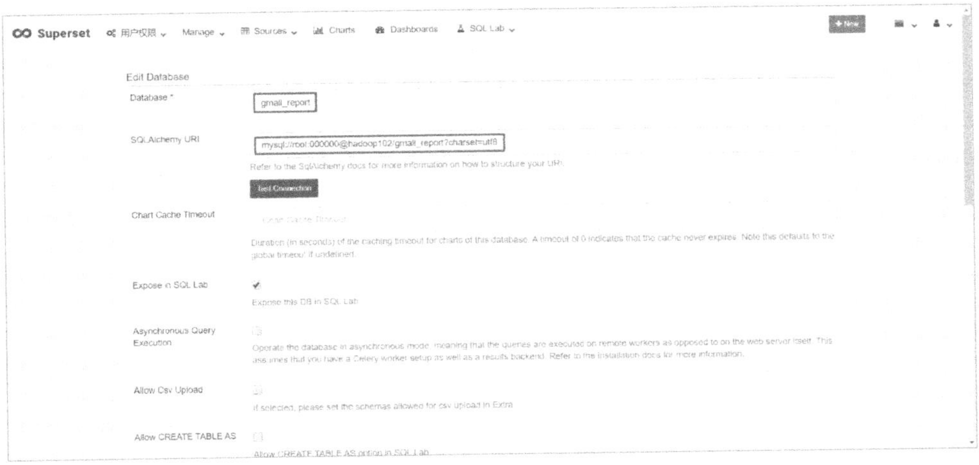

图 7-8　编辑 Database 相关配置

"SQLAlchemy URI"的编写规范为 mysql://账号:密码@服务器 IP 地址/数据库名称?charset=utf8。

Step4：单击"Test Connection"按钮，弹出"Seems Ok！"提示，表示连接成功，如图 7-9 所示。

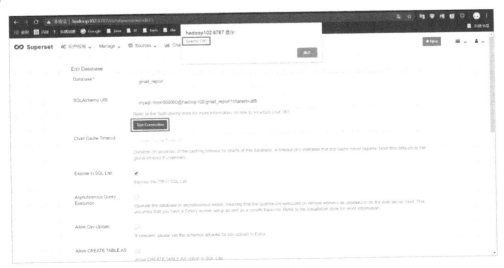

图 7-9　测试是否连接成功

Step5：保存配置，如图 7-10 所示。

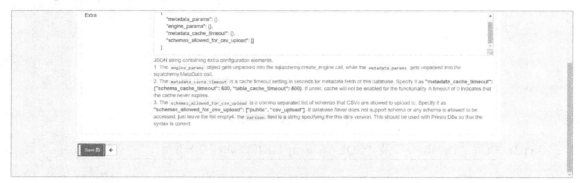

图 7-10　保存配置

2）Table 配置

Step1：选择"Sources"→"Tables"选项，如图 7-11 所示。

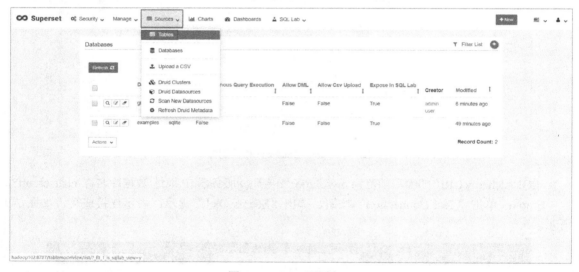

图 7-11　Table 配置入口

Step2：单击 ⊕ 图标添加 Table，如图 7-12 所示。

图 7-12　添加 Table 操作

Step3：配置 Table，如图 7-13 所示。

图 7-13　配置 Table

7.3.2　制作仪表盘

1．创建空白仪表盘

（1）选择"Dashboards"选项并单击 图标，如图 7-14 所示。

图 7-14　创建空白仪表盘入口

（2）配置仪表盘，如图 7-15 所示。

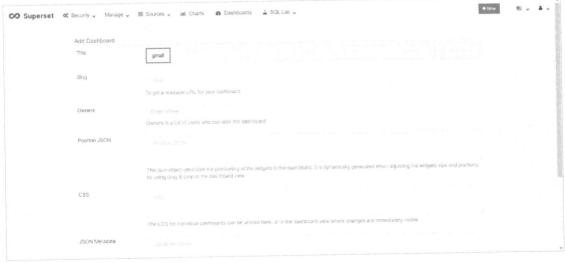

图 7-15　配置仪表盘

（3）保存仪表盘，如图 7-16 所示。

图 7-16　保存仪表盘

2. 创建图表

（1）选择"Charts"选项并单击图标，如图 7-17 所示。

图 7-17　创建图表入口

（2）选择数据源及图表类型，如图 7-18 所示。

图 7-18　选择数据源及图表类型

（3）选择合适的图表样式，如图 7-19 所示。

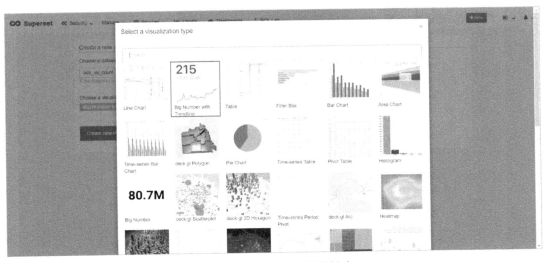

图 7-19　选择合适的图表样式

（4）创建图表，如图 7-20 所示。

图 7-20　创建图表

（5）可将语言修改为中文，以方便配置，如图 7-21 所示。

图 7-21　语言修改入口

（6）按照说明配置图表，如图 7-22 所示。

图 7-22　配置图表

（7）配置完成后，单击"Run Query"按钮，执行查询操作，如图 7-23 所示。

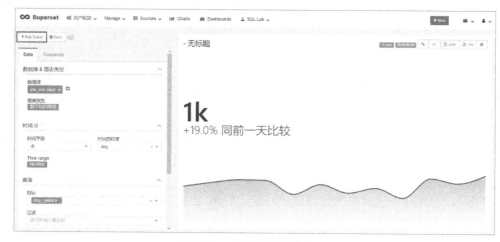

图 7-23　执行查询操作入口

（8）保存图表并将其添加到仪表盘中，如图7-24所示。

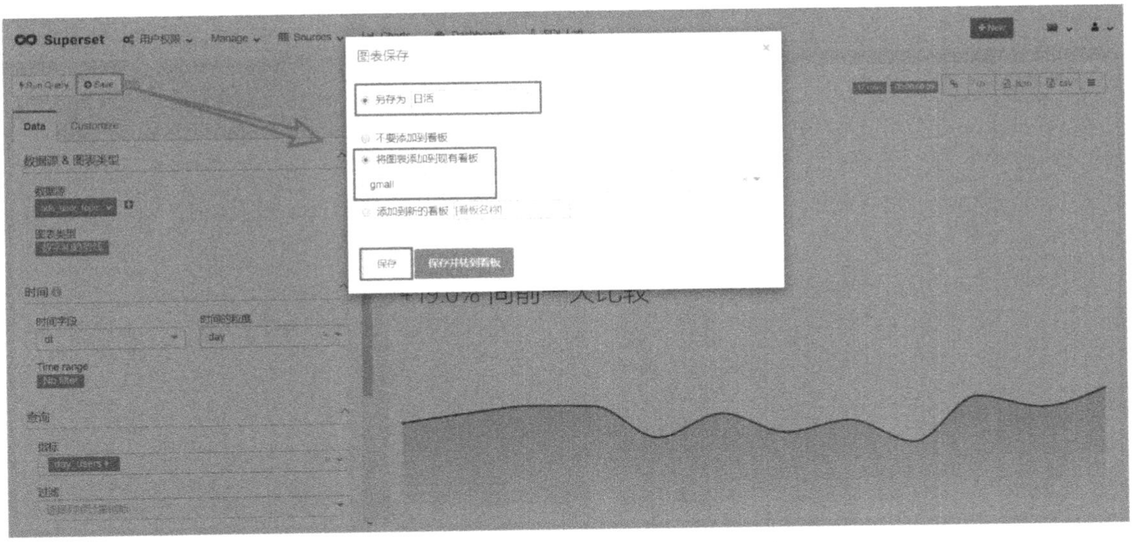

图7-24　保存图表并将其添加到仪表盘中

3．编辑仪表盘

（1）单击"Edit dashboard"按钮，编辑仪表盘，如图7-25所示。

图7-25　编辑仪表盘入口

（2）拖动图表，调整仪表盘布局，如图7-26所示。

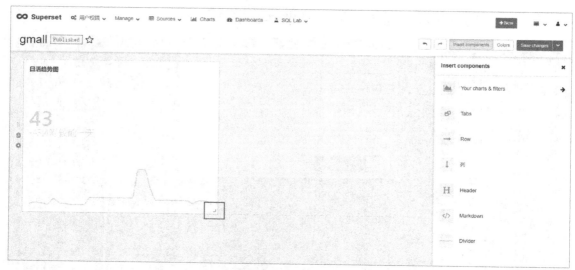

图 7-26　调整仪表盘布局

（3）单击"Edit dashboard"按钮右侧的下三角按钮，在弹出的下拉列表中选择"Auto_refresh dashboard"选项，可调整仪表盘的自动刷新时间，如图 7-27 所示。

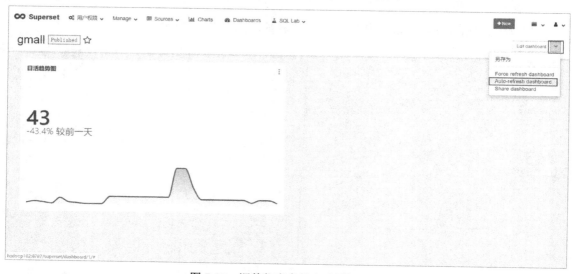

图 7-27　调整仪表盘的自动刷新时间

7.4　本章总结

本章对会员主题和地区主题 2 个主题指标进行了可视化，使读者对数据仓库的具体作用有了更形象的体会。数据仓库在实现数据需求，获取最终的结果数据之余，有的时候，也需要承担数据可视化的责任。业务实现代码不属于大数据范畴，此处不再赘述。数据仓库关键在于为可视化提供正确的结果数据。

第 8 章

即席查询模块

除了前几章讲解的需求实现，数据仓库系统还需要满足简单的即席查询需求。本章为读者讲解 3 个即席查询的框架，分别是 Presto、Druid 和 Kylin，这 3 个框架在大数据领域应用十分广泛，性能上各有千秋，下面分别讲解。

8.1 Presto

Presto 是 Facebook 推出的一个开源的分布式 SQL 查询引擎，数据规模可以支持 GB 到 PB 级，主要应用于处理秒级查询的场景。

注意：虽然 Presto 可以解析 SQL，但它不是一个标准的数据库，不是 MySQL、Oracle 的代替品，也不能用来处理在线事务（OLTP）。

8.1.1 Presto 特点

Presto 是一个 master-slave 架构，由一个 Coordinator 节点、一个 Discovery Server 节点、多个 Worker 节点组成。Discovery Server 通常内嵌于 Coordinator 节点中；Coordinator 节点负责解析 SQL 语句，生成执行计划，分发执行任务给 Worker 节点；Worker 节点负责执行查询任务。Worker 节点启动后向 Discovery Server 服务注册，Coordinator 节点从 Discovery Server 节点获得可以正常工作的 Worker 节点。如果配置了 Hive Connector，则需要配置一个 Hive Metastore 服务，为 Presto 提供 Hive 元信息，Worker 节点与 HDFS 交互读取数据，Presto 架构如图 8-1 所示。

Presto 基于内存运算，减少了磁盘 I/O 操作，计算速度更快。它能够连接多个数据源，并跨数据源进行连表查询，如从 Hive 中查询大量网站访问记录，然后从 MySQL 中匹配出设备信息。

虽然 Presto 能够处理 PB 级的海量数据，但 Presto 并不是把 PB 级的数据都放在内存中进行计算，而是根据场景，如 count、AVG 等聚合运算，边读数据边计算，再清内存，再读数据再计算的，这种方式消耗的内存并不高。如果多表连接，就可能产生大量的临时数据，因此，计算速度会变慢，此时使用 Hive 反而更好。

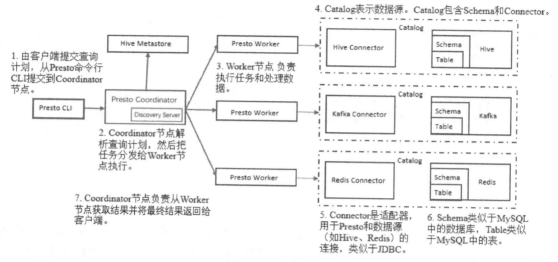

图 8-1 Presto 架构

8.1.2 Presto 安装

1. Presto Server 的安装

（1）下载 Presto Server 安装包。

（2）将下载的 presto-server-0.196.tar.gz 导入 hadoop102 的 /opt/software 目录下，并解压到 /opt/module 目录下。

```
[atguigu@hadoop102 software]$ tar -zxvf presto-server-0.196.tar.gz -C /opt/module
```

（3）修改 presto-server-0.196 的名称为 presto。

```
[atguigu@hadoop102 module]$ mv presto-server-0.196/ presto
```

（4）进入 /opt/module/presto 目录，创建存储数据的文件夹。

```
[atguigu@hadoop102 presto]$ mkdir data
```

（5）进入 /opt/module/presto 目录，创建存储配置文件的文件夹。

```
[atguigu@hadoop102 presto]$ mkdir etc
```

（6）在 /opt/module/presto/etc 目录下创建 jvm.config 配置文件。

```
[atguigu@hadoop102 etc]$ vim jvm.config
```

在配置文件中添加如下内容。

```
-server
-Xmx16G
-XX:+UseG1GC
-XX:G1HeapRegionSize=32M
-XX:+UseGCOverheadLimit
-XX:+ExplicitGCInvokesConcurrent
-XX:+HeapDumpOnOutOfMemoryError
```

```
-XX:+ExitOnOutOfMemoryError
```

（7）Presto 可以支持多个数据源，在 Presto 中称为 Catalog，这里我们配置支持 Hive 的数据源，配置一个 Hive 的 Catalog，需要新建一个目录 catalog，在 catalog 目录下新建文件 hive.properties。

```
[atguigu@hadoop102 etc]$ mkdir catalog
[atguigu@hadoop102 catalog]$ vim hive.properties
```

在文件中添加如下内容。

```
connector.name=hive-hadoop2
hive.metastore.uri=thrift://hadoop102:9083
```

（8）将 hadoop102 上的 presto 分发到 hadoop103、hadoop104。

```
[atguigu@hadoop102 module]$ xsync presto
```

（9）分发之后，分别进入 hadoop102、hadoop103、hadoop104 三台节点服务器的/opt/module/presto/etc 目录，配置 node 属性，每台节点服务器中 node 属性的 id 都不一样。

```
[atguigu@hadoop102 etc]$vim node.properties
node.environment=production
node.id=ffffffff-ffff-ffff-ffff-ffffffffffff
node.data-dir=/opt/module/presto/data

[atguigu@hadoop103 etc]$vim node.properties
node.environment=production
node.id=ffffffff-ffff-ffff-ffff-fffffffffffe
node.data-dir=/opt/module/presto/data

[atguigu@hadoop104 etc]$vim node.properties
node.environment=production
node.id=ffffffff-ffff-ffff-ffff-fffffffffffd
node.data-dir=/opt/module/presto/data
```

（10）Presto 是由一个 Coordinator 节点和多个 Worker 节点组成的。在 hadoop102 上配置为 Coordinator，在 hadoop103、hadoop104 上配置为 Worker。

① 在 hadoop102 上配置 Coordinator 节点。

```
[atguigu@hadoop102 etc]$ vim config.properties
```

添加内容如下。

```
coordinator=true
node-scheduler.include-coordinator=false
http-server.http.port=8881
query.max-memory=50GB
discovery-server.enabled=true
discovery.uri=http://hadoop102:8881
```

② 在 hadoop103 上配置 Worker 节点。

```
[atguigu@hadoop103 etc]$ vim config.properties
```

添加内容如下。

```
coordinator=false
http-server.http.port=8881
query.max-memory=50GB
discovery.uri=http://hadoop102:8881
```

在 hadoop104 上配置 Worker 节点。

```
[atguigu@hadoop104 etc]$ vim config.properties
```

添加内容如下。

```
coordinator=false
http-server.http.port=8881
query.max-memory=50GB
discovery.uri=http://hadoop102:8881
```

（11）在 hadoop102 的/opt/module/hive 目录下，以 atguigu 角色启动 Hive Metastore。

```
[atguigu@hadoop102 hive]$
nohup bin/hive --service metastore >/dev/null 2>&1 &
```

（12）分别在 hadoop102、hadoop103、hadoop104 上启动 Presto Server。

① 在前台启动 Presto，控制台显示的日志如下。

```
[atguigu@hadoop102 presto]$ bin/launcher run
[atguigu@hadoop103 presto]$ bin/launcher run
[atguigu@hadoop104 presto]$ bin/launcher run
```

② 在后台启动 Presto。

```
[atguigu@hadoop102 presto]$ bin/launcher start
[atguigu@hadoop103 presto]$ bin/launcher start
[atguigu@hadoop104 presto]$ bin/launcher start
```

（13）查看日志的路径为/opt/module/presto/data/var/log。

2. Presto 命令行客户端的安装

（1）下载 Presto 的客户端。

（2）将 presto-cli-0.196-executable.jar 上传到 hadoop102 的/opt/module/presto 目录下。

（3）修改文件名称。

```
[atguigu@hadoop102 presto]$ mv presto-cli-0.196-executable.jar  prestocli
```

（4）增加执行权限。

```
[atguigu@hadoop102 presto]$ chmod +x prestocli
```

（5）执行 prestocli 命令，启动 Presto 客户端。

```
[atguigu@hadoop102 presto]$ ./prestocli --server hadoop102:8881 --catalog hive --schema default
```

（6）Presto 命令行操作。

Presto 的命令行操作相当于 Hive 的命令行操作。每张表必须加上 Schema。

例如：

```
select * from schema.table limit 100
```

3. Presto 可视化客户端安装

（1）将 yanagishima-18.0.zip 上传到 hadoop102 的/opt/module 目录下。

（2）解压 yanagishima-18.0.zip。

```
[atguigu@hadoop102 module]$ unzip yanagishima-18.0.zip
cd yanagishima-18.0
```

（3）进入/opt/module/yanagishima-18.0/conf 目录，创建 yanagishima.properties 配置文件。

```
[atguigu@hadoop102 conf]$ vim yanagishima.properties
```

在配置文件中添加如下内容。

```
jetty.port=7080
presto.datasources=atiguigu-presto
presto.coordinator.server.atiguigu-presto=http://hadoop102:8881
catalog.atiguigu-presto=hive
schema.atiguigu-presto=default
sql.query.engines=presto
```

（4）在/opt/module/yanagishima-18.0 目录下执行以下命令启动 Presto。

```
[atguigu@hadoop102 yanagishima-18.0]$ nohup bin/yanagishima-start.sh >y.log 2>&1 &
```

（5）启动 Web 页面（http://hadoop102:7080），即可查询相关信息。

（6）查看 Presto 表结构，如图 8-2 所示。

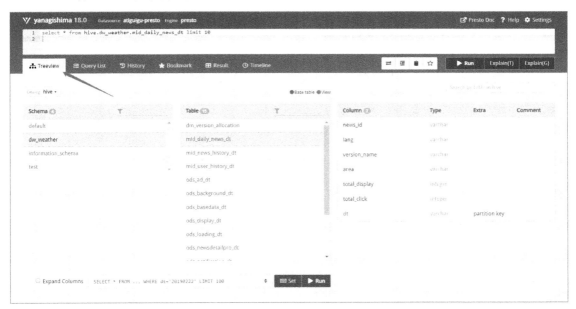

图 8-2　查看 Presto 表结构

在"Treeview"页面下可以查看所有表的结构，包括 Schema、Table、Column 等。

比如，执行 select * from hive.dw_weather.mid_daily_news_dt limit 10。

每张表后面都有一个复制图标，单击此图标会复制完整的表名，然后在上面的文本框中输入 SQL 语句即可，如图 8-3 所示。

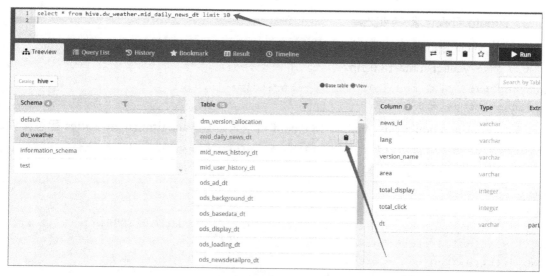

图 8-3　Presto 复制表名输入 SQL 语句

还可以查询列表中其他的表格，比如，执行 select * from hive.dw_weather.tmp_news_click limit 10，按 Ctrl+Enter 组合键显示查询结果，如图 8-4 所示。

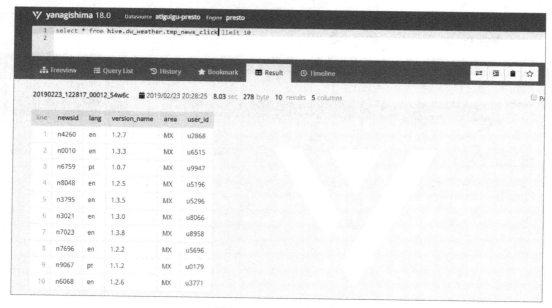

图 8-4　Presto 查询结果

8.1.3　Presto 优化之数据存储

想要使用 Presto 更高效地查询数据，需要在数据存储方面利用一些优化手段。

1．合理设置分区

与 Hive 类似，Presto 会根据元数据信息读取分区数据，合理地设置分区能减少 Presto 数据读取量，提升查询性能。

2．使用 ORC 格式存储

Presto 对 ORC 文件读取进行了特定优化，因此，在 Hive 中创建 Presto 使用的表时，建议采用 ORC 格式存储。相对于 Parquet 格式，Presto 对 ORC 格式支持得更好。

3．使用压缩

对数据进行压缩可以减少节点服务器间的数据传输对 I/O 带宽产生的压力，即席查询需要快速解压，建议采用 Snappy 格式压缩。

8.1.4　Presto 优化之查询 SQL

想要使用 Presto 更高效地查询数据，需要在编写查询 SQL 语句方面利用一些优化手段。

1．只选择需要的字段

由于采用列式存储，所以只选择需要的字段可加快字段的读取速度，减少数据量。避免采用*读取所有字段。

```
[GOOD]: select time, user, host from tbl

[BAD]:  select * from tbl
```

2．过滤条件必须加上分区字段

对于有分区的表，where 语句中优先使用分区字段进行过滤。acct_day 是分区字段，visit_time 是具体访问时间。

```
[GOOD]: select time, user, host from tbl where acct_day=20171101

[BAD]:  select * from tbl where visit_time=20171101
```

3．group by 语句优化

合理安排 group by 语句中字段的执行顺序，对性能有一定的提升。将 group by 语句中的字段按照每个字段去重后的数据量进行降序排列。

```
[GOOD]: select group by uid, gender

[BAD]:  select group by gender, uid
```

4．order by 时使用 limit

order by 时需要扫描数据到单个 Worker 节点进行排序，导致单个 Worker 节点需要占用大量内存。如果是查询 Top *N* 或者 Bottom *N*，则使用 limit 可减少排序计算时间和内存压力。

```
[GOOD]: select * from tbl order by time limit 100

[BAD]:  select * from tbl order by time
```

5．使用 join 语句时将大表放在左边

Presto 中 join 的默认算法是 broadcast join，即将 join 左边表的数据分割到多个 Worker 节

点，然后将 join 右边表的数据整个复制一份发送到每个 Worker 节点进行计算。如果右边的表数据量太大，则可能会报内存溢出错误。

```
[GOOD] select ... from large_table l join small_table s on l.id = s.id
[BAD]  select ... from small_table s join large_table l on l.id = s.id
```

8.1.5 Presto 注意事项

使用 Presto 需要注意如下几项。

1．字段名引用

避免字段名与关键字冲突：MySQL 对与关键字冲突的字段名加反引号，Presto 对与关键字冲突的字段名加双引号。当然，如果字段名不与关键字冲突，则可以不加双引号。

2．时间函数

在 Presto 中比较时间的时候，需要添加 timestamp 关键字，而在 MySQL 中可以直接进行比较。

```
/*MySQL 中的写法*/
select t from a where t > '2017-01-01 00:00:00';

/*Presto 中的写法*/
select t from a where t > timestamp '2017-01-01 00:00:00';
```

3．不支持 insert overwrite 语法

Presto 中不支持 insert overwrite 语法，只能先删除数据，然后插入数据。

4．Parquet 格式

Presto 目前支持 Parquet 格式，支持查询，但不支持 insert。

8.2　Druid

Druid 是一个高性能的实时分析数据库。它在 PB 级数据处理、毫秒级查询、数据实时处理方面，比传统的 OLAP 系统有显著的性能提升。

8.2.1　Druid 简介

Druid 具有如下技术特点。
- 列式存储格式。Druid 使用列式存储格式，它只需要加载特定查询所需要的列。查询速度快。
- 可扩展的分布式系统。Druid 通常部署在数十到数百台服务器的集群中，并且提供数百万条/秒的摄取率，保留数百万条记录，以及具有亚秒到几秒的查询延迟。
- 大规模的并行查询。Druid 可以在整个集群中进行大规模的并行查询。

- 实时摄取或批量处理。Druid 可以实时摄取数据（实时获取的数据可立即用于查询）或批量处理数据。
- 自愈、自平衡、易操作。集群的扩展和缩小，只需添加或删除服务器，集群将在后台自动重新平衡，无须停机。
- 数据有效地进行了预聚合或预计算，查询速度快。
- 结果数据应用了 Bitmap 压缩技术。

Druid 适用于如下几种场景。
- 适用于将清洗好的记录实时录入，但不需要更新操作的场景。
- 适用于支持宽表，不用 join 的场景（换句话说就是一张单表）。
- 适用于实时性要求高的场景。
- 适用于对数据质量的敏感度不高的场景。

8.2.2 Druid 框架原理

Druid 总体架构如图 8-5 所示。

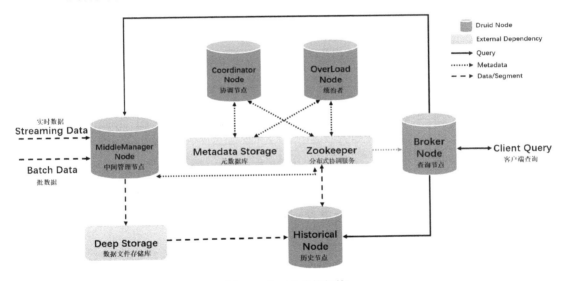

图 8-5　Druid 总体架构

Druid 的总体架构包含如下四类节点。
- 中间管理节点（MiddleManager Node）：及时摄入实时数据，并生成 Segment 数据文件。
- 历史节点（Historical Node）：加载已生成的数据文件，以供用户查询数据。
- 查询节点（Broker Node）：对外提供数据查询服务，并同时从中间管理节点与历史节点中查询数据，合并后返回调用方。
- 协调节点（Coordinator Node）：负责历史节点的数据负载均衡，以及通过规则（Rule）管理数据的生命周期。

集群还包含如下三类外部依赖。
- 元数据库（Metadata Storage）：存储 Druid 集群的元数据信息，比如，Segment 的相关信息，一般使用 MySQL 或 PostgreSQL 存储。
- 分布式协调服务（Zookeeper）：为 Druid 集群提供一致性协调服务的组件，通常为 Zookeeper。
- 数据文件存储库（Deep Storage）：存放生成的 Segment 数据文件，并提供历史服务器下载功能，对于单节点集群，可以是本地磁盘，而对于分布式集群，一般是 HDFS 或 NFS。

8.2.3 Druid 数据结构

基于 DataSource 和 Segment 的 Druid 数据结构与 Druid 架构相辅相成，它们共同成就了 Druid 的高性能优势。

DataSource 相当于关系型数据库中的表（Table）。DataSource 的结构如下。
- 时间列：表明每行数据的时间值，默认使用 UTC 时间格式且精确到毫秒级。
- 维度列：维度来自 OLAP 的概念，用来标识数据行的各个类别信息。
- 指标列：用于聚合和计算的列。通常是一些数字，计算操作包括 Count、Sum 等。

DataSource 结构如表 8-1 所示。

表 8-1 DataSource 结构

时间列（Timestamp）	维度列		指标列（click）
	publisher	country	
2019-01-01T01:01:35Z	www.atguigu.com	China	0
2019-01-01T01:03:35Z	www.atguigu.com	China	1
2019-01-01T02:05:35Z	www.baidu.com	China	1
2019-01-01T02:07:35Z	www.baidu.com	China	0
2019-01-01T03:09:35Z	www.google.com	USA	1
2019-01-01T03:10:35Z	www.google.com	USA	0

无论是实时摄取数据还是批量处理数据，Druid 在基于 DataSource 结构存储数据时，可选择对任意的指标列进行聚合操作。该聚合操作主要基于维度列与时间列。表 8-2 显示的是 DataSource 聚合后的数据结构。

表 8-2 DataSource 聚合后的数据结构

时间列（Timestamp）	维度列		指标列（click）
	publisher	country	
2019-01-01T01:03:35Z	www.atguigu.com	China	1
2019-01-01T02:05:35Z	www.baidu.com	China	1
2019-01-01T03:09:35Z	www.google.com	USA	1

在数据存储时便可对数据进行聚合操作是 Druid 的特点，该特点使得 Druid 不仅能够节省存储空间，而且能够提高聚合查询的效率。

DataSource 是一个逻辑概念，Segment 是数据的实际物理存储格式。Druid 将不同时间范围内的数据存储在不同的 Segment 数据块中，这便是所谓的数据横向切割。按照时间横向切割数据，避免了全表查询，极大地提高了效率。

在 Segment 中，采用列式存储格式对数据进行压缩存储（Bitmap 压缩技术），这便是所谓的数据纵向切割。

8.2.4 Druid 安装（单机版）

1. 安装部署

从 imply 页面下载 Druid 最新版本的安装包，imply 集成了 Druid，提供了 Druid 从部署到配置再到各种可视化工具的完整解决方案。

（1）将 imply-2.7.10.tar.gz 上传到 hadoop102 的/opt/software 目录下，并解压到/opt/module 目录下。

```
[atguigu@hadoop102 software]$ tar -zxvf imply-2.7.10.tar.gz -C /opt/module
```

（2）修改 imply-2.7.10 的名称为 imply。

```
[atguigu@hadoop102 module]$ mv imply-2.7.10 imply
```

（3）修改配置文件。

① 修改 Druid 的 ZK 配置。

```
[atguigu@hadoop102 _common]$ vim /opt/module/imply/conf/druid/_common/common.runtime.properties
```

需要修改的内容如下。

```
druid.zk.service.host=hadoop102:2181,hadoop103:2181,hadoop104:2181
```

② 修改启动命令参数，使其不校验、不启动内置 ZK。

```
[atguigu@hadoop102 supervise]$ vim /opt/module/imply/conf/supervise/quickstart.conf
```

需要修改的内容如下。

```
:verify bin/verify-java
#:verify bin/verify-default-ports
#:verify bin/verify-version-check
:kill-timeout 10

#!p10 zk bin/run-zk conf-quickstart
```

（4）启动。

① 启动 Zookeeper。

```
[atguigu@hadoop102 imply]$ zk.sh start
```

② 通过 bin 目录下的 supervise 命令启动 imply。

```
[atguigu@hadoop102 imply]$ bin/supervise -c conf/supervise/quickstart.conf
```

说明：每启动一个服务均会打印出一条日志。读者可以在/opt/module/imply/var/sv/目录下查看服务启动时的日志信息。

③ 启动采集 Flume 和 Kafka（主要是为了节省内存开销，同时将 hadoop102 的内存调整为 8GB）。

```
[atguigu@hadoop102 imply]$ fl.sh start
[atguigu@hadoop102 imply]$ kf.sh start
```

按 Ctrl+C 组合键中断监督进程，如果想在中断服务后重新启动，则需要删除/opt/module/imply/var 目录。

2．Web 页面使用

（1）启动日志生成程序（延时 1s 发送一条日志）。

```
[atguigu@hadoop102 server]$ lg.sh 1000 5000
```

（2）在浏览器地址栏中输入 hadoop102:9095/datasets/，打开 imply Web UI 页面，如图 8-6 所示。

图 8-6　imply Web UI 页面

（3）单击"Load data"按钮，然后单击"Apache Kafka"按钮，如图 8-7 所示。

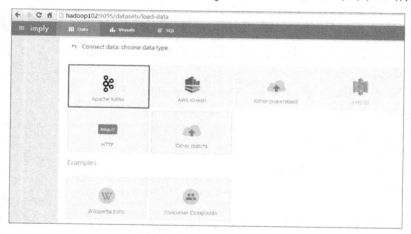

图 8-7　加载 Kafka 源数据

（4）添加 Kafka 集群和主题信息，如图 8-8 所示。

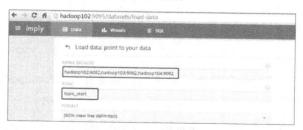

图 8-8　添加 Kafka 集群和主题信息

（5）确认数据样本格式，如图 8-9 所示。

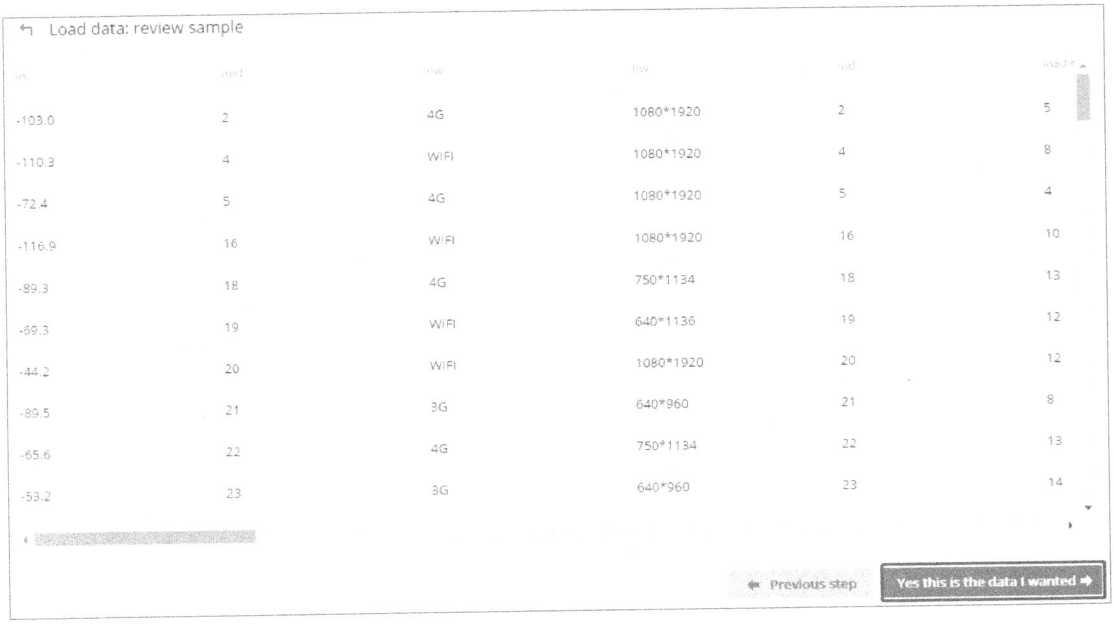

图 8-9　确认数据样本格式

（6）加载数据，必须有时间字段，如图 8-10 所示。

图 8-10　加载数据设置

（7）配置要加载的列，如图 8-11 所示。

图 8-11 配置要加载的列

（8）配置 Kafka 数据的主题名称，如图 8-12 所示。

图 8-12 配置 Kafka 数据的主题名称

（9）确认加载数据的配置，如图 8-13 所示。

图 8-13 确认加载数据的配置

（10）连接 Kafka 的 topic_start，如图 8-14 和图 8-15 所示。

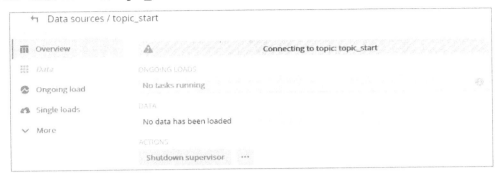

图 8-14　连接 Kafka 中

图 8-15　连接 Kafka 成功

（11）选择"SQL"，如图 8-16 所示。

图 8-16　选择"SQL"选项

（12）查询指标，如图 8-17 所示。

图 8-17　查询指标

8.3　Kylin

Apache Kylin 是一个开源的分布式分析引擎，提供 Hadoop/Spark 之上的 SQL 查询接口及多维分析功能，支持超大规模数据，最初由 eBay 开发并贡献至开源社区。它能在亚秒内查询巨大的 Hive 表。

8.3.1　Kylin 简介

Kylin 架构如图 8-18 所示。

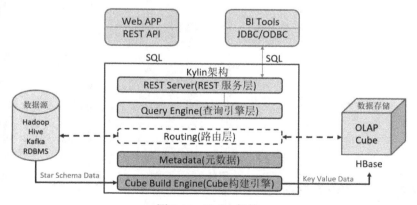

图 8-18　Kylin 架构

Kylin 具有如下几个关键组件。

1. REST Server

REST Server 是面向应用程序开发的入口点,旨在完成针对 Kylin 平台的应用开发工作,可以提供查询、获取结果、触发 Cube 构建任务、获取元数据及获取用户权限等功能,另外可以通过 REST API 接口实现 SQL 查询。

2. Query Engine

当 Cube 准备就绪后,查询引擎即可获取并解析用户所查询的问题。它随后会与系统中的其他组件进行交互,从而向用户返回对应的结果。

3. Routing

在最初设计时,设计者曾考虑将 Kylin 不能执行的查询引导到 Hive 中继续执行,但在实践后发现,Hive 与 Kylin 的查询速度差异过大,导致用户无法对查询的速度有一致的期望,大多数查询几秒内就返回结果了,而有些查询则要等几分钟到几十分钟,因此,用户体验非常糟糕。最后这个路由功能在发行版中被默认关闭。

4. Metadata

Kylin 是一款元数据驱动型应用程序。元数据管理工具是一大关键性组件,用于对保存在 Kylin 中的所有元数据进行管理,其中包括最为重要的 Cube 元数据。其他组件的正常运作都需要以元数据管理工具为基础。Kylin 的元数据存储在 HBase 中。

5. Cube Build Engine

Cube 构建引擎的设计目的在于处理所有离线任务,其中包括 Shell 脚本、Java API、Map Reduce 任务等。Cube Build Engine 对 Kylin 中的全部任务加以管理与协调,从而确保每项任务都能得到切实执行并解决期间出现的故障。

Kylin 的主要特点及说明如下。

- 支持标准 SQL 接口:Kylin 以标准的 SQL 作为对外服务的接口。
- 支持超大规模数据集:Kylin 对大数据的支撑能力是目前所有技术中较为领先的。早在 2015 年 eBay 的生产环境中就能支持百亿条记录的秒级查询,之后在移动的应用场景中又有了支持千亿条记录的秒级查询的案例。
- 亚秒级响应:Kylin 拥有优异的查询响应速度,这点得益于预计算,很多复杂的计算,比如,连接、聚合,在离线的预计算过程中就已经完成,这大大降低了查询时刻所需的计算量,提高了响应速度。
- 高伸缩性和高吞吐率:单节点 Kylin 可实现每秒 70 个查询,还可以搭建 Kylin 的集群。
- BI 工具集成。Kylin 可以与现有的 BI 工具集成。

Kylin 开发团队还贡献了 Zeppelin 的插件,用户可以使用 Zeppelin 访问 Kylin 服务。

8.3.2 HBase 安装

在安装 Kylin 前需要先安装部署好 Hadoop、Hive、Zookeeper 和 HBase,并且需要在/etc/profile 目录下配置 HADOOP_HOME、HIVE_HOME、HBASE_HOME 环境变量,注意执行 source/etc/profile 命令使其生效。

（1）保证 Zookeeper 集群的正常部署，并启动它。

```
[atguigu@hadoop102 zookeeper-3.4.10]$ bin/zkServer.sh start
[atguigu@hadoop103 zookeeper-3.4.10]$ bin/zkServer.sh start
[atguigu@hadoop104 zookeeper-3.4.10]$ bin/zkServer.sh start
```

（2）保证 Hadoop 集群的正常部署，并启动它。

```
[atguigu@hadoop102 hadoop-2.7.2]$ sbin/start-dfs.sh
[atguigu@hadoop103 hadoop-2.7.2]$ sbin/start-yarn.sh
```

（3）解压 HBase 安装包到指定目录。

```
[atguigu@hadoop102 software]$ tar -zxvf hbase-1.3.1-bin.tar.gz -C /opt/module
```

（4）修改 HBase 对应的配置文件。

① hbase-env.sh 文件的修改内容如下。

```
export JAVA_HOME=/opt/module/jdk1.8.0_144
export HBASE_MANAGES_ZK=false
```

② hbase-site.xml 文件的修改内容如下。

```xml
<configuration>
 <property>
  <name>hbase.rootdir</name>
  <value>hdfs://hadoop102:9000/hbase</value>
 </property>

 <property>
  <name>hbase.cluster.distributed</name>
  <value>true</value>
 </property>

<!-- 0.98 版本后新变动的内容,之前版本没有.port,默认端口为 60000 -->
 <property>
  <name>hbase.master.port</name>
  <value>16000</value>
 </property>

 <property>
  <name>hbase.zookeeper.quorum</name>
     <value>hadoop102,hadoop103,hadoop104</value>
 </property>

 <property>
  <name>hbase.zookeeper.property.dataDir</name>
     <value>/opt/module/zookeeper-3.4.10/zkData</value>
 </property>
</configuration>
```

③ 在 regionservers 文件中增加如下内容。

```
hadoop102
```

```
hadoop103
hadoop104
```

④ 软连接 Hadoop 配置文件到 HBase。

```
[atguigu@hadoop102 module]$ ln -s /opt/module/hadoop-2.7.2/etc/hadoop/core-site.xml
/opt/module/hbase/conf/core-site.xml
[atguigu@hadoop102 module]$ ln -s /opt/module/hadoop-2.7.2/etc/hadoop/hdfs-site.xml
/opt/module/hbase/conf/hdfs-site.xml
```

⑤ 将 HBase 远程发送到其他集群。

```
[atguigu@hadoop102 module]$ xsync hbase/
```

（5）启动 HBase 服务。

① 启动方式 1，语句如下。

```
[atguigu@hadoop102 hbase]$ bin/hbase-daemon.sh start master
[atguigu@hadoop102 hbase]$ bin/hbase-daemon.sh start regionserver
```

提示：如果集群之间的节点时间不同步，会导致 regionserver 无法启动，并抛出 ClockOutOfSyncException 异常。读者可参考 3.3.1 节中的相关内容。

② 启动方式 2，语句如下。

```
[atguigu@hadoop102 hbase]$ bin/start-hbase.sh
```

（6）停止 HBase 服务，语句如下。

```
[atguigu@hadoop102 hbase]$ bin/stop-hbase.sh
```

（7）查看 HBase 页面。

HBase 服务启动成功后，可以通过"host:port"的方式来访问 HBase 页面：http://hadoop102:16010。

8.3.3 Kylin 安装

（1）下载 Kylin 安装包。

（2）解压 apache-kylin-2.5.1-bin-hbase1x.tar.gz 到/opt/module/目录下。

```
[atguigu@hadoop102 sorfware]$ tar -zxvf apache-kylin-2.5.1-bin-hbase1x.tar.gz -C /opt/module/
```

注意：启动前需检查/etc/profile 目录中的 HADOOP_HOME、HIVE_HOME 和 HBASE_HOME 是否配置完毕。

（3）启动。

① 在启动 Kylin 之前，需要先启动 Hadoop（HDFS、YARN、JobHistoryServer）、Zookeeper 和 HBase。需要注意的是，要同时启动 Hadoop 的历史服务器，对 Hadoop 集群配置进行如下修改。

- 配置 mapred-site.xml 文件。

```
[atguigu@hadoop102 hadoop]$ vim mapred-site.xml
<property>
    <name>mapreduce.jobhistory.address</name>
    <value>hadoop102:10020</value>
</property>
<property>
    <name>mapreduce.jobhistory.webapp.address</name>
    <value>hadoop102:19888</value>
</property>
```

- 配置 yarn-site.xml 文件。

```
[atguigu@hadoop102 hadoop]$ vim yarn-site.xml
<!-- 日志聚集功能开启 -->
<property>
    <name>yarn.log-aggregation-enable</name>
    <value>true</value>
</property>
<!-- 日志保留时间设置为 7 天 -->
<property>
    <name>yarn.log-aggregation.retain-seconds</name>
    <value>604800</value>
</property>
```

- 修改配置后，分发配置文件，重启 Hadoop 的 HDFS 和 YARN 的所有进程。
- 启动 Hadoop 的历史服务器。

```
[atguigu@hadoop102 hadoop-2.7.2]$ sbin/mr-jobhistory-daemon.sh start historyserver
```

② 启动 Kylin。

```
[atguigu@hadoop102 kylin]$ bin/kylin.sh start
```

启动之后查看各台节点服务器的进程。

```
-------------------- hadoop102 ----------------
3360 JobHistoryServer
31425 HMaster
3282 NodeManager
3026 DataNode
53283 Jps
2886 NameNode
44007 RunJar
2728 QuorumPeerMain
31566 HRegionServer
-------------------- hadoop103 ----------------
5040 HMaster
2864 ResourceManager
9729 Jps
2657 QuorumPeerMain
4946 HRegionServer
```

```
2979 NodeManager
2727 DataNode
-------------------- hadoop104 ----------------
4688 HRegionServer
2900 NodeManager
9848 Jps
2636 QuorumPeerMain
2700 DataNode
2815 SecondaryNameNode
```

在浏览器地址栏中输入 hadoop102:7070/kylin/login，查看 Web 页面，如图 8-19 所示。

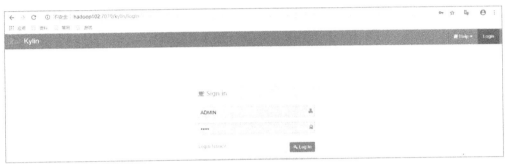

图 8-19　Kylin 的 Web 页面

用户名为 ADMIN，密码为 KYLIN（系统已填）。

（4）关闭 Kylin。

```
[atguigu@hadoop102 kylin]$ bin/kylin.sh stop
```

8.3.4　Kylin 使用

以 gmall 数据库中的 dwd_fact_payment_info 表作为事实表，以 dwd_dim_base_province、dwd_dim_user_info_his 表作为维度表，构建星形模型，并演示如何使用 Kylin 进行 OLAP 分析。

1．创建工程

（1）单击 ■ 图标创建工程，如图 8-20 所示。

图 8-20　创建工程入口

(2）填写工程名称和描述信息，并单击"Submit"按钮提交，如图 8-21 所示。

图 8-21　填写工程名称和描述信息并提交

2．获取数据源

（1）选择"Data Source"选项卡，如图 8-22 所示。

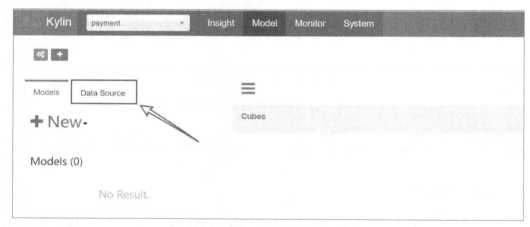

图 8-22　选择"Data Source"选项卡

（2）单击图 8-23 中箭头所指的图标，导入 Hive 表。

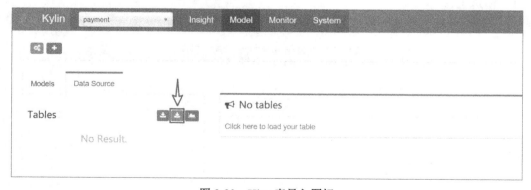

图 8-23　Hive 表导入图标

（3）选择所需数据表，并单击"Sync"按钮，如图 8-24 所示。

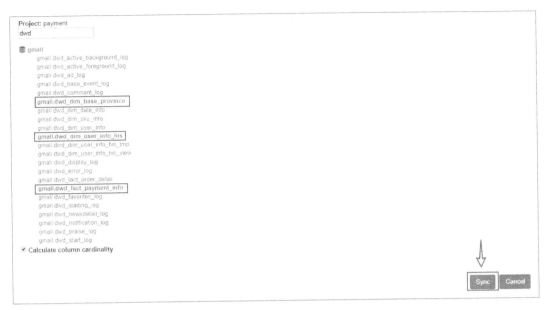

图 8-24 选择表格并同步

3．创建 Model

（1）选择"Models"选项卡，然后单击"New"按钮，接着单击"New Model"按钮，如图 8-25 所示。

图 8-25 创建 Model 示意

（2）填写 Model 信息，然后单击"Next"按钮，如图 8-26 所示。

图 8-26 填写 Model 信息

（3）选择事实表，如图 8-27 所示。

图 8-27　选择事实表

（4）选择维度表，并指定事实表和维度表的连接条件，然后单击"OK"按钮，如图 8-28 所示。

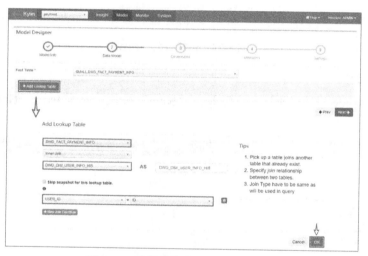

图 8-28　选择维度表并指定连接条件

维度表添加完毕之后，单击"Next"按钮，如图 8-29 所示。

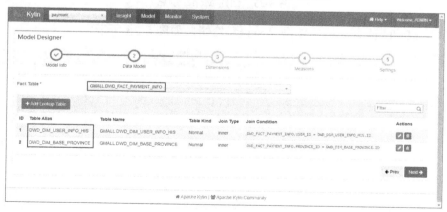

图 8-29　维度表添加完毕

(5)指定维度字段,并单击"Next"按钮,如图 8-30 所示。

图 8-30　指定维度字段

(6)指定度量字段,并单击"Next"按钮,如图 8-31 所示。

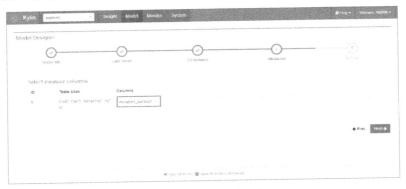

图 8-31　指定度量字段

(7)指定事实表分区字段(仅支持时间分区),并单击"Save"按钮,如图 8-32 所示,Model 创建完毕。

图 8-32　指定事实表分区字段

4．构建 Cube

（1）单击"New Cube"按钮，如图 8-33 所示。

图 8-33　构建 Cube 入口

（2）填写 Cube 信息，选择 Cube 所依赖的 Model，并单击"Next"按钮，如图 8-34 所示。

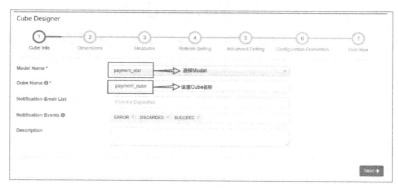

图 8-34　选择 Model 并设置 Cube 名称

（3）选择 Cube 所需的维度，如图 8-35 所示。

图 8-35　选择 Cube 所需的维度

（4）选择 Cube 所需的度量值，如图 8-36 所示。

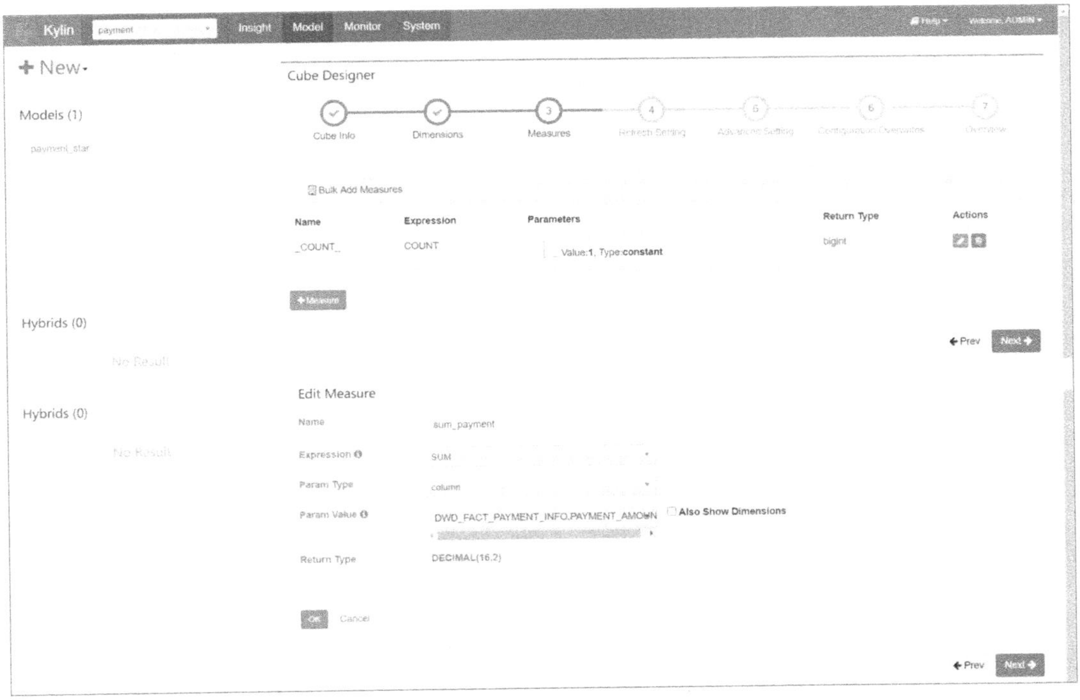

图 8-36　选择 Cube 所需的度量值

（5）Cube 自动合并设置。每天 Cube 需按照日期分区字段进行构建，每次构建的结果会保存到 HBase 的一张表中，为提高查询效率，需将每日的 Cube 进行合并，此处可设置合并周期，如图 8-37 所示。

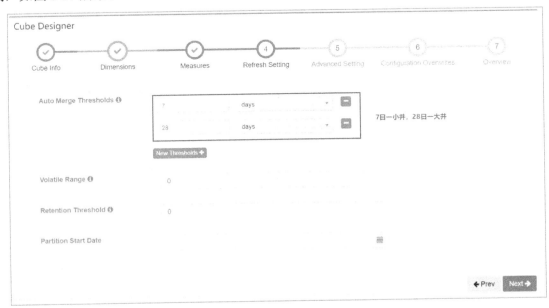

图 8-37　设置合并周期

（6）Kylin 高级配置（优化相关配置，暂时跳过），如图 8-38 所示。

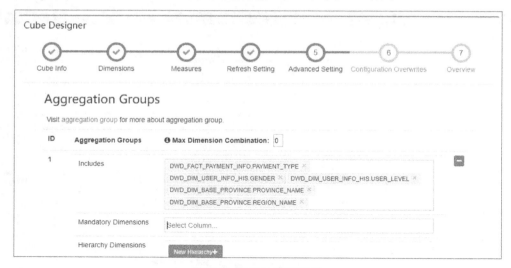

图 8-38　Kylin 高级配置

（7）Kylin 属性值覆盖相关配置，如图 8-39 所示。

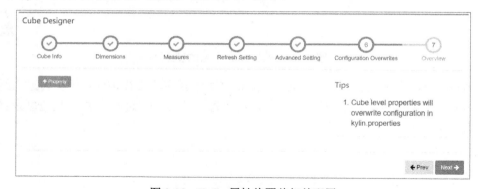

图 8-39　Kylin 属性值覆盖相关配置

（8）Cube 设计信息总览，如图 8-40 所示，单击"Save"按钮，Cube 创建完成。

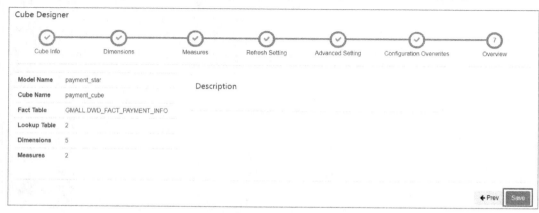

图 8-40　Cube 设计信息总览

（9）构建 Cube（计算），单击对应 Cube 的"Action"下拉按钮，选择"Build"选项，如图 8-41 所示。

图 8-41　构建 Cube（计算）

（10）选择要构建的时间区间，并单击"Submit"按钮，如图 8-42 所示。

图 8-42　选择要构建的时间区间

（11）选择"Monitor"，查看构建进度，如图 8-43 所示。

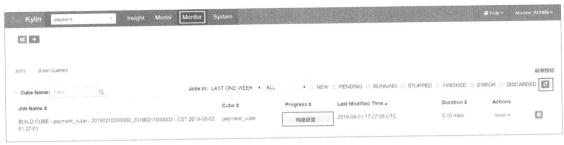

图 8-43　查看构建进度

5．使用进阶

1）如何处理每日全量维度表

如果按照上述流程构建 Cube，则会出现如图 8-44 所示的错误。

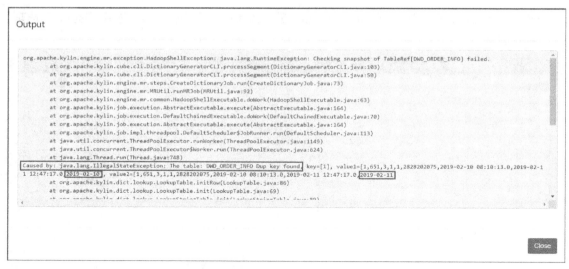

图 8-44 构建流程报错

错误原因分析如下。

出现上述错误的原因是 Model 中的维度表 dwd_dim_user_info_his 为拉链表，所以使用整张表作为维度表，必然会出现同一个 user_id 对应多条数据的问题。针对上述问题，有以下两种解决方案。

方案一：在 Hive 中创建维度表的临时表，并在该临时表中存放前一天的分区数据，在 Kylin 中创建模型时选择该临时表作为维度表。

方案二：与方案一思路相同，但不使用物理临时表，而使用视图（view）实现相同的功能。

此处采用方案二。

（1）创建维度表视图（使用视图获取前一天的分区数据）。

```
-- 拉链维度表视图
create view dwd_dim_user_info_his_view as select * from dwd_dim_user_info_his
where end_date='9999-99-99';

-- 全量维度表视图
create view dwd_dim_sku_info_view as select * from dwd_dim_sku_info where
dt=date_add(current_date,-1);
```

（2）在 DataSource 中导入新创建的视图，如图 8-45 所示，可选择将之前的维度表删除。

图 8-45 导入新创建的视图

（3）重新创建 Model、Cube。
（4）重新查询结果，如图 8-46 所示。

图 8-46　重新查询结果

2）如何实现每日自动构建 Cube

Kylin 提供了 REST API，因此，我们可以将构建 Cube 的命令写到脚本中，然后将脚本交给 Azkaban 或者 Oozie 调度工具，以实现定时调度的功能。

脚本如下。

```bash
#! /bin/bash
cube_name=payment_view_cube
do_date=`date -d '-1 day' +%F`

# 获取 00:00 的时间戳
start_date_unix=`date -d "$do_date 08:00:00" +%s`
start_date=$(($start_date_unix*1000))

# 获取 24:00 的时间戳
stop_date=$(($start_date+86400000))

curl -X PUT -H "Authorization: Basic QURNSU46S11MSU4=" -H 'Content-Type: application/json' -d '{"startTime":'$start_date', "endTime":'$stop_date', "buildType":"BUILD"}' http://hadoop102:7070/kylin/api/cubes/$cube_name/build
```

8.3.5　Kylin Cube 构建原理

Apache Kylin 的工作原理本质上是 MOLAP（Multidimension On-Line Analysis Processing）Cube，也就是多维立方体分析，是数据分析中非常经典的理论，下面对其进行简要介绍。

维度：观察数据的角度。比如，关于员工数据，可以从性别角度来观察，也可以更加细化，从入职时间或者地区的角度来观察。维度是一组离散的值，比如，性别中的男和女，或者时间维度上的每个独立的日期。因此，在统计时可以将维度值相同的记录聚合在一起，然后应用聚合函数进行累加，以及求平均值、最大值和最小值等聚合计算。

度量：被聚合（观察）的统计值，也就是聚合运算的结果。比如，员工数据中不同性别员工的人数，又如，在同一年入职的员工数。

有了维度和度量，就可以对一张数据表或者一个数据模型上的所有字段进行分类了，它们要么是维度，要么是度量（可以被聚合）。于是就有了根据维度和度量进行预计算的 Cube 理论，如图 8-47 所示。

给定一个数据模型，我们可以对其上的所有维度进行聚合，对于 N 个维度来说，组合的所有可能性共有 2^N 种。对每种维度组合的度量值进行聚合计算，然后将结果保存为一个物化视图，称为 Cuboid。所有维度组合的 Cuboid 作为一个整体，称为 Cube，如图 8-48 所示。

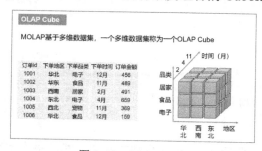

图 8-47　OLAP Cube

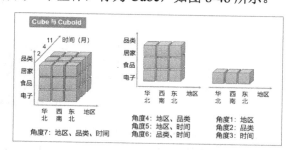

图 8-48　Cube 与 Cuboid

接下来介绍 Kylin Cube 的构建算法，主要分为两种：逐层构建算法和快速构建算法。

1．逐层构建算法

我们知道，一个 N 维的 Cube 是由 1 个 N 维子立方体、N 个 $N-1$ 维子立方体、$N\times(N-1)/2$ 个 $N-2$ 维子立方体、…、N 个 1 维子立方体和 1 个 0 维子立方体构成的，一共由 2^N 个子立方体构成，在逐层构建算法中，按维度数逐层减少来计算，每个层级的结果（除了第一层，第一层是从原始数据聚合而来的）是基于它上一层级的结果聚合得出的。比如，[Group by A, B]的结果，可以基于[Group by A, B, C]的结果，通过去掉 C 后聚合得出，这样可以减少重复计算。当 0 维度 Cuboid 计算出来的时候，整个 Cube 的计算也就完成了，计算过程如图 8-49 所示。

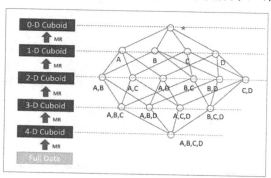

图 8-49　Kylin Cube 逐层构建算法计算过程

每轮的计算都是一个 MapReduce 任务，且串行执行；一个 N 维的 Cube，至少需要执行 N 次 MapReduce 任务，如图 8-50 所示。

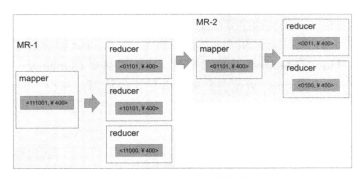

图 8-50 MapReduce 任务构建 Cube

逐层构建算法的优点如下。

（1）此算法充分利用 MapReduce 的优点，处理了中间复杂的排序和 Shuffle 工作，所以构建算法的代码清晰简单、易于维护。

（2）受益于 Hadoop 的日趋成熟，此算法非常稳定，即便在集群资源紧张时，也能保证工作最终完成。

逐层构建算法的缺点如下。

（1）当 Cube 有较多维度的时候，所需要的 MapReduce 任务也会相应增加；由于 Hadoop 的任务调度需要耗费额外资源，特别是集群较庞大的时候，反复递交任务造成的额外开销会相当大。

（2）由于 mapper 逻辑中并未进行聚合操作，所以每轮 MR 的 Shuffle 工作量都很大，从而导致效率低下。

（3）对 HDFS 的读/写操作较多：由于每层计算的输出会被用作下一层计算的输入，所以这些 key-value 需要写到 HDFS 上；当所有计算都完成后，Kylin 还需要执行额外的一轮任务将这些文件转成 HBase 的 HFile 格式，以导入 HBase 中。

总体而言，该算法的效率较低，尤其是当 Cube 维度数较多的时候。

2．快速构建算法

快速构建算法也被称作"逐段"（By Segment）或"逐块"（By Split）算法，Kylin Cube 从 1.5.x 版本开始引入该算法，该算法的主要思想是每个 mapper 将其所分配到的数据块计算成一个完整的小 Cube 段（包含所有的 Cuboid）。每个 mapper 将计算完成的 Cube 段输出给 reducer 并进行合并，生成大 Cube，也就是最终结果。Kylin 快速构建算法的计算过程如图 8-51 所示。

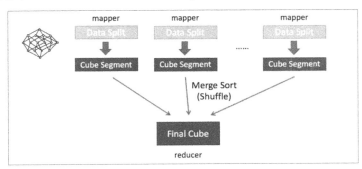

图 8-51 Kylin 快速构建算法的计算过程

与逐层构建算法相比，快速构建算法主要有两点不同。

（1）mapper 会利用内存进行预聚合，算出所有组合；mapper 输出的每个 key 都是不同的，这样会减少输出到 Hadoop MapReduce 的数据量，也不再需要 Combiner。

（2）执行一轮 MapReduce 任务便会完成所有层次的计算，减少了 Hadoop 任务的调配，如图 8-52 所示。

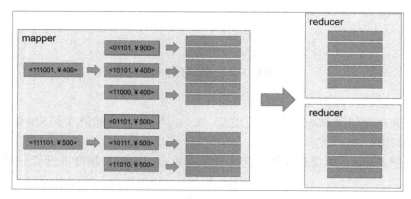

图 8-52　快速构建算法 MapReduce 示意

8.3.6　Kylin Cube 构建优化

1．衍生维度

衍生维度用于在有效维度内将维度表中的非主键维度排除，并使用维度表中的主键（其实是事实表中相应的外键）来代替它们。Kylin 会在底层记录维度表主键与维度表其他维度之间的映射关系，以便在查询时能够动态地将维度表中的主键"翻译"成非主键维度，并进行实时聚合，如图 8-53 所示。

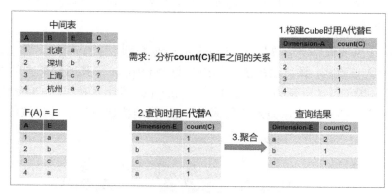

图 8-53　衍生维度示意

虽然衍生维度具有非常大的优势，但并不是说所有维度表中的维度都需要变成衍生维度，如果从维度表主键到维度表某个维度所需要的聚合工作量非常大，则不建议使用衍生维度。

2．聚合组

聚合组（Aggregation Group）是一个强大的剪枝工具。聚合组假设一个 Cube 的所有维度均可以根据业务需求划分成若干组（当然也可以是一个组），由于同一个分组内的维度更可能同时被同一个查询用到，因此会表现出更加紧密的内在关联。每个分组的维度集合均是 Cube 所有维度的一个子集，不同的分组各自拥有一个维度集合，它们可能与其他分组有相同的维度，也可能没有相同的维度。每个分组各自独立地根据自身的规则贡献出一批需要被物化的 Cuboid，所有分组贡献的 Cuboid 的并集就成了当前 Cube 中所有需要物化的 Cuboid 的集合。不同的分组有可能会贡献出相同的 Cuboid，构建引擎会察觉到这一点，并保证每个 Cuboid 无论在多少个分组中出现，它们都只会被物化一次。

对于每个分组内部的维度，用户可以使用如下三种可选的方式定义它们之间的关系。

（1）强制维度：如果一个维度被定义为强制维度，那么在这个分组产生的所有 Cuboid 中每个 Cuboid 都会包含该维度。每个分组中都可以有 0 个、1 个或多个强制维度。如果根据这个分组的业务逻辑，该维度一定会在过滤条件或分组条件中，则可以在该分组中把该维度设置为强制维度，如图 8-54 所示。

（2）层级维度：每个层级包含两个或多个维度。假设一个层级中包含 $D_1, D_2, \cdots, D_n$ n 个维度，那么在该分组产生的任何 Cuboid 中，这 n 个维度只会以（ ），(D_1)，(D_1, D_2)，$\cdots$，$(D_1, D_2, \cdots, D_n)$ $n+1$ 种形式中的一种出现。每个分组中可以有 0 个、1 个或多个层级，不同的层级之间不应有共享的维度。如果根据这个分组的业务逻辑，多个维度之间存在层级关系，则可以在该分组中把这些维度设置为层级维度，如图 8-55 所示。

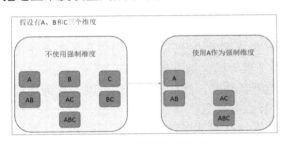
图 8-54　强制维度示意

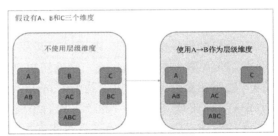
图 8-55　层级维度示意

（3）联合维度，每个联合中包含两个或多个维度，如果某些列形成一个联合，那么在该分组产生的任何 Cuboid 中，这些联合维度要么一起出现，要么都不出现。每个分组中可以有 0 个或多个联合，但是不同的联合之间不应有共享的维度（否则它们将合并成一个联合）。如果根据这个分组的业务逻辑，多个维度在查询中总是同时出现，则可以在该分组中把这些维度设置为联合维度，如图 8-56 所示。

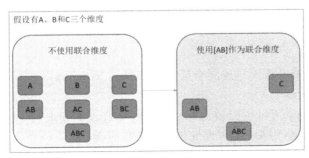
图 8-56　联合维度示意

上述操作可以在 Cube Designer 的 Advanced Setting 中的 Aggregation Groups 区域完成，如图 8-57 所示。

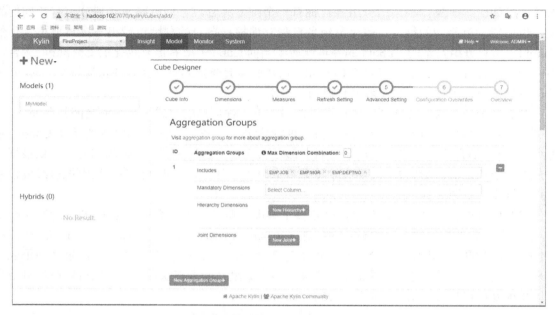

图 8-57 使用聚合组构建优化的示意

聚合组的设计非常灵活，甚至可以用来描述一些极端的设计。假设我们的业务需求非常单一，只需要某些特定的 Cuboid，那么可以创建多个聚合组，每个聚合组代表一个 Cuboid。具体的方法是在聚合组中先包含某个 Cuboid 所需的所有维度，然后把这些维度都设置为强制维度。这样当前的聚合组就只能产生我们想要的那个 Cuboid 了。

有的时候，Cube 中有一些基数非常大的维度，如果不进行特殊处理，它们就会和其他的维度进行各种组合，从而产生一大堆包含它们的 Cuboid。包含高基数维度的 Cuboid 在行数和体积上往往非常庞大，这会导致整个 Cube 的膨胀率变大。如果根据业务需求知道这个高基数的维度只会与若干个维度（而不是所有维度）同时被查询到，那么就可以通过聚合组对这个高基数维度进行一定的"隔离"。我们把这个高基数的维度放入一个单独的聚合组中，再把所有可能会与这个高基数维度一起被查询到的其他维度也放进来。这样，这个高基数的维度就被"隔离"在一个聚合组中了，所有不会与它一起被查询到的维度都没有和它一起出现在任何一个分组中，因此也就不会有多余的 Cuboid 产生。这点也大大减少了包含该高基数维度的 Cuboid 的数量，可以有效地控制 Cube 的膨胀率。

3. RowKey 优化

Kylin 会把所有的维度按照顺序组合成一个完整的 RowKey，并且按照这个 RowKey 升序排列 Cuboid 中所有的行。

设计良好的 RowKey 可以更有效地完成数据的查询过滤和定位，减少 I/O 次数，提高查询速度，RowKey 中的维度次序，对查询性能有显著的影响。

RowKey 的设计原则如下。

（1）被用作 where 过滤的维度放在没有被用作 where 过滤的维度前边，如图 8-58 所示。

（2）基数大的维度放在基数小的维度前边，如图 8-59 所示。二维 Cuboid，由三维 Cuboid 计算而来，此处，三维 Cuboid-1110/1101，均可通过计算得到二维 Cuboid-1100，Kylin 的规则是选择 Cuboid 小的，即选择三维 Cuboid-1101。我们应保证 Kylin 所选的 Cuboid-1101 为数据量较小的一个，而三维 Cuboid-1110/1101 的数据量大小实际是由维度 C 和 D 的基数决定的。为保证 Cuboid-1101 的数据量小于 Cuboid-1110，需保证维度 D 的基数小于维度 C 的基数。

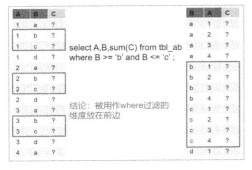

图 8-58 被用作 where 过滤的维度放在前边

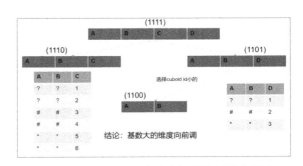

图 8-59 基数大的维度放在基数小的维度前边

4．并发粒度优化

当 Segment 中某一个 Cuboid 的大小超出一定的阈值时，系统会将该 Cuboid 的数据分配到多个分区中，以实现 Cuboid 数据读取的并行化，从而优化 Cube 的查询速度。具体的实现方式如下：构建引擎根据 Segment 估计的大小，以及参数"kylin.hbase.region.cut"的设置决定 Segment 在存储引擎中需要的分区数量，如果存储引擎是 HBase，那么分区的数量就对应于 HBase 中的 Region 数量。"kylin.hbase.region.cut"的默认值是 5.0，单位是 GB，也就是说，对于一个大小是 50GB 的 Segment，构建引擎会给它分配 10 个分区。用户还可以通过设置"kylin.hbase.region.count.min"（默认值为 1）和"kylin.hbase.region.count.max"（默认值为 500）两个参数来决定每个 Segment 最少或最多被划分成几个分区。

由于每个 Cube 的并发粒度控制不同，因此建议读者在 Cube Designer 的 Configuration Overwrites 中进行配置，如图 8-60 所示，可以通过单击"Property"按钮，为每个 Cube 量身定制控制并发粒度的参数。假设将当前 Cube 的"kylin.hbase.region.count.min"的参数值设置为 2，"kylin.hbase.region.count.max"的参数值设置为 100。这样无论 Segment 的大小如何变化，它的分区数量最小都不会小于 2，最大都不会大于 100。相应地，Segment 背后的存储引擎（HBase）为了存储 Segment，也不会使用小于 2 个或大于 100 个的分区。我们还调整了"kylin.hbase.region.cut"参数的默认值，这样 50GB 的 Segment 基本上会被分配到 50 个分区中，相比默认设置，Cuboid 最多会获得 5 倍的并发量。

大数据分析——数据仓库项目实战

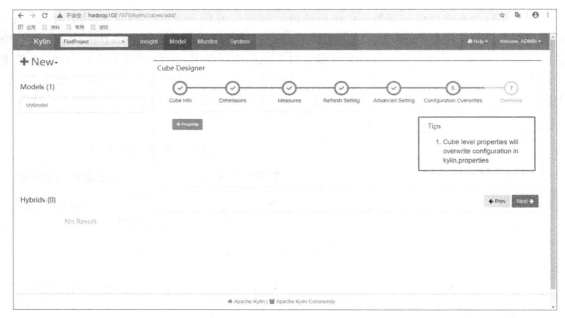

图 8-60 Kylin Cube 属性设置

8.3.7 Kylin BI 工具集成

可以与 Kylin 结合使用的可视化工具有很多，如下所示。
- ODBC：与 Tableau、Excel、Power BI 等工具集成。
- JDBC：与 Saiku、BIRT 等 Java 工具集成。
- REST API：与 JavaScript、Web 网页集成。

1．JDBC

（1）新建项目并导入依赖。

```
<dependencies>
    <dependency>
        <groupId>org.apache.kylin</groupId>
        <artifactId>kylin-jdbc</artifactId>
        <version>2.5.1</version>
    </dependency>
</dependencies>
```

（2）编码，语句如下。

```
package com.atguigu;

import java.sql.*;

public class TestKylin {

    public static void main(String[] args) throws Exception {
```

```java
// Kylin_JDBC 驱动
String KYLIN_DRIVER = "org.apache.kylin.jdbc.Driver";

// Kylin_URL
String KYLIN_URL = "jdbc:kylin://hadoop102:7070/FirstProject";

// Kylin 的用户名
String KYLIN_USER = "ADMIN";

// Kylin 的密码
String KYLIN_PASSWD = "KYLIN";

// 添加驱动信息
Class.forName(KYLIN_DRIVER);

// 获取连接
Connection connection = DriverManager.getConnection(KYLIN_URL, KYLIN_USER, KYLIN_PASSWD);

// 预编译 SQL 语句
PreparedStatement ps = connection.prepareStatement("SELECT sum(sal) FROM emp group by deptno");

// 执行查询
ResultSet resultSet = ps.executeQuery();

// 遍历打印
while (resultSet.next()) {
    System.out.println(resultSet.getInt(1));
}
    }
}
```

(3) 结果展示，如图 8-61 所示。

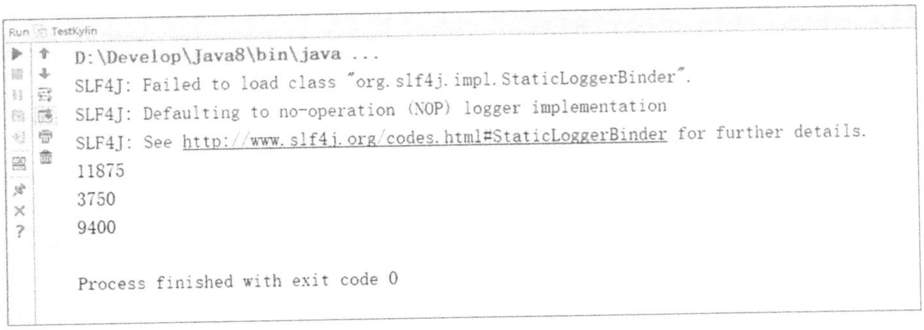

图 8-61 JDBC 结果展示

2. Zeppelin

1）Zeppelin 安装与启动

（1）将 zeppelin-0.8.0-bin-all.tgz 上传至 Linux。

（2）解压 zeppelin-0.8.0-bin-all.tgz 到 /opt/module 目录下。

```
[atguigu@hadoop102 sorfware]$ tar -zxvf zeppelin-0.8.0-bin-all.tgz -C /opt/module/
```

（3）修改名称。

```
[atguigu@hadoop102 module]$ mv zeppelin-0.8.0-bin-all/ zeppelin
```

（4）启动 Zeppelin。

```
[atguigu@hadoop102 zeppelin]$ bin/zeppelin-daemon.sh start
```

读者可登录 Zeppelin 网页（http://hadoop102:8080/#/）进行查看，Web 默认端口为 8080，如图 8-62 所示。

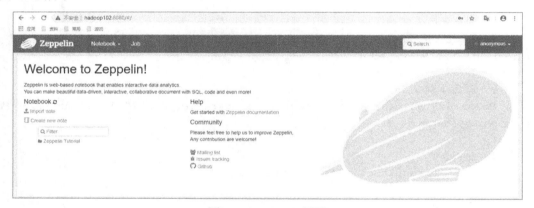

图 8-62　Zeppelin 网页

2）配置 Zeppelin 支持 Kylin

（1）选择"anonymous"→"Interpreter"选项，配置解释器，如图 8-63 所示。

（2）搜索 Kylin 插件并修改相应的配置，如图 8-64 所示。

（3）修改完成后，单击"Save"按钮保存修改内容，如图 8-65 所示。

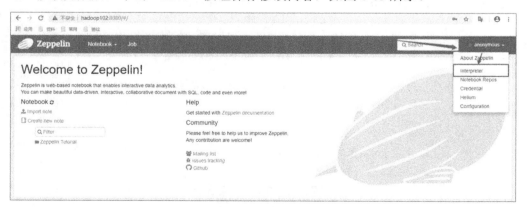

图 8-63　配置解释器

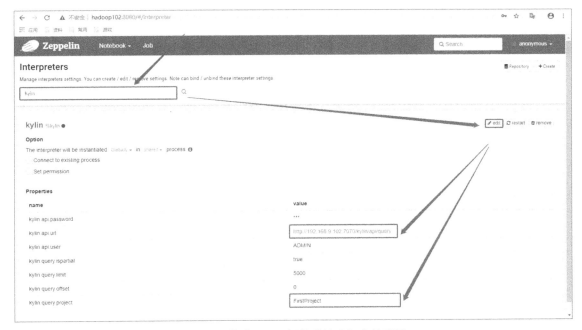

图 8-64　搜索 Kylin 插件并修改相应的配置

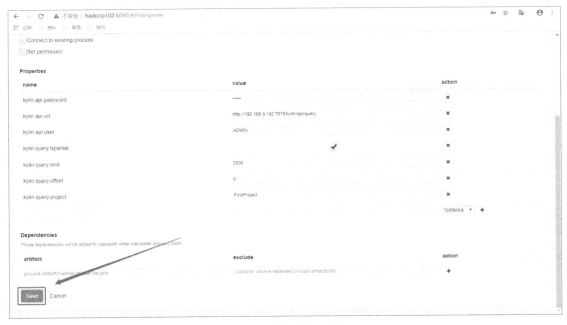

图 8-65　保存修改内容

3）案例实操

需求：查询员工的详细信息，并使用各种图表进行展示。

（1）选择"Notebook"→"Creat new note"选项，创建新的 note，如图 8-66 所示。

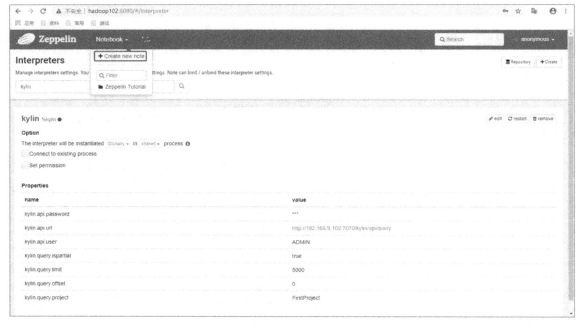

图 8-66　创建新 note 入口

（2）在"Note Name"文本框中输入"test_kylin"并单击"Create"按钮，如图 8-67 所示，note 创建成功的页面如图 8-68 所示。

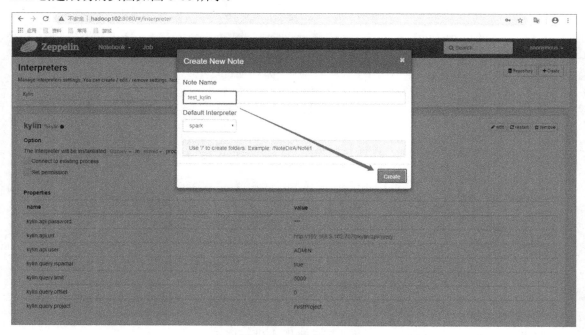

图 8-67　创建新 note

图 8-68　note 创建成功的页面

（3）执行查询操作，如图 8-69 所示。

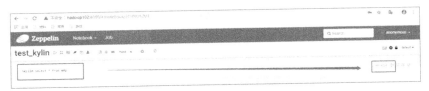

图 8-69　执行查询操作

（4）展示结果，如图 8-70 所示。

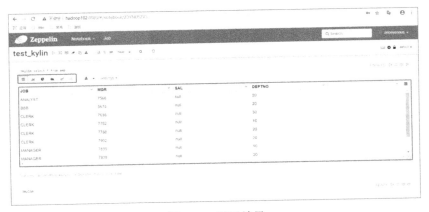

图 8-70　展示结果

（5）其他图表格式，如图 8-71～图 8-75 所示。

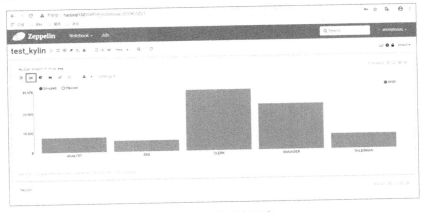

图 8-71　柱状图表示意

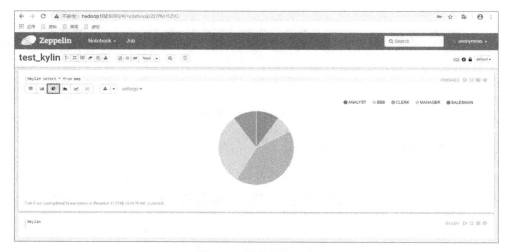

图 8-72　饼状图表示意

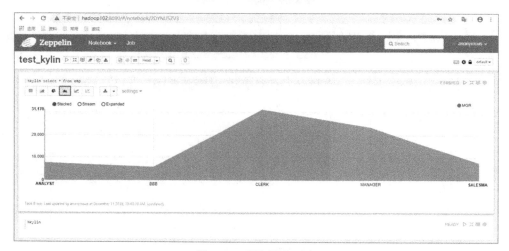

图 8-73　面积图表示意

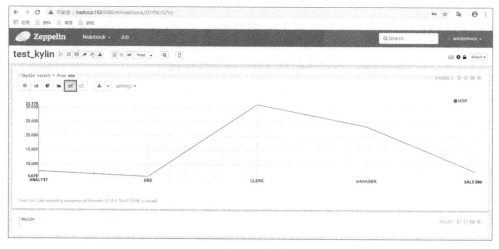

图 8-74　折线图表示意

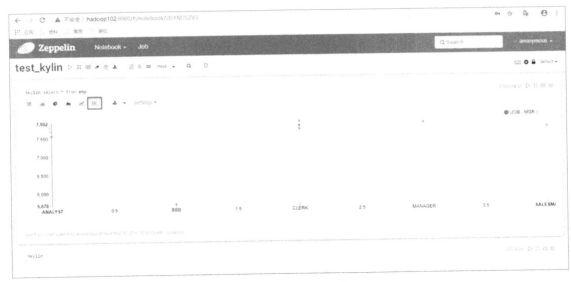

图 8-75　散点图表示意

8.4　即席查询框架对比

目前应用比较广泛的几种即席查询框架有 Druid、Kylin、Presto、Impala、Spark SQL 和 Elasticsearch，针对响应时间、数据支持、技术特点等方面的对比如表 8-3 所示。

表 8-3　即席查询框架对比

对比项目	Druid	Kylin	Presto	Impala	Spark SQL	Elasticsearch
亚秒级响应	Y	Y	N	N	N	N
百亿数据集	Y	Y	Y	Y	Y	Y
SQL 支持	Y	Y	Y	Y	Y	N
离线	Y	Y	Y	Y	Y	Y
实时	Y	Y	N	N	N	Y
精确去重	N	Y	Y	Y	Y	N
多表 join	N	Y	Y	Y	Y	N
DBC for BI	N	Y	Y	Y	Y	N

针对上表的对比情况，分析汇总如下。
- Druid：实时处理时序数据的 OLAP 数据库，因为它的索引首先按照时间进行分片，查询的时候也是按照时间线去路由索引的。
- Kylin：核心是 Cube，Cube 是一种预计算技术，基本思路是预先对数据进行多维索引，查询时只扫描索引而不访问原始数据，从而提高查询速度。
- Presto：它没有使用 MapReduce，大部分场景下比 Hive 快一个数量级，其中的关键是所有的处理都在内存中完成。
- Impala：基于内存运算，速度快，支持的数据源没有 Presto 多。
- Spark SQL：基于 Spark 平台上的一个 OLAP 框架，基本思路是增加机器以实现并行计算，从而提高查询速度。
- Elasticsearch：最大的特点是使用倒排索引解决了索引存在的问题。根据研究，Elasticsearch 在数据获取和聚集时用的资源比在 Druid 中高。

框架选型如下。
- 从超大数据的查询效率来看：Druid > Kylin > Presto > Spark SQL。
- 从支持的数据源种类来看：Presto > Spark SQL > Kylin > Druid。

8.5 本章总结

本章主要对三个应用比较广泛的即席查询框架进行了讲解。对于数据仓库系统来说，即席查询是不可或缺的环节。本章对三个即席查询框架的特点、安装部署等方面进行了说明，并对目前比较流行的几个即席查询框架进行了对比，在实际应用中，读者可以根据自己项目的具体情况进行选取。

第9章 元数据管理模块

国内企业进行元数据管理的方向有三个：一是基于数据平台进行元数据管理，由于大数据平台的兴起，目前企业逐步开始针对 Hadoop 环境进行元数据管理；二是基于企业数据整体管理规划开展对元数据的管理，也是企业数据资产管理的基础；三是将元数据作为某个平台的组件进行此平台特有的元数据管理，它作为一个中介或中转互通平台各组件间的数据。基于数据平台的元数据管理相对成熟，也是业界最早进行元数据管理的切入点或者说是数据平台建设的必备。社区中开源的元数据管理系统方案，常见的有 Hortonworks 主推的 Apache Atlas，本书将以 Atlas 为例对元数据管理进行介绍。

9.1 Atlas 入门

Atlas 为组织提供开放式元数据管理和治理功能，用以构建组织的数据资产目录，对这些资产进行分类和管理，并为数据科学家、数据分析师和数据治理团队提供围绕这些数据资产的协作功能。

9.1.1 Atlas 概述

Atlas 的整体设计侧重于数据血缘关系的采集，以及表格维度的基本信息和业务属性信息的管理。为了达到这个目的，Atlas 设计了一套通用的 Type 体系来描述这些信息。Type 的主要基础类型包括 DataSet 和 Process，前者用来描述各种数据源本身，后者用来描述一个数据处理的流程，比如，一个 ETL 任务。

Atlas 现有的 Bridge 实现，从数据源的角度来看，主要覆盖了 Hive、HBase、HDFS 和 Kafka，此外，还有适配于 Sqoop、Storm 和 Falcon 的 Bridge，不过这三者更多是从 Process 的角度入手的，最后落地的数据源还是上述四种。

9.1.2 Atlas 架构原理

Atlas 架构如图 9-1 所示。

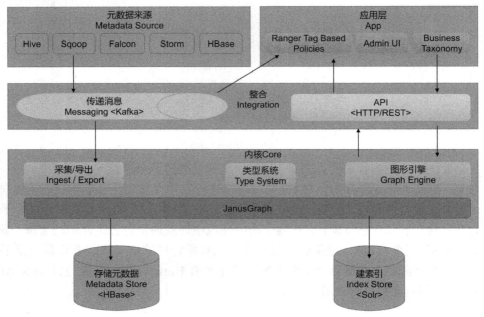

图 9-1　Atlas 架构

Atlas 架构具有如下关键组件。

- Metadata Store<HBase>：采用 HBase 存储元数据。
- Index Store<Solr>：采用 Solr 建索引。
- Ingest/Export：采集组件允许将元数据添加到 Atlas。同样地，导出组件将 Atlas 检测到的元数据更改并公开为事件。
- Type System：用户为他们想要管理的元数据对象定义模型。Type System 称为实体的类型实例，表示受管理的实际元数据对象。
- Graph Engine：Atlas 在内部使用 Graph 模型持久保存它管理的元数据对象。
- API<HTTP/REST>：Atlas 的所有功能都通过 REST API 向最终用户展示，该 API 允许创建、更新和删除类型和实体。它也是查询和发现 Atlas 管理的类型和实体的主要机制。
- Messaging<Kafka>：除了 API，用户还可以选择使用基于 Kafka 的消息传递接口与 Atlas 集成。
- Metadata Sources：目前，Atlas 支持从以下来源提取和管理元数据：HBase、Hive、Sqoop、Storm 和 Falcon。
- Admin UI：该组件是一个基于 Web 的应用程序，允许数据管理员和科学家发现和注释元数据。这里最重要的是搜索界面和类似 SQL 的查询语言，可用于查询 Atlas 管理的元数据类型和对象。
- Ranger Tag Based Policies：权限管理模块。
- Business Taxonomy：业务分类。

9.2 Atlas 安装及使用

Atlas 的安装及使用需要基于 Hadoop、Zookeeper、Kafka、HBase、Solr、Hive、Azkaban 等，所以需要先对以上框架进行安装部署。

9.2.1 安装前环境准备

1．安装 JDK8、Hadoop 2.7.2

（1）安装 JDK、Hadoop 集群，可参考 3.3 节中的相关内容。
（2）启动 Hadoop 集群。

```
[atguigu@hadoop102 hadoop-2.7.2]$ sbin/start-dfs.sh
[atguigu@hadoop103 hadoop-2.7.2]$ sbin/start-yarn.sh
```

2．安装 Zookeeper 3.4.10

（1）安装 Zookeeper 集群，可参考 4.3.1 节中的相关内容。
（2）启动 Zookeeper 集群。

```
[atguigu@hadoop102 zookeeper-3.4.10]$ zk.sh start
```

3．安装 Kafka 0.11.0.2

（1）安装 Kafka 集群，可参考 4.3.3 节中的相关内容。
（2）启动 Kafka 集群。

```
[atguigu@hadoop102 kafka]$ kf.sh start
```

4．安装 HBase 1.3.1

（1）安装 HBase 集群，可参考 8.3.2 节中的相关内容。
（2）启动 HBase 集群。

```
[atguigu@hadoop102 hbase]$ bin/start-hbase.sh
```

5．安装 Solr 5.2.1

（1）Solr 的版本必须是 5.2.1，参见官方网站。
（2）下载 Solr 安装包。
（3）把 solr-5.2.1.tgz 上传到 hadoop102 的/opt/software 目录下。
（4）解压 solr-5.2.1.tgz 到/opt/module/目录下。

```
[atguigu@hadoop102 software]$ tar -zxvf solr-5.2.1.tgz -C /opt/module/
```

（5）修改 solr-5.2.1 的名称为 solr。

```
[atguigu@hadoop102 module]$ mv solr-5.2.1/ solr
```

（6）进入 solr/bin 目录，修改 solr.in.sh 文件。

```
[atguigu@hadoop102 solr]$ vim bin/solr.in.sh
# 添加下列指令
ZK_HOST="hadoop102:2181,hadoop103:2181,hadoop104:2181"
```

```
SOLR_HOST="hadoop102"
# Sets the port Solr binds to, default is 8983
# 可修改端口
SOLR_PORT=8983
```

（7）分发 Solr，进行 Cloud 模式部署。

```
[atguigu@hadoop102 module]$ xsync solr
```

提示：分发完成后，将每台机器/opt/module/solr/bin 目录下的 solr.in.sh 文件的 "SOLR_HOST=" 修改为对应主机名即可。

（8）在三台节点服务器上分别启动 Solr，即 Cloud 模式。

```
[atguigu@hadoop102 solr]$ bin/solr start
[atguigu@hadoop103 solr]$ bin/solr start
[atguigu@hadoop104 solr]$ bin/solr start
```

提示：启动 Solr 前，需要先启动 Zookeeper 集群。

（9）通过 Web 网页访问 8983 端口，可指定三台节点服务器中任意一台节点服务器的 IP 地址（http://hadoop102:8983/solr/#/），打开 Solr UI 界面，如图 9-2 所示。

图 9-2 Solr UI 界面

提示：只有 UI 界面出现 Cloud 菜单栏时，Solr 的 Cloud 模式才算部署成功。

（10）编写 Solr 启动、停止脚本。

① 在 hadoop102 的/home/atguigu/bin 目录下创建脚本 s.sh。

```
[atguigu@hadoop102 bin]$ vim s.sh
```

在脚本中编写如下内容。

```
#!/bin/bash

case $1 in
```

```
"start"){
    for i in hadoop102 hadoop103 hadoop104
    do
        ssh $i "/opt/module/solr/bin/solr start"
    done
};;
"stop"){
    for i in hadoop102 hadoop103 hadoop104
    do
        ssh $i "/opt/module/solr/bin/solr stop"
    done
};;
esac
```

② 增加脚本执行权限。

```
[atguigu@hadoop102 bin]$ chmod 777 s.sh
```

③ Solr 集群启动脚本。

```
[atguigu@hadoop102 module]$ s.sh start
```

④ Solr 集群停止脚本。

```
[atguigu@hadoop102 module]$ s.sh stop
```

6. 安装 Hive 1.2.1

安装 Hive 1.2.1，可参考 6.2.2 节中的相关内容。

7. 安装 Azkaban 2.5.0

安装 Azkaban 2.5.0，可参考 6.9.1 节中的相关内容。

8. 安装 Atlas 0.8.4

（1）将 apache-atlas-0.8.4-bin.tar.gz 上传到 hadoop102 的/opt/software 目录下。

（2）解压 apache-atlas-0.8.4-bin.tar.gz 到/opt/module/目录下。

```
[atguigu@hadoop102 software]$ tar -zxvf apache-atlas-0.8.4-bin.tar.gz -C /opt/module/
```

（3）修改 apache-atlas-0.8.4 的名称为 atlas。

```
[atguigu@hadoop102 module]$ mv apache-atlas-0.8.4/ atlas
```

9.2.2 集成外部框架

Atlas 的安装方式分为集成自带的 HBase+Solr 和集成外部的 HBase+Solr。通常在企业开发中选择集成外部的 HBase+Solr 安装方式，方便项目整体进行集成操作。

1. Atlas 集成 HBase

（1）进入/opt/module/atlas/conf 目录，修改配置文件 atlas-application.properties。

```
[atguigu@hadoop102 conf]$ vim atlas-application.properties

# 修改 Atlas 存储数据的主机
atlas.graph.storage.hostname=hadoop102:2181,hadoop103:2181,hadoop104:2181
```

（2）进入/opt/module/atlas/conf/hbase 目录，添加 HBase 集群的配置文件到${Atlas_Home}。

```
[atguigu@hadoop102 hbase]$ ln -s /opt/module/hbase/conf/ /opt/module/atlas/conf/hbase/
```

（3）在/opt/module/atlas/conf 目录下的 atlas-env.sh 文件中添加 HBASE_CONF_DIR。

```
[atguigu@hadoop102 conf]$ vim atlas-env.sh

# 添加 HBase 配置文件路径
export HBASE_CONF_DIR=/opt/module/atlas/conf/hbase/conf
```

2. Atlas 集成 Solr

（1）进入/opt/module/atlas/conf 目录，修改配置文件 atlas-application.properties。

```
[atguigu@hadoop102 conf]$ vim atlas-application.properties

# 修改如下配置
atlas.graph.index.search.solr.zookeeper-
url=hadoop102:2181,hadoop103:2181,hadoop104:2181
```

（2）将 Atlas 配置的 solr 文件夹复制到外部 Solr 集群的各节点中。

```
[atguigu@hadoop102 conf]$ cp -r /opt/module/atlas/conf/solr /opt/module/solr/
```

（3）进入/opt/module/solr 目录，修改复制过来的 solr 文件夹名称为 atlas_conf。

```
[atguigu@hadoop102 solr]$ mv solr atlas_conf
```

（4）在 Cloud 模式下，启动 Solr（先启动 Zookeeper 集群），并创建 collection。

```
[atguigu@hadoop102 solr]$ bin/solr create -c vertex_index -d
/opt/module/solr/atlas_conf -shards 3 -replicationFactor 2

[atguigu@hadoop102 solr]$ bin/solr create -c edge_index -d
/opt/module/solr/atlas_conf -shards 3 -replicationFactor 2

[atguigu@hadoop102 solr]$ bin/solr create -c fulltext_index -d
/opt/module/solr/atlas_conf -shards 3 -replicationFactor 2
```

注意：如果需要删除 vertex_index、edge_index、fulltext_index 等 collection，则可以执行如下语句。

```
[atguigu@hadoop102 solr]$ bin/solr delete -c ${collection_name}
```

（5）验证 collection 是否创建成功

登录 Solr Web 控制台（http://hadoop102:8983/solr/#/~cloud），显示的内容如图 9-3 所示。

图 9-3　Solr 控制台显示的内容

3. Atlas 集成 Kafka

（1）进入/opt/module/atlas/conf 目录，修改配置文件 atlas-application.properties。

```
[atguigu@hadoop102 conf]$ vim atlas-application.properties

######### 通知配置 #########
atlas.notification.embedded=false
atlas.kafka.zookeeper.connect=hadoop102:2181,hadoop103:2181,hadoop104:2181
atlas.kafka.bootstrap.servers=hadoop102:9092,hadoop103:9092,hadoop104:9092
atlas.kafka.zookeeper.session.timeout.ms=4000
atlas.kafka.zookeeper.connection.timeout.ms=2000

atlas.kafka.enable.auto.commit=true
```

（2）启动 Kafka 集群，并创建 Topic。

```
[atguigu@hadoop102 kafka]$ bin/kafka-topics.sh --zookeeper hadoop102:2181,hadoop103:2181,hadoop104:2181 --create --replication-factor 3 --partitions 3 --topic _HOATLASOK

[atguigu@hadoop102 kafka]$ bin/kafka-topics.sh --zookeeper hadoop102:2181,hadoop103:2181,hadoop104:2181 --create --replication-factor 3 --partitions 3 --topic ATLAS_ENTITIES
```

4. Atlas 的其他配置

（1）进入/opt/module/atlas/conf 目录，修改配置文件 atlas-application.properties。

```
[atguigu@hadoop102 conf]$ vim atlas-application.properties

######### 服务器属性 #########
atlas.rest.address=http://hadoop102:21000
# If enabled and set to true, this will run setup steps when the server starts
atlas.server.run.setup.on.start=false

######### 实体审核配置 #########
atlas.audit.hbase.zookeeper.quorum=hadoop102:2181,hadoop103:2181,hadoop104:2181
```

(2)记录性能指标,进入/opt/module/atlas/conf 目录,修改 atlas-log4j.xml 文件。

```
[atguigu@hadoop102 conf]$ vim atlas-log4j.xml

# 去掉如下代码的注释
<appender name="perf_appender" class="org.apache.log4j.
DailyRollingFileAppender">
    <param name="file" value="${atlas.log.dir}/atlas_perf.log" />
    <param name="datePattern" value="'.'yyyy-MM-dd" />
    <param name="append" value="true" />
    <layout class="org.apache.log4j.PatternLayout">
        <param name="ConversionPattern" value="%d|%t|%m%n" />
    </layout>
</appender>

<logger name="org.apache.atlas.perf" additivity="false">
    <level value="debug" />
    <appender-ref ref="perf_appender" />
</logger>
```

5. Atlas 集成 Hive

(1)进入/opt/module/atlas/conf 目录,修改配置文件 atlas-application.properties。

```
[atguigu@hadoop102 conf]$ vim atlas-application.properties

######### Hive Hook Configs #######
atlas.hook.hive.synchronous=false
atlas.hook.hive.numRetries=3
atlas.hook.hive.queueSize=10000
atlas.cluster.name=primary
```

(2)将 atlas-application.properties 配置文件加入 atlas-plugin-classloader-0.8.4.jar 中。

```
[atguigu@hadoop102 hive]$ zip -u /opt/module/atlas/hook/hive/atlas-plugin-classloader-0.8.4.jar /opt/module/atlas/conf/atlas-application.properties

[atguigu@hadoop102 hive]$ cp /opt/module/atlas/conf/atlas-application.properties /opt/module/hive/conf/
```

注意:这个配置不能参照官方网站的做法:将配置文件复制到 Hive 的 conf 目录中。参照官方网站的做法,一直读取不到 atlas-application.properties 配置文件,笔者查看源码发现这个配置文件是在 classpath 中读取的,所以将它解压到 jar 包中。

(3)在/opt/module/hive/conf 目录下的 hive-site.xml 文件中设置 Atlas hook。

```
[atguigu@hadoop102 conf]$ vim hive-site.xml
<property>
    <name>hive.exec.post.hooks</name>
    <value>org.apache.atlas.hive.hook.HiveHook</value>
</property>
```

```
[atguigu@hadoop102 conf]$ vim hive-env.sh
```

#在 Tez 引擎依赖的 jar 包后面追加与 Hive 插件相关的 jar 包
```
export HIVE_AUX_JARS_PATH=/opt/module/hadoop-2.7.2/share/hadoop/common/hadoop-lzo-0.4.20.jar$TEZ_JARS,/opt/module/atlas/hook/hive/atlas-plugin-classloader-0.8.4.jar,/opt/module/atlas/hook/hive/hive-bridge-shim-0.8.4.jar
```

9.2.3 集群启动

（1）启动 Hadoop 集群。
```
[atguigu@hadoop102 hadoop-2.7.2]$ sbin/start-dfs.sh
[atguigu@hadoop103 hadoop-2.7.2]$ sbin/start-yarn.sh
```

（2）启动 Zookeeper 集群。
```
[atguigu@hadoop102 zookeeper-3.4.10]$ zk.sh start
```

（3）启动 Kafka 集群。
```
[atguigu@hadoop102 kafka]$ kf.sh start
```

（4）启动 HBase 集群。
```
[atguigu@hadoop102 hbase]$ bin/start-hbase.sh
```

（5）启动 Solr 集群。
```
[atguigu@hadoop102 solr]$ bin/solr start
[atguigu@hadoop103 solr]$ bin/solr start
[atguigu@hadoop104 solr]$ bin/solr start
```

（6）进入/opt/module/atlas 目录，重新启动 Atlas 服务。
```
[atguigu@hadoop102 atlas]$ bin/atlas_stop.py

[atguigu@hadoop102 atlas]$ bin/atlas_start.py
```
错误信息查看路径为/opt/module/atlas/logs/*.out 和/opt/module/atlas/logs/application.log。
访问地址：http://hadoop102:21000。
账户：admin。
密码：admin。

9.2.4 导入 Hive 元数据到 Atlas

（1）配置 Hive 环境变量。
```
[atguigu@hadoop102 hive]$ sudo vim /etc/profile

# 配置 Hive 环境变量
export HIVE_HOME=/opt/module/hive
export PATH=$PATH:$HIVE_HOME/bin/
```

```
[atguigu@hadoop102 hive]$ source /etc/profile
```

（2）启动 Hive。

```
[atguigu@hadoop102 hive]$ hive
```

（3）进入/opt/module/atlas 目录，将 Hive 元数据导入 Atlas 中。

```
[atguigu@hadoop102 atlas]$ bin/import-hive.sh

Using Hive configuration directory [/opt/module/hive/conf]
Log file for import is /opt/module/atlas/logs/import-hive.log
log4j:WARN No such property [maxFileSize] in org.apache.log4j.PatternLayout.
log4j:WARN No such property [maxBackupIndex] in org.apache.log4j.PatternLayout.
```

输入用户名（admin）和密码（admin）。

```
Enter username for atlas :- admin
Enter password for atlas :-
Hive Meta Data import was successful!!!
```

9.3 Atlas 界面查看及使用

9.3.1 查看基本信息

1. 登录

（1）在浏览器地址栏中输入 http://hadoop102:21000/login.jsp，登录 Atlas，如图 9-4 所示。

图 9-4 Atlas 登录

（2）账号、密码默认都为 admin，登录成功的页面如图 9-5 所示。

图 9-5　Atlas 登录成功的页面

2. 查看 Hive 库

查看相应的 Hive 库，类型选择"hive_db"，如图 9-6 所示。

图 9-6　查看 Hive 库

3. 查看 Hive 进程

查看 Hive 进程，类型选择"hive_process"，如图 9-7 所示。

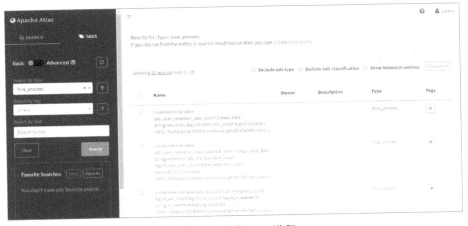

图 9-7　查看 Hive 进程

4．查看 Hive 表

查看 Hive 表，类型选择"hive_table"，如图 9-8 所示。

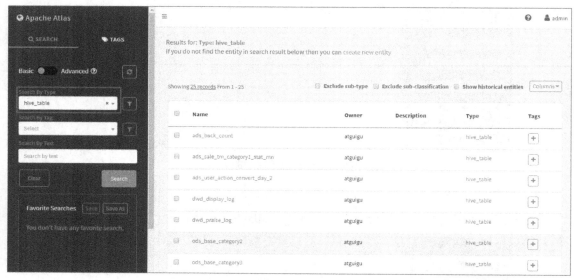

图 9-8　查看 Hive 表

5．查看 Hive 列

查看 Hive 列，类型选择"hive_column"，如图 9-9 所示。

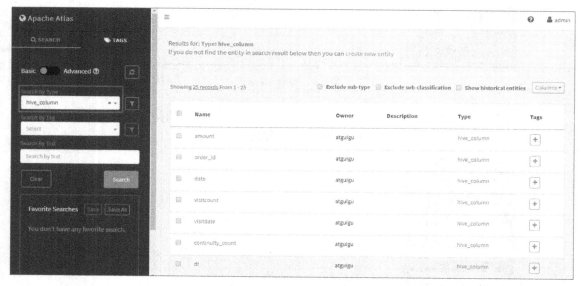

图 9-9　查看 Hive 列

6．筛选查询条件

比如，要查询 name 为 id 的列，则在"Search By Text"文本框中输入"where name="id"", 其他选项筛选条件的写法一样，如图 9-10 所示。

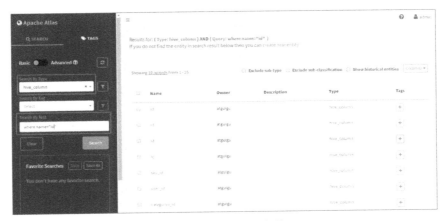

图 9-10　筛选查询条件

7. 查看具有血缘依赖列的数据

查看类型选择"hive_column_lineage",如图 9-11 所示。

图 9-11　查看具有血缘依赖列的数据

9.3.2　查看血缘依赖关系

1. 第一次查看表血缘依赖关系

（1）先选择"hive_db"类型，然后单击"Type"列的"hive_db"，查看对应的数据库。例如，查看 gmall 数据库，如图 9-12 所示。

图 9-12　查看指定数据库

(2)单击"Tables"按钮,查看 gmall 数据库中的所有表,如图 9-13 所示。

图 9-13　查看 gmall 数据库中的所有表

(3)选择"Properties"选项卡,查看表详情,如图 9-14 所示。

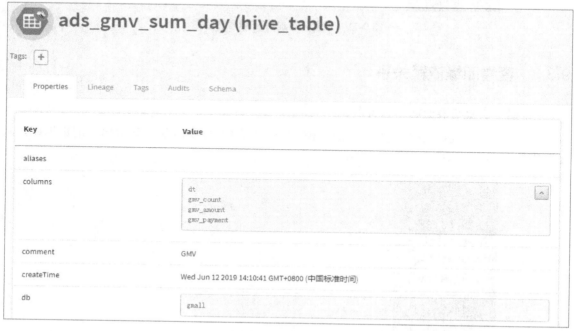

图 9-14　查看表详情

(4)选择"Lineage"选项卡,查看表血缘依赖关系,如图 9-15 所示。

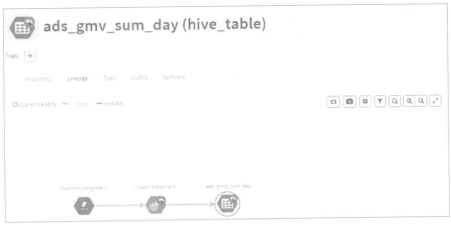

图 9-15　查看表血缘依赖关系

（5）选择"Audits"选项卡，查看表修改操作对应的时间和详情，如图 9-16 所示。

图 9-16　查看表修改操作对应的时间和详情

（6）选择"Schema"选项卡，查看表字段信息，如图 9-17 所示。

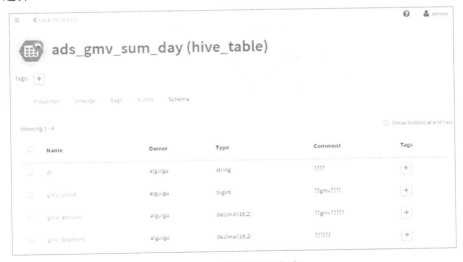

图 9-17　查看表字段信息

2. 第一次查看字段血缘依赖关系

（1）单击"gmv_count"字段查看相应信息，如图 9-18 所示。

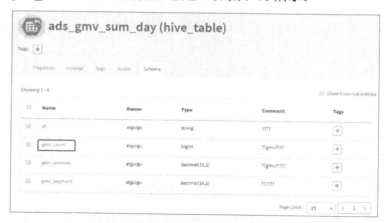

图 9-18　查看 gmv_count 字段

（2）选择"Properties"选项卡，显示字段详情，如图 9-19 所示。

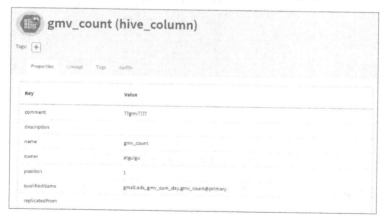

图 9-19　查看字段详情

（3）选择"Lineage"选项卡，查看字段血缘关系，如图 9-20 所示。

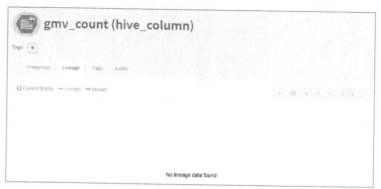

图 9-20　查看字段血缘关系

（4）选择"Audits"选项卡，查看字段修改操作对应的时间和详情，如图 9-21 所示。

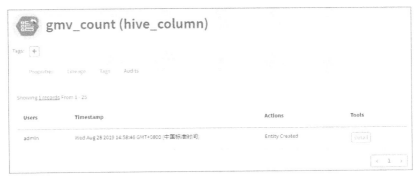

图 9-21　查看字段修改操作对应的时间和详情

3. 启动 GMV 全流程任务

（1）启动 Azkaban。

① 启动 Executor 服务器。在 Executor 服务器目录下执行启动命令。

```
[atguigu@hadoop102 executor]$ pwd
/opt/module/azkaban/executor
[atguigu@hadoop102 executor]$ bin/azkaban-executor-start.sh
```

② 启动 Web 服务器。在 Azkaban Web 服务器目录下执行启动命令。

```
[atguigu@hadoop102 server]$ pwd
/opt/module/azkaban/server
[atguigu@hadoop102 server]$ bin/azkaban-web-start.sh
```

③ 查看 Web 页面：https://hadoop102:8443。

（2）上传任务。参考 6.9.3 节中的相关内容。

（3）等待 Azkaban 运行结束，查看结果。

（4）查看 Atlas 表血缘依赖关系，如图 9-22 所示。

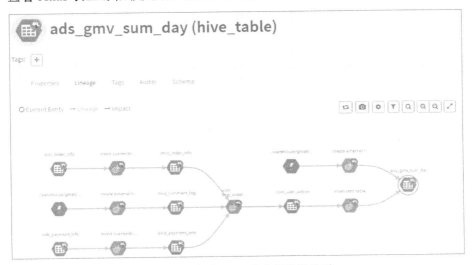

图 9-22　查看 Atlas 表血缘依赖关系

(5) 查看 Atlas 字段血缘依赖关系，如图 9-23 所示。

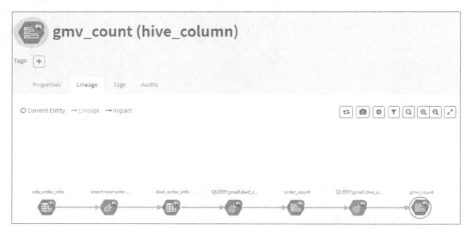

图 9-23　查看 Atlas 字段血缘依赖关系

9.4　本章总结

本章以 Atlas 为例，主要从 Atlas 的概述和架构原理、Atlas 安装前的环境准备、与外部框架的集成、界面的查看及使用几个方面，为读者讲解了大数据的元数据管理系统。Atlas 只是为大数据的元数据管理提供了一种解决方案，其本身也存在一定的局限性，读者如果感兴趣，可以对 Atlas 进行深入了解或者探索学习其他的元数据管理框架。